1992

Proceedings of the Conference on Technology in Collegiate Mathematics

The Twilight of the Pencil and Paper

First Annual Conference

Proceedings of the Conference on Technology in Collegiate Mathematics

The Twilight of the Pencil and Paper

First Annual Conference

The Ohio State University
Department of Mathematics
Columbus, Ohio
October 27–29, 1988

Conference Organizers: Franklin Demana, The Ohio State University
Thomas Ralley, The Ohio State University
Alan Osborne, The Ohio State University
Bert K. Waits, The Ohio State University
John Harvey, The University of Wisconsin-Madison

Edited by: Franklin Demana, The Ohio State University
John Harvey, The University of Wisconsin-Madison

ADDISON-WESLEY PUBLISHING COMPANY
Reading, Massachusetts • Menlo Park, California
New York • Don Mills, Ontario • Wokingham, England
Amsterdam • Bonn • Sydney • Singapore • Tokyo
Madrid • San Juan

Library of Congress Cataloging-in-Publication Data

Conference on Technology in Collegiate Mathematics (1st : 1988 :
    Ohio State University)
        Proceedings of the Conference on Technology in Collegiate
    Mathematics : the twilight of the pencil and paper / edited by
    Franklin Demana and Bert K. Waits.
            p.   cm.
        Bibliography: p.
        ISBN 0-201-50049-3 :
        1. Mathematics– Study and teaching (Higher)–Congresses.
    2. Mathematics–Data processing–Study and teaching (Higher)–
    Congresses. I. Demana, Franklin D., 1938– .II. Waits, Bert K.
    QA11.A1C657   1989
    510'.71'1–dc20   89-31611
    CIP

ABCDEFGHIJ–MU–89876543210

Table of Contents

CONFERENCE ON TECHNOLOGY IN COLLEGIATE MATHEMATICS
OCTOBER 27–29, 1988
THE OHIO STATE UNIVERSITY

Thursday, October 27, 1988

1:00 pm - 1:25 pm   Welcome

Dean William Kern
College of Mathematical
and Physical Sciences

1:30 pm - 2:20 pm   The Twilight of Paper-and-Pencil: Undergraduate Mathematics at the End of the Century

Professor Thomas Tucker
Colgate University

2:25 pm - 3:15 pm   Adapting the Maple Computer Algebra System to the Mathematics Curriculum

Profesor Stan Devitt
University of Saskatchewan

3:15 pm - 3:30 pm   Break

3:30 pm - 4:45 pm   Changes in Pedagogy and Testing When Using Technologies in College-Level Mathematics Courses

Professor John Harvey
University of Wisconsin-Madison

Reaction   *Professor Alan Osborne*
The Ohio State University

5:00 pm - 7:00 pm   Contributed Paper Session I (See separate listing for schedule.)

Room 1: Conference Theater
Room 2: Ohio Suites A-B-C
Room 3: Failer Lounge (3rd floor)

7:00 pm - 10:00 pm   Open software sharing session in the **Stecker Lounge** area third floor of the Ohio Union.

Friday, October 28, 1989

8:00 am - 8:30 am   Coffee outside the Ohio Union Conference Theater

8:30 am - 9:20 am   The Effect of Technology on Teaching College Mathematics

Professor Anthony Ralston
SUNY at Buffalo

9:25 am - 10:15 am   Using Computer Graphing to Enhance the Teaching and Learning of Calculus and Precalculus Mathematics

Professor Franklin Demana
The Ohio State University

10:15 am - 10:45 am   Break

10:45 am - Noon   Contributed Paper Session II (See separate listing for schedule.)

     Room 1: Conference Theater
     Room 2: Ohio Suites A-B-C
     Room 3: Buckeye Suites E-F-G

Noon - 1:15 pm   Break

1:15 pm - 2:05 pm   *Mathematica*™: A System for Doing Mathematics by Computer

Professor Stephen Wolfram
Wolfram Research Inc.

2:10 pm - 3:00 pm   Algebraic, Graphical, and Numerical Computing in Elementary Calculus: Report of a Project at St. Olaf College

Professor Paul Zorn
St. Olaf College

3:00 pm - 3:30 pm   Break

3:30 pm - 4:20 pm   Calculus and *Mathematica*™: An Electronic Calculus Textbook

Professor Jerry Uhl
University of Illinois - Urbana

Friday, October 28, 1988 (cont.)

4:20 pm - 4:30 pm   Break

4:30 pm - 6:30 pm   Contributed Paper Session III (See separate listing for schedule.)

        Room 1: Conference Theater
        Room 2: Ohio Suites A-B-C
        Room 3: Buckeye Suites E-F-G

7:00 pm - 8:00 pm   Reception in the Ohio State Faculty Club (South Oval) hosted by <u>Addison-Wesley Publishing Company</u>

8:00 pm - 10:00 pm   Open software sharing session in the **Stecker Lounge** area - third floor of the Ohio Union.

Saturday, October 29, 1988

8:00 am - 8:45 am   Coffee outside the Ohio Union Conference Theater

9:00 am - 11:00 am   **Workshops - By Ticket Only**

1. An Introduction to Maple on CMS

Professor Jeanette Palmiter
Kenyon College

2. Maple - An Electronic Blackboard on the Macintosh

Professor Stan Devitt
University of Saskatchewan

3. Using Computer Graphing to Enhance the Teaching of Calculus and Precalculus Mathematics

Professor Bert K. Waits
The Ohio State University

4. Linear Algebra and MATLAB

Professor Thomas Ralley
The Ohio State University

Saturday, October 29, 1988 (cont.)

9:00 am - 11:00 am **Demonstrations**

    1. Using the HP28C CAS Calculator to Enhance the Teaching of Calculus (Presented in the Conference Theater.)

             Professor Thomas Tucker
             Colgate University

    2. Using Hand-Held Graphing Computers in College Mathematics (Presented in Ohio Suites A-B-C.)

             Professor Gregory Foley
             The Ohio State University

    3. Calculus and *Mathematica* ™ : An Electronic Calculus Textbook (Presented in The Memorial Room.)

             Professor Jerry Uhl
             University of Illinois-Urbana

11:00 am - 11:30 am Break

11:30 am - Noon Getting Started with Symbolic Computation

             Professor Zaven Karian
             Denison University

12:10 pm - 1:10 pm Calculus, Technology, the National Science Foundation, and the Future

             Professor John Kenelly
             National Science Foundation
             (on leave from Clemson University)

Contributed Papers

(Manuscripts in Alphabetical Order)

Hollie Adams
William C. Bauldry
Charlene E. Beckmann
Michael J. Beeson
Dona V. Boccio
Christine Browning
James Burgmeier and Larry Kost
Chris K. Caldwell
Pamela Cemen and Jerry Johnson
Stephen D. Comer
Jere Confrey and Erick Smith
J. M. Anthony Danby
Abdi Darai
Robert Decker and John Williams
Gloria S. Dion
J. A. Eidswick
Arnold Feldman and Marjolein de Wit
John Frederick Fink, Margret Joft, David James
Michael B. Fiske
Charles G. Fleming and Judy D. Halchin
W. J. Hardell and J. J. Malone
Andrew Hugine, Jr. and Manuel Keepler
Jerry Johnson
Mark J. Kiemele
Larry E. Knop
Ernest J. Manfred and George Rezendes
James E. Mann, Jr.
John Masterson
James M. Meehan
Emily H. Moore
T. D. Morley
Umesh P. Nagarkatte and Shailaja U. Nagarkatte
Henry C. Nixt
Arnold Ostebee
Jerry W. Phillips
Boris D. Rakover
V. S. Ramamurthi
Peter Rice

John Selden and Annie Selden
Lester Senechal
Peter Shenkin
Larry E. Sherwood
Samuel W. Spero
Alan Stickney
Marvin Stotz
David L. Stout
Paul Thompson
Frederic W. Tufte
Charles Vonder Embse
Russell C. Walker
Gerald L. White
Agnes Wieschenberg
Daniel S. Yates
Lee L. Zia

Using Computer Graphing to Enhance the Teaching and Learning of Calculus and Precalculus Mathematics

Franklin Demana and Bert K. Waits

The Ohio State University

Today's technology is dramatically changing the way mathematics is valued and used in the "real world". Corresponding change that recognizes how technology can be used to enhance the teaching and learning of mathematics is needed. The technology based approach to the teaching and learning of mathematics described in this paper was piloted for two years and field tested for one year in The Ohio State University Calculator and Computer Precalculus (C²PC) Project [9]. The C²PC teachers are using two important technology driven instructional models. Students participate in an interactive lecture-demonstration instructional model in a classroom containing a single computer. Computer laboratories and classrooms where students have graphing calculators provide a setting for a guided-discovery instructional model. Teachers use a carefully prepared sequence of questions and activities to help students understand or discover important mathematical concepts.

The C²PC project was supported in pilot by the Ohio Board of Regents and British Petroleum and was supported in field test by the NSF [1]. Besides the authors, Alan Osborne and Gregory Foley from the College of Education are part of the C²PC project team. The C²PC approach and textbook, *College Algebra and Trigonometry, A Graphing Approach* [13] will be used in all college algebra and trigonometry courses at Ohio State beginning Autumn Quarter, 1989. Ohio State has been on the leading edge of using technology in freshman mathematics instruction for over 15 years [19].

Computer Based Graphing

The standard traditional approach uses arithmetic and algebraic information to produce graphs of functions and relations and to develop geometric intuition important in the study of calculus and advanced mathematics. The C²PC approach uses computers or graphing calculators (really pocket

computers) to quickly obtain accurate graphs to provide many more examples and further strengthen geometric understanding and foreshadow the study of calculus.

The graphing technology is under student control. Students can choose the viewing window or rectangle in which to display a graph. The *viewing rectangle* $[L, R]$ by $[B, T]$ is the rectangular portion of the coordinate plane determined by $L \leq x \leq R$ and $B \leq y \leq T$ (Figure 1). The $[-10, 10]$ by $[-10, 10]$ viewing rectangle is called the *standard viewing rectangle*.

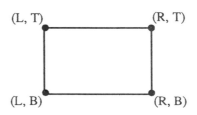

Figure 1. The Viewing Rectangle $[L, R]$ by $[B, T]$

Graphing calculators and the graphing software *Master Grapher* [21] used by C²PC students has important zoom-in and zoom-out features. *Master Grapher* contains powerful function, conic, polar, parametric, and two variable surface graphing utilities. Versions are available for the IBM, Apple IIe, c or GS, and the Macintosh computer. The graphs in this article were created using the Macintosh version of *Master Grapher*.

Zoom-in is a process of framing a small rectangular area within a given viewing rectangle, making it the new viewing rectangle, and then quickly replotting the graph in this new viewing rectangle. This feature permits the user to create a sequence of nested rectangles that "squeeze down" on a key point of a graph. Zoom-in is very useful for solving equations, systems of equations, inequalities, and for determining maximum and minimum values of functions. The graphing zoom-in process yields answers as accurate as any numerical method.

Zoom-out is a process of increasing the absolute value of the viewing rectangle parameters.

Preparation of this paper was supported in part by grants from the Ohio Board of Regents, British Petroleum of Ohio, and NSF grant number TPE-8751353. Conclusions and findings are those of the authors and do not necessarily represent the views of the funding agencies.

The zoom-out process is useful for examining limiting, end behavior of functions and relations and for determining "complete" graphs. A *complete graph* is the entire graph displayed in an appropriate viewing rectangle, for example, $x^2 + y^2 = 16$ in $[-10, 10]$ by $[-10, 10]$; or a *portion* of a graph displayed in an appropriate viewing rectangle which shows all of the important behavior and features of the graph, for example, $f(x) = x^3 - x + 15$ in $[-10, 10]$ by $[-10, 30]$. Of course, it is possible to create a function for which you cannot determine *one* viewing rectangle that gives a complete graph. Thus, several viewing rectangles may be needed to describe a complete graph.

The Role of Graphing in Calculus

Calculus textbook authors assume that students have control of graphing. Graphs of functions are often used to illustrate the definition of limit. For example, the following excerpt taken from a standard calculus textbook appeared right after the definition of limit of a function.

> "The function f defined by $f(x) = \frac{1}{x}$ provides an illustration in which no limit exists as x approaches 0. If x is assigned values closer and closer to 0 (but $x \neq 0$), $f(x)$ increases without bound numerically as illustrated by Figure 2."

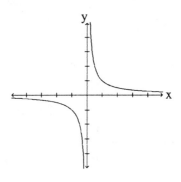

Figure 2. The graph of $f(x) = \frac{1}{x}$

Here the author assumes that students have enough understanding and control of the graph of $f(x) = \frac{1}{x}$ to use it to help understand the concept of limit. In reality, many calculus students are not able to produce a correct sketch of this graph. This important subtle notion of limit is further confounded by lack of understanding about graphs of functions.

The same textbook uses Figure 3 to illustrate the meaning of $\lim_{x \to a} f(x) = L$. Notice the depth of understanding about graphs required by this figure. Many entering calculus students are not even able to correctly produce the graph of a quadratic function.

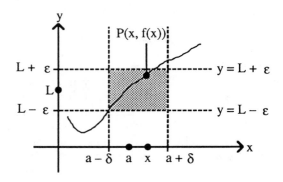

Figure 3. A Geometric Illustration of $\lim_{x \to a} f(x) = L$

The following test item (Figure 4) appeared on the Second International Mathematics Study (SIMS) 12th grade test [18]. United States 12th grade calculus students scored 29% on the pretest and 44% on the posttest on this item. United States precalculus students scored 22% on the pretest and 31% on the posttest on this item. The international posttest average score on this item was 58%.

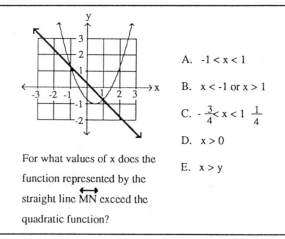

For what values of x does the function represented by the straight line $\overleftrightarrow{MN}$ exceed the quadratic function?

A. $-1 < x < 1$

B. $x < -1$ or $x > 1$

C. $-\frac{3}{4} < x < 1\frac{1}{4}$

D. $x > 0$

E. $x > y$

Figure 4. A SIMS Test Item.

All students must acquire a better understanding about graphs of functions in precalculus if more students are to be successful in calculus. It is our position that the proper use of technology in precalculus courses can significantly enhance student understanding and facility with graphing. This in turn will lead to better understanding of important concepts in calculus.

Changes in Mathematics as a Consequence of Technology

The role of algebraic manipulation. Some leaders in mathematics education call for drastic reduction in time spent on algebraic manipulation. We are convinced that the amount of time on this topic should be reduced, but are not ready to completely abandon algebraic manipulation. First hand observations have convinced us that the use of technology helps student gain new understanding about and provides motivation for important algebraic processes. Graphing gives a geometric interpretation to algebraic procedures. We have found that students are willing, even eager, to perform both arithmetic and algebraic procedures when those procedures answer questions generated by graphs.

Example 1. Determine the real zeros, the end behavior, and draw a complete graph of

$$f(x) = \frac{x^3 - 7x^2 - 12x + 54}{x - 1}.$$

Solution. It can be shown that the graph of f in Figure 5 is complete. One important connection students need to make is that the zeros of f are the same as the x-intercepts of the graph of f. Because the graph is complete, we can be sure that there are three real zeros. Zooming in around an x-intercept to find a zero helps establish and solidify this connection. There appears to be a zero near $x = -3$. We can use this geometric observation to motivate students to divide the numerator of f by $x+3$ or to compute $f(-3)$. Thus, arithmetic and algebraic ideas can be motivated by a graph.

If we zoom in around the zero of f between 2 and 3 a few times we obtain the graph in Figure 6 and can read that 2.354 is a reasonable approximation. We say that 2.354 is a zero of f with error at most 0.01, the distance between the horizontal scale marks in Figure 6.

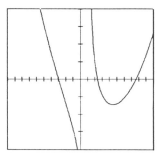

Figure 5. A Complete Graph of $f(x) = \frac{x^3 - 7x^2 - 12x + 54}{x - 1}$ in $[-10, 10]$ by $[-40, 40]$

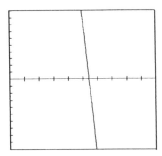

Figure 6. A Zoom-In View of a Zero of $f(x) = \frac{x^3 - 7x^2 - 12x + 54}{x - 1}$ in $[2.3, 2.4]$ by $[-0.1, 0.1]$

In general, the *error* in using a point (x, y) in the viewing rectangle $[L, R]$ by $[B, T]$ to approximate any point (a, b) in the viewing rectangle is *at most* $R - L$ for x and $T - B$ for y. Of course, there are better error bounds possible by overlaying a lattice in a viewing rectangle or by using scale marks appearing in a viewing rectangle. We can use zoom-in to find that the other positive zero of f is 7.645 with error at most 0.01.

The ability to quickly obtain a graph of $y = f(x)$ makes it very natural to discuss the geometric interpretation of solving the equation $f(x) = 0$ or the inequality $f(x) > 0$. Solving equations and inequalities using a zoom-in procedure soon becomes an easy geometric problem of finding x-intercepts, or when one graph is above or below another, or when one graph is above or below the x-axis.

If we zoom out a few times we can obtain the graph in Figure 7. Notice this graph looks very much like the graph of $y = x^2$. In fact, if we overlay the graph of $y = x^2$ the two graphs will appear coincident. This is the geometric meaning of end behavior; the behavior of a function for large $|x|$. The graph of $y = x^2$ is called an *end behavior model* of the rational function f. With selected examples of rational functions as a guide, students can be led to conjecture the end behavior of a rational function. Our students quizzed us for a way to determine, without using zoom-out, the end behavior of such functions. This discussion led to the introduction of the *end behavior asymptote* of a rational function. Their attention was held as we used *long division* to rewrite f as follows:

$$f(x) = x^2 - 6x - 18 + \frac{36}{x - 1}$$

Our students were then able to use this form to draw a correct rough sketch of f by replacing f by the end behavior *asymptote* $y = x^2 - 6x - 18$ for values of x away from $x = 1$ and the hyperbola $y = \frac{36}{x - 1}$ for x near 1. This algebraic procedure and added insight was due to the ability of students to produce large numbers of graphs in a short period of time. We have found that we can do more with algebraic manipulation when it is *not* the focus of a lesson.

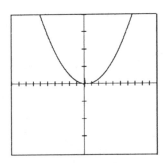

Figure 7. A Zoom-Out View of $f(x) = \frac{x^3 - 7x^2 - 12x + 54}{x - 1}$ in $[-100, 100]$ by $[-4000, 4000]$

Establishing connections among problem situations, algebraic representations, and geometric representations. The easy availability of geometric representations gives students and teachers the opportunity to explore and exploit the connections between algebraic and geometric representations and makes multiple representations of problem situations possible. Analyzing the problem situation through both algebraic and geometric representations deepens student understanding about the problem situation. Instead of the usual negative attitude about word problems, students gain more confidence about problems with the added technique of analyzing and solving them graphically. Word problems seem less mysterious and not as formidable with the addition of a geometric representation and powerful graphic problem solving methods.

Example 2. Squares of side length x are removed from a 8.5 inch by 11 inch piece of cardboard (Figure 8). A box with no top is formed by folding along the dashed lines in Figure 8.

(a) Express the volume V of the box as a function of x.

(b) Draw a complete graph of the algebraic model V.

(c) Which portion of the geometric model (graph) in (b) represents the problem situation.

(d) Determine x so that the box has maximum possible volume and find this maximum volume.

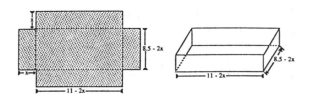

Figure 8. The Box Problem

Solution.

(a) The formula $V = LWH$ can be applied to obtain the volume V as a function of x. The height is x, the length is $11 - 2x$, and the width is $8.5 - 2x$. Thus, $V(x) = x(8.5 - $

$2x)(11-2x)$ is an *algebraic representation* of the volume as a function of x.

(b) A complete graph of $y = V(x) = x(8.5 - 2x)(11 - 2x)$ is shown in Figure 9. Students will need to experiment with different viewing rectangles until a complete graph is determined. Students would be expected to have had considerable computer based experience graphing cubic polynomials before investigating this problem.

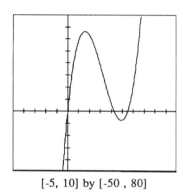

[-5, 10] by [-50 , 80]

Figure 9. A Complete Graph of $V(x) = x(8.5 - 2x)(11 - 2x)$

(c) The physical limitations inherent in removing a square of side length x implies that x must be positive. Because the smaller side of the rectangular piece of cardboard is 8.5 , $2x$ must be less than 8.5 , or x must be less than 4.25 . Thus, the values of x that make sense in this problem situation are $0 < x < 4.25$. This means that only the portion of the graph in Figure 9 in the first quadrant that is above the x-axis with $x < 4.25$ represents the problem situation. Therefore, the graph in Figure 10 is a *complete graph of the problem situation.*

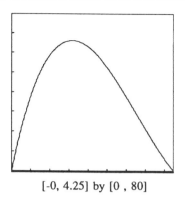

[-0, 4.25] by [0 , 80]

Figure 10. A Complete Graph of the Box Problem

(d) Figure 9 strongly suggests that there is a maximum value of V of about 66 and it occurs when x is about 1.6 . We find that considerable discussion is necessary for students to readily associate the coordinates of the "maximum point" with a solution to this real world "maximization" problem. First, the connection between the coordinate representation of points (a,b) of the graph of V and $b = V(a)$ must be established. That is, in (a,b), a represents a possible side length of a removed square and b the corresponding volume of the resulting box only for certain values of a and b. Such discussion helps establish the connections among the graphical representation, the algebraic representation $y = V(x)$, and the problem situation representation. Now, if (a,b) are the coordinates of the highest point, students can see that the maximum volume is $b = V(a)$ and that a is the side length of the removed square. Such connections must be carefully developed with many examples during the school year. Once this kind of activity is well established, it is easy to move to zoom-in as a procedure for determining very accurate solutions to these types of problems. Figure 11 illustrates the last viewing rectangle used in a zoom-in process. The figure shows that the volume is 66.14823 with error at most 0.0001 and the associated value of the side length of the removed square is 1.5854 with error at most 0.001.

6

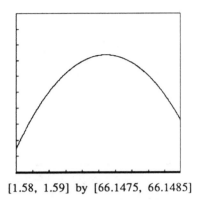

[1.58, 1.59] by [66.1475, 66.1485]

Figure 11. A Zoom-In View of the Relative Maximum of V

Example 2 illustrates how graphing can be used by precalculus students to foreshadow the study of calculus. Furthermore, graphing technology removes the barriers imposed by limited algebraic techniques available to precalculus and calculus students.

Problems need no longer be contrived. Realistic problems are accessible to students much earlier in their study of mathematics through the use of technology. Lack of familiarity or facility with algebraic techniques need no longer be a barrier to quality problem solving activity by students.

Example 3. A couple can afford to pay $600 per month for a 25 year home loan. What APR (annual percentage rate) interest rate will permit them to purchase a $65,000 home?

Solution. Let x be the *monthly* interest rate. Then $12x$ is the APR rate. It is not difficult to establish that x is given by [20]

$$65,000 = 600\frac{1 - (1 + x)^{-300}}{x}.$$

Because there is no closed form solution to this equation, a numerical method is required to find a solution. In fact, a graphing based method is quite natural. One way to solve $f(x) = g(x)$ graphically is to simply graph $y = g(x)$ and $y = f(x)$ in the same viewing rectangle and then look for points common to both graphs (points of intersection).

Let $f(x)$ be the left-hand side and $g(x)$ the right-hand side of the above equation. In this problem

it is particularly important to choose a reasonable first viewing rectangle. The problem situation indicates that we need only graph f and g in the first quadrant. (Why?) It must be established that the y values represents possible dollar amounts for the loan. Thus, the maximum y value for a viewing rectangle must be greater than 65,000. Because x is a monthly interest rate, it is reasonable to assume x is less than 0.1 (10% per month). Figure 12 shows complete graphs of

$$f(x) = 65,000 \quad \text{and} \quad g(x) = 600\frac{1 - (1 + x)^{-300}}{x}$$

in the $[0, 0.1]$ by $[0, 100,000]$ viewing rectangle. That there is only one solution is readily apparent. The graph in Figure 12 suggest x is about 0.01. Zoom–in can be used to determine that the monthly interest rate x is 0.008503 with error less than 0.00001 as shown in Figure 13. Thus, the desired APR rate of the home loan is 10.20%.

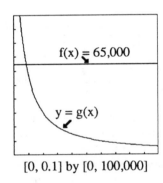

[0, 0.1] by [0, 100,000]

Figure 12. The graphs of $f(x) = 65,000$ and $g(x) = 600\frac{1-(1+x)^{-300}}{x}$

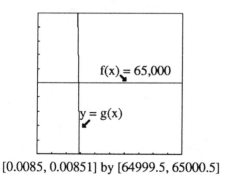

[0.0085, 0.00851] by [64999.5, 65000.5]

Figure 13. A Zoom-In View of the Loan Problem

Graphing surfaces (functions of two variables) is easily accessible to precalculus and calculus students when technology is used. Obtaining graphs of surfaces by hand is a difficult task for both student and teacher. Students have a good bit of trouble visualizing in three dimensions. Teachers have a hard time producing quick, accurate graphs of functions of two variables. The graphing software *Master Grapher* used in C²PC has a powerful utility which allows the user to obtain accurate graphs of functions of two variables. The user can obtain the graphs for $a \leq x \leq b$, $c \leq y \leq d$, and $e < z < f$, and then choose an arbitrary point in three dimensional space from which to view the graph. Once the first graph is drawn the points are stored in an array so that the graph can be redrawn quickly from different views. The user can choose any point in three dimensional space from which to view the graph. The resolution of a graph is under user control.

This three dimensional grapher allows the user to interactively explore the behavior of surfaces. Local maximum and minimum values of functions of two variables can be investigated graphically. The grapher can help students deepen understanding and intuition about functions of two variables. It can provide a geometric representation of multidimensional problem situations to go along with an algebraic representation. The connections between these two representations can be also explored and exploited to gain better understanding about problem situations in a manner similar to using a single variable function grapher.

First the user chooses a region of three dimensional space in which to draw a graph of a function of two variables. The set $\{(x,y,z) \mid a \leq x \leq b,\ c \leq y \leq d,\ e \leq z \leq f\}$ is called the *viewing box* $[a, b]$ by $[c, d]$ by $[e, f]$.

Next, the user decides how to view the graph contained in the selected viewing box. Two points can be selected. The point at which the user places his/her "eye" is called the *viewing point*. The point at which the view of the eye is directed is called the *aiming point*.

Example 4. A box with no lid has volume 6 ft³. Determine the dimensions of a box with minimum surface area.

Solution. Let x be the width of the box and y the length. The height h of the box is given by $h = \frac{6}{xy}$. If z is the surface area of the box, then

$$z = 2hx + 2hy + xy \quad \text{or} \quad z = \frac{12}{x} + \frac{12}{y} + xy.$$

Because x, y, and z must be positive, we need to investigate the graph only in the first octant. Figure 14 gives the graph of z in the viewing box $[0, 10]$ by $[0, 10]$ by $[0, 20]$ with aiming point $(5, 5, 10)$ and viewing point with spherical coordinates with respect to the aiming point of ρ (distance) $= 50$, θ (rotation) $= 30°$, and ϕ (elevation) $= 90°$.

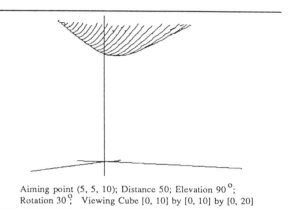

Aiming point (5, 5, 10); Distance 50; Elevation 90°; Rotation 30°; Viewing Cube [0, 10] by [0, 10] by [0, 20]

Figure 14. The graph of $z = \frac{12}{x} + \frac{12}{y} + xy$

Notice the graph in Figure 14 suggests the existence of a relative minimum. *Master Grapher* allows the user to interrupt the graphing process and read the coordinates of a point. We can use this technique to estimate the coordinates of the lowest point in Figure 14 to be $(2.3, 2.3, 15.7)$. The actual answer can be shown to be $(\sqrt[3]{12}, \sqrt[3]{12}, 3\sqrt[3]{144}) = (2.289\ldots, 2.289\ldots, 15.724\ldots)$.

Motion Simulation. In an article in *The American Mathematical Monthly*, Neal Koblitz [17] discussed four complicated real-world problems that are not typically solved in calculus textbooks. One problem is especially intriguing to us because it can be simulated and studied with a parametric equation graphing utility. Furthermore, an elementary, non-calculus geometric solution can be obtained with a function graphing utility. The solution suggested by Koblitz involves determining a *derivative to minimize* an expression and then involves *solving a complicated equation iteratively using Newton's method.*

Example 5. You are standing on the ground at point B (Figure 15), a distance of 75 feet from the bottom of a ferris wheel 20 feet in radius. Your arm is at the same level as the bottom of the ferris wheel. Your friend is on the ferris wheel, which makes one revolution (counterclockwise) every 12 seconds. At the instant when she is at point A you throw a ball to her at 60 ft/sec at an angle of 60°

above the horizontal. Take $g = -32$ ft/sec^2, and neglect air resistance. Find the closest distance the ball gets to your friend ... accurate to within $\frac{1}{2}$ foot. [17, p. 256]

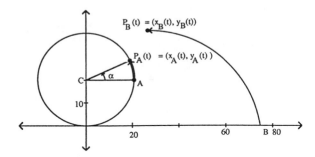

Figure 15. The Ferris Wheel Problem

Solution. The problem situation can be nicely simulated using the parametric equation graphing utility of *Master Grapher* or by using a graphing calculator with a short program to graph parametric equations [7]. The minimization problem can also be easily solved by a non-calculus, graphing zoom–in procedure. The ferris wheel is placed in a rectangular coordinate system with a diameter along the y-axis, the bottom at the origin, and the top at the point $(0, 40)$ (Figure 15). The ferris wheel is a circle with center $C(0, 20)$. Let t be the time in seconds the ball is in flight, $P_A(t) = (x_A(t), y_A(t))$ the position of the friend on the ferris wheel at time t, and $P_B(t) = (x_B(t), y_B(t))$ the position of the ball at time t. Notice that $P_A(0) = A = (20, 20)$ and $P_B(0) = B = (75, 0)$. It is easy to show, using only right triangle trigonometry and high school physics, that $P_A(t)$ and $P_B(t)$ are given by

$$x_A(t) = 20 \cos\left(\frac{\pi t}{6}\right)$$

$$y_A(t) = 20 + 20 \sin\left(\frac{\pi t}{6}\right)$$

and

$$x_B(t) = 75 - 30t$$

$$y_B(t) = 30\sqrt{3}t - 16t^2.$$

Our parametric graphing utility allows the student to graph any relation $(x(t), y(t))$ defined parametrically by specifying a t interval as [tmin, tmax], and a *viewing rectangle* $[a, b]$ by $[c, d]$. To simulate the problem situation, set tmin $= 0$ and let tmax take on different values and observe the paths of the friend and the ball. The two paths are plotted *simultaneously* producing an excellent simulation of the problem situation. Figure 16 show actual screen dumps from a Macintosh computer of the four simulations given by tmax $1, 1.5, 2,$ and 3. The same results can be obtained using *Master Grapher* on an IBM PC or Apple II computer or a graphing calculator [14]. Each simulation takes less than 5 seconds!

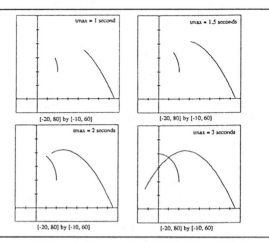

Figure 16. Simulations of the Ferris Wheel Problem for t max $1, 1.5, 2$, and 3.

Figure 16 indicates that the two *paths* have a common point. However, the values of t (time) that produces the common point are *different* for each set of parametric equations. It is really the *endpoints* of both curves that are of interest. The solution to the problem can be found by determining a value of t that *minimizes* the distance between the ball and the position of the friend on the ferris wheel. A parametric equation graphing utility that produces simultaneous graphs can be used to approximate the solution using "guess and check." It is easily shown that a value of tmax between 2.1 and 2.3 seems to yield the minimum distance.

The speed of computer graphing makes a "guess and check" simulation method possible and appropriate (some students even say "fun!") for mathematical exploration. The next figure gives a closer view of the solution. Each graph is drawn in the $[5, 15]$ by $[33, 42]$ viewing rectangle. The t range for the graphs are $[0, 2.1]$, $[0, 2.15]$, $[0, 2.2]$, and $[0, 2.3]$, respectively. It is easy to see from these graphs that the common point of the two paths is reached at different times. These static figures fail to do justice to the insight gained by observing the dynamic, "real time" computer generated simulation. That is, by observing simultaneously the position of the ball and the friend on the ferris wheel as t (time) increases.

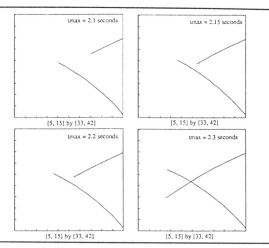

Figure 17. Simulations of the Ferris Wheel Problem for t max $2.1, 2.15, 2.2$, and 2.3

By estimating the distance between the end points of the two paths, a student can quickly determine that the minimum distance occurs when t is near 2.2 seconds, and that the actual minimum distance is probably less that 2 feet. Notice that the scale marks in Figure 17 are one unit in length.

The distance formula can be used to write the distance D between $P_A(t)$ and $P_B(t)$ as a function of time t.

$$D(t) = \sqrt{(x_A(t) - x_B(t))^2 + (y_A(t) - y_B(t))^2}$$
$$= [(20\cos\left(\frac{\pi t}{6}\right) - 75 + 30t)^2$$
$$+ (20 + 20\sin\left(\frac{\pi t}{6}\right) - 30\sqrt{3}\,t + 16t^2)^2]^{\frac{1}{2}}$$

Solving the equation involving the derivative ($D'(t) = 0$) is very difficult (try it "by hand"). However, the minimum value of the function D can be found easily and quickly by drawing a graph of $y = D(t)$ and using a graphing *zoom–in* process to determine the coordinates of the minimum. Figure 18 gives the graph of $y = D(t)$ for $0 \le t \le 3$. Figure 19 is the result after several iterations of the graphic zoom–in procedure.

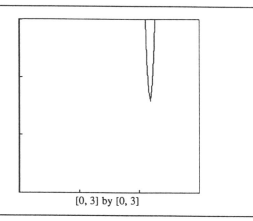

[0, 3] by [0, 3]

Figure 18. The Graph of $y = D(t)$

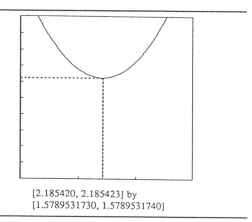

[2.185420, 2.185423] by
[1.5789531730, 1.5789531740]

Figure 19. A Zoom-In View of the Graph of $y = D(t)$

The coordinates $(2.1854214, 1.5789531736)$ of the local minimum of $y = D(t)$ can be read from Figure 19. The error in the first coordinate is at most $0.000001 = 10^{-6}$, the distance between horizontal scale marks. The error in the second coordinate is at most 10^{-10}, the distance between vertical scale marks. Thus, the minimum distance is 1.5789531736 feet with error at most 10^{-10} feet, and occurs when t is 2.1854214 feet with error at most 10^{-6} seconds.

Other interesting problems can be posed and solved using this computer simulation approach. For example, how could the angle of elevation be adjusted so that the ball comes within *very* easy catching distance (say 6 inches)? How close together will two balls come if thrown at the same time by two people facing each other (vary the distance between the two people, the angles of elevation, and the initial velocities)?

Students are often exposed for the first time to important topics such as parametric and polar equations and graphing 3 dimensional surfaces in calculus. This makes the task of a calculus teacher even more ominous because students must quickly learn and apply these ideas. Little or no attention is given to these topics in precalculus and calculus because of the difficulties of graphing such curves by hand. Technology permits students to quickly determine a graph and to discover the role of a parameter by experimentation. More realistic and interesting problems are possible because of the speed and power of technology and the fact that algebraic complication is not a factor when technology is used. Important new approaches (such as the computer simulation demonstrated in the previous example) are possible with technology. Students need a rich intuitive background prior to the study of calculus and other advanced mathematics and science courses. Technology deepens the level of student understanding and reduces the time necessary to acquire such understanding.

References

1. Demana, Franklin and Bert K. Waits. "Enhancing Problem Solving Skills in Mathematics Through Microcomputers." *Collegiate Microcomputer* 5 (February 1987), 72–75.

2. Demana, Franklin and Bert K. Waits. "Problem Solving Using Microcomputers." *The College Mathematics Journal* 18 (May 1987) 236–241.

3. Demana, Franklin and Bert K. Waits. "Solving Problems Graphically using Microcomputers." *The UMAP Journal* 8 (Spring 1987) 1–7.

4. Demana, Franklin and Bert K. Waits. "Microcomputer Graphing–A Microscope for the Mathematics Student." *School Science and Mathematics* 88 (March 1988), 218–224.

5. Demana, Franklin and Bert K. Waits. "Pitfalls in Graphical Computation, or Why a Single Graph Isn't Enough." *The College Mathematics Journal* 19 (March 1988) 177.

6. Demana, Franklin and Bert K. Waits. "Manipulative Algebra–The Culprit or the Scapegoat?" *The Mathematics Teacher* 81 (May 1988), 332.

7. Demana, Franklin and Bert K. Waits. "Using Graphing Calculators to Enhance the Teaching and Learning of Precalculus Mathematics: Teacher Inservice Worksheets for the CASIO Graphing Calculator." Department of Mathematics. Columbus (August, 1988).

8. Demana, Franklin and Bert K. Waits. "An Application of Programming and Mathematics: Writing a Computer Graphing Program." *The Journal of Computers in Mathematics and Science Teaching* 7 (Summer 1988), 47–48.

9. Demana, Franklin and Bert K. Waits. "The Ohio State University Calculator and Computer Precalculus Project: The Mathematics of Tomorrow Today!" *The AMATYC Review* 10 (Fall 1988), 46–55.

10. Demana, Franklin and Bert K. Waits. "Using Computer Graphing to Enhance the Teaching and Learning of Calculus and Precalculus Mathematics". MAA Minicourse Monograph (Atlanta, January, 1988); (Phoenix, January, 1989).

11. Demana, Franklin and Bert K. Waits. "Computers and the Rational Root Theorem–Another View." *The Mathematics Teacher* 80 (February, 1989), 124–125.

12. Demana, Franklin and Bert K. Waits. "A Computer Graphing Based Approach to Solving Inequalities." *The Mathematics Teacher*, (May, 1989), 327–331.

13. Demana, Franklin and Bert K. Waits. *College Algebra and Trigonometry, A Graphing Approach.* Addison-Wesley Publishing Company, Reading, MA, 1990.

14. Demana, Franklin, Bert K. Waits, Greg Foley, and Alan Osborne (1990) *Graphing Calculator and Computer Graphing Laboratory Manual* to accompany *College Algebra and Trigonometry, A Graphing Approach* Addison–Wesley. Reading, MA, 1990.

15. Demana, Franklin and Bert K. Waits. "A Hidden Case of Negative Amortization." In press (*The College Mathematics Journal*).

16. Demana, Franklin and Bert K. Waits. "Around the Sun in a Graphing Calculator." In press (*The Mathematics Teacher*).

17. Koblitz, Neal. "Problems That Teach the Obvious but Difficult," *The American Mathematical Monthly*, Volume 95, Number 3 (March 1988) 254–257.

18. Travers, Kenneth J., Director. *Second International Mathematics Study, Detailed National Report United States.* Champaign, Illinois: Stipes Publishing Company, 1985.

19. Waits, Bert K. and Joan Leitzel. "Hand Held Calculators in the Freshman Mathematic Classroom". *American Mathematical Monthly* 83. November, 1976.

19. Waits, Bert K. and Joan Leitzel. "Hand Held Calculators in the Freshman Mathematic Classroom". *American Mathematical Monthly* 83. November, 1976.

20. Waits, Bert K. "The Mathematics of Finance Revisited Through the Hand Calculator." In *Applications of Mathematics, 1979 Yearbook of the National Council of Mathematics*, The National Council of Teachers of Mathematics, Washington, D.C., 1987.

21. Waits, Bert K., and Franklin Demana. *Master Grapher* (Computer Software: IBM, Apple II, and Macintosh Versions). Addison–Wesley Publishing Company, Reading, MA, 1990.

Adapting the Maple Computer Algebra System to the Mathematics Curriculum*

J. S. Devitt

The University of Saskatchewan

Introduction

This year has seen considerable excitement in the mathematical community over the increased availability of symbolic algebra software for use in the classroom. It is not that the software has just emerged. In fact, such software has been under development since the early 1960's [7,6]. Rather, it is a combination of issues including the evolution of the software and hardware that combine to make sufficiently powerful versions of the software available to individual students and instructors.

Numerous attempts to include computation in the curriculum have been made over the past decade using a variety of software and hardware platforms. However, the ready availability of computational support for algebra is a key step towards providing an adequate computational environment for doing college and research level mathematics. It is not intended as a replacement for the computational mainstays of graphics or numerics any more than algebra is intended as a replacement for numerical computation or visual representation. The point is that algebraic computation plays a crucial role in modelling and in analyzing and setting up problems. In describing our visual and numerical observations we often rely heavily on algebra to state and analyze the underlying properties of the model? With symbolic algebra we provide the computational support necessary to derive and manipulate this underlying model.

Much of the past development effort for computer algebra systems has gone into extending them to carry out large and difficult algebraic computations. Considerable progress is being made in this direction but there is more to doing mathematics than completing large computations. As these systems mature it is possible to place some emphasis on using them to do mathematics. This places different demands on them.

For example, we may want to transform an indefinite integral using integration by parts rather than evaluating it. Computer algebra systems must allow us to manipulate and observe algebraic entities from a variety of points of view.

This article raises some of the issues about the style of use of computer algebra systems through a variety of examples based on the use of Maple^{TM} to solve particular problems arising from first year calculus. [1] These examples run using Maple version 4.2 on, for example, a standard Macintosh Plus with one megabyte of memory. [2] They emphasize how the Maple system can and has been adapted to the needs which arise in talking about and teaching mathematics. The examples do not come close to describing the full computational power of Maple. Our aim is to illustrate a philosophy or style of use appropriate for such tools.

Daily use in the classroom

Ideally we should be able to use a symbolic algebra system in an interactive session much like we would the blackboard and chalk. A live symbolic algebra session projected by means of an overhead projector can be the focus of every lecture. At this point the black board may still be used to sketch rough diagrams motivating concepts and to state theorems. However, almost all computation can be done by machine.

- Students observe first hand how to carry out certain basic tasks within the system.

- The instructor role model supports the overall goals of applying the available tools to the problem at hand. We don't imply that "Computers aren't much use for what I do!" by avoiding the machine at every possible turn.

- The focus of the lectures is on the structure of problem solving rather than on the details of every computation.

---

[1] Maple is a trademark of the University of Waterloo.

[2] Maple: Symbolic computation for the Macintosh is now available from Brooks/Cole Publishing Co., Monterey, CA. Macintosh is a registered trademark of Apple Computer Inc.

Through the use of online documents containing a mixture of explanations and system commands, we can extend this classroom experience to the laboratory.

When is a solution a solution?

A question which has been forcefully raised by the "Four Colour Theorem" [8] is "What is the nature of a mathematical solution?".

Symbolic algebra systems cause a real dilemma. Consider the solutions to the following problems proposed in the American Mathematical Monthly.

Compare the following sum and product [9]

```
#--> Sum(
#--> (-1)^(k-1)*binomial(N,k)*k/(1+(k-1)*a),
#--> #--> k=1..N)
#--> #--> =
#--> #--> Product( (k+1) / (k+1/a), k=1..N-1);
```

```
   N
 -----          (k - 1)
  \       (-1)           binomial(N, k) k
   )      ----------------------------------
  /               1 + (k - 1) a
 -----
 k = 1

          N - 1
         --------,
      ,  | |    k + 1
    =    | | ---------
         | | k + 1/a
         | |
         k = 1
```

Here the Sum and Product deliberately returned *unevaluated* though of course the algebra system may be unable to complete a computation and so would return unevaluated of its own accord.

We next force Maple to try evaluating the above expression using a routine which effectively replaces "Sum" by "sum", etc. [3]

```
#--> Eval(");
```

```
          1 + a           1 + a
N GAMMA(-------)  GAMMA(------- - 1/a + N - 1)
          a               a
---------------------------------------------
       1 + a                 1 + a
 GAMMA(------- - 1/a) GAMMA(------- + N - 1)
         a                     a
      GAMMA(N + 1) GAMMA(1 + 1/a)
  = ------------------------------
           GAMMA(N + 1/a)
```

---

[3] There is a convention that uppercase equivalents of lowercase commands are inert. **Eval()** is a function in the student package.

We appear to be successful and the answer can be further simplified further as:

```
#--> simplify(");

        1 + a
 GAMMA(-------) GAMMA(N + 1)
        a
 ---------------------------
               1 + N a
         GAMMA(---------)
                  a

               1 + a
         GAMMA(-------) GAMMA(N + 1)
                a
   = ---------------------------
                 1 + N a
           GAMMA(---------)
                    a
```

The name """ stands for "the previous expression".
Or consider this ordinary differential equations problem.[10]

```
#--> ODE := y3 = y1*(3*y2^ 2 - y3*y1);

                             2
             ODE := y3 = y1 (3 y2 - y3 y1)

#--> y1 := diff(y(x),x);

                       d
             y1 := ---- y(x)
                       dx

#--> y2 := diff(y(x),x,x);

                        2
                        d
             y2 := ----- y(x)
                        2
                       dx

#--> y3 := diff(y(x),x,x,x);

                        3
                        d
             y3 := ----- y(x)
                        3
                       dx

#--> #--> dsolve(ODE,y(x));

y(x)

        2      2           2       2   2      1/2
   = ((x  exp(C) + 2 x exp(C) C1 + C1 exp(C) - 1)

          2
       / exp(C)

       - C1
```

```
   ln(2

     exp(C)

         2       2          2        2       2
        (x  exp(C)  + 2 x exp(C)  C1 + C1  exp(C)  - 1)

             ^ (1/2)
                    2          2
        + 2 x exp(C)  + 2 exp(C)  C1)

       1 / exp(C))

      1/2
  (-1)    exp(C)

 + ln(2

     exp(C)

         2      2         2       2       2
        (x exp(C)  + 2 x exp(C) C1 + C1 exp(C) - 1)

           (1/2)
                  2           2
        + 2 x exp(C)  + 2 exp(C)  C1)

        1/2
   (-1)    C1

 + C2
```

Though the final answer is not as elegant as we might like, it can be simplified.

`#--> subs( exp(C)=K,");`

```
y(x)
          2 2      2        2 2   1/2
         (x K + 2 x K C1 + C1 K - 1)
   = (----------------------------------
                      2
                      K

    - C1

          2 2     2        2 2    1/2       2
   ln(2 K (x K + 2 x K  C1 + C1 K   - 1)    + 2 x K
         2
       + 2 K C1)

      1 / K)

      1/2
  (-1)    K

          2 2     2       2 2    1/2       2
 + ln(2 K (x K + 2 x K C1 + C1 K    - 1)   2+ 2 x K
         2
       + 2 K C1)
```

```
        1/2
  (-1)     C1
```

```
 + C2
```

In both these examples, we observe that:
- The user has control over what evaluation is attempted and when it takes place.
- The system supports a high level of abstraction providing a near 1-1 correspondence with the tasks at hand.
- The system is fairly powerful.
- We will need to develop verification techniques if we are to find these answers acceptable.

The verifications refered to above will benefit from machine computation as well. For example, we could substitute the solution to the differential equation back into the original equation using the command `subs(",ODE);`.

Some sample symbolic computations

We begin by reviewing some of the basic capabilities of Maple. All the examples you see here run on the basic Mac Plus with only one megabyte of memory. The actual Maple commands used are indicated here by use of the prompt `#-->`.

The overall capabilities of a symbolic algebra system can be quite impressive. First noticed may be the freedom from concern about machine specific restrictions such as limits on the size of integers and refusals to represent rationals exactly.

```
#--> 2/3;
```

```
                2/3
```

```
#--> 100!;
```

```
93326215443944152681699238856266700490715968264381621468592\
96389521759999322991560894146397615651828625369792082722375\
825118521091686400000000000000000000000000000
```

```
#--> evalf(Pi,100);
```

```
.3141592653589793238462643383279502884197169399375105820974\
9445923078164062862089986280348253421170680
   * 10 ^ 1
```

Essential operations such as integer greatest common divisors and factoring are made available interactively.

```
#--> 3^ 500;
```

```
36360291795869936842385267079543319118023385026001623040346\
03583258060019158389548419850826297938878330817970253440385\
57528559315170130661429924309165620257800217712478476434501\
25342836565813209972590371590152578728008385990139795377610\
001
```

```
#--> igcd(",5^ 300);
```

```
                1
```

Our computational experience might have convinced us that we should never leave a variable unassigned, but here the variable may well be the data.

```
#--> s1 := Sum(i^ 20,i=1..n);
```

$$
s1 := \sum_{i=1}^{n} i^{20}
$$

```
#--> s2 := Sum(1/i^ 3,i=1..infinity);
```

$$
s2 := \sum_{i=1}^{infinity} \frac{1}{i^3}
$$

Unevaluated expressions like these can be evaluated. exactly by Maple.[4]

```
#--> Eval(s1);
```

```
             21              20              19
1/21 (n + 1)   - 1/2 (n + 1)   + 5/3 (n + 1)

              17   1292          15           13
   - 19/2 (n + 1)  + ------ (n + 1)   - 323 (n + 1)
                      21

     41990          11   223193          9              7
   + ------- (n + 1)   - -------- (n + 1) + 6460 (n + 1)
      33                 63

     68723         5   219335          3   174611
   - ------- (n + 1) + -------- (n + 1) - -------- n
      10                63                 330

     174611
   - --------
      330
```

```
#--> Eval(s2);
```

```
                    Zeta(3)
```

The Maple system generally assumes it is working with polynomials having exact rational number coefficents. However, we can choose to work with algebraic extensions of the rationals.

```
#--> f := expand( (x^ 2 - (3*alpha-1)*x + alpha^ 2)
#--> * (x-2*alpha)^ 2 );

          4     3              2    2          3    3
f := x - 7 x alpha + 17 x alpha - 16 alpha x + x

              2        2         4
   - 4 alpha x + 4 x alpha + 4 alpha
```

---

[4] the Eval() command is a command from the student package which is part of the basic maple system and is used to convert from unevaluated forms of expression to ones which evaluate.

```
#--> g := expand( (x^ 2+x+1)^ 2 * (x-2*alpha)^ 2 );
       6     5          4     2    5     4
g := x - 4 x alpha + 4 x alpha + 2 x - 8 x alpha

         3     2     4       3            2       2      3
   + 8 x alpha + 3 x - 12 x alpha + 12 x alpha + 2 x

           2            2   2               2
   - 8 alpha x + 8 x alpha + x - 4 alpha x + 4 alpha
#--> alpha := RootOf(x^ 5+x^ 3+1,x);
                              5    3
                alpha := RootOf(_Z + _Z + 1)
#--> evala(Gcd(f,g));
    2           5    3            5    3    2
   x - 4 RootOf(_Z + _Z + 1) x + 4 RootOf(_Z + _Z + 1)
```

The library of about a thousand routines organized into man subdirectories. subdirectories. Most of these are automatically loaded when needed. However, some mechanism must be provided for focussing on a specific subject area such as linear algebra, first year calculus, or number theory. An environment for linear algebra can be quickly constructed by the `with()` command which defines (but does not load) a library of routines on this specific topic.

```
#--> with(linalg);
Warning: new definition for trace
[jacobian, band, smith, add, vectdim, trace, gausselim,

   orthog, laplacian, transpose, cond, rowspace, leastsqrs,

   scalarmul, adj, genmatrix, hadamard, dotprod, mulcol,

   swaprow, vandermonde, submatrix, det, swapcol,

    singularvals, bezout, definite, hilbert, range, mdet,

   kernel, linsolve, indexfunc, diverge, hessian, addrow,

   sylvester, multiply, adjoint, nullspace, mulrow,

   inverse, rank, subvector, rowdim, toeplitz, angle,

   ismith, eigenvals, colspace, addcol, crossprod, grad,

   norm, coldim, curl, permanent]
```

Once defined the routines will be automatically loaded when needed. These generally apply to matrices with symbolic entries.

```
#--> jacobian([1/sin(x+y)],[x,y]);
array ( 1 .. 1, 1 .. 2,

           cos(x + y)        cos(x + y)
   [- --------------, - --------------]
              2                 2
           sin(x + y)        sin(x + y)

)
```

Much of the work in designing computer algebra systems has gone into developing sophisticated and powerful methods to attack large computational problems. Indefinite integration has received considerable attention. The Risch integration algorithm is used for many classes of problems.

```
#--> f1 := x / (exp(x) + 1);
```

$$f1 := \frac{x}{\exp(x) + 1}$$

```
#--> int(f1,x);
```

$$\int \frac{x}{\exp(x) + 1}\, dx$$

Though it is not obvious in this example, Maple has actually *proved* that the above integral cannot be expressed in terms of elementary functions.[5] Again we are faced with the question "What constitutes a proof?". Our ability to comprehend and verify solutions becomes essential. The following slightly different integral can be expressed in terms of elementary functions;

```
#--> f2 := 1 / (exp(x) + 1);
```

$$f2 := \frac{1}{\exp(x) + 1}$$

```
#--> int(f2,x);
```

$$- \ln(\exp(x) + 1) + \ln(\exp(x))$$

as can the following integral.

```
#--> num := x*(x+1) * ( (x^ 2*exp(x^ 2)^ 2 - log(x+1)^ 2)^ 2
#--> + 2*x*exp(x^ 2)^ 3 * (x - (2*x^ 3+2*x^ 2+x+1)*log(x+1)) ):
#--> den := ( (x+1)*log(x+1)^ 2 - (x^ 3+x^ 2)*exp(x^ 2)^ 2 )^ 2:
#--> f3 := num/den;
```

```
f3 :=

   x (x + 1)

         2    2 2           2 2
     ((x  exp(x ) - ln(x + 1) )

       + 2

               2 3         3     2
         x exp(x ) (x - (2 x + 2 x + x + 1) ln(x + 1))

       )

                    2       3    2       2 2 2
   / ((x + 1) ln(x + 1) - (x + x ) exp(x ) )
```

---

[5] An unevaluated return in Maple generally means that Maple has been unable to solve the problem. This integral happens to fall into the class of integrals for which Maple has a complete algorithm so we can make the stronger claim.

```
#--> #--> int(f3,x);

                        2
               x exp(x ) ln(x + 1)
x - ln(x + 1) + ---------------------------
                  2      2 2         2
                 x exp(x ) - ln(x + 1)

                           2
   + 1/2 ln(ln(x + 1) + x exp(x ))

                           2
   - 1/2 ln(ln(x + 1) - x exp(x ))
```

For definite integration problems, we can automatically invoke numerical techniques in situations that warrant it.

We first represent the problem without even attempting to evaluate it.

```
#--> r1 := Int( exp(-t) / sqrt(1-t^ 2),
#--> t = -1..1 );

                   1
                  /
                  |      exp(- t)
          r1 :=   |   ------------- dt
                  |           2 1/2
                  /       (1 - t )
                 -1
```

We next force maple to attempt the evaluation (ie. to use `int()`).

```
#--> Eval(");

              1
             /
             |    exp(- t)
             |  ------------- dt
             |          2 1/2
             /      (1 - t )
            -1
```

Finally, as this did not evaluate we apply numerical techniques directly to the unevaluated result. The algebraic representation of the integrand allows for a sophisticated analysis of the numerical problem and the use of various tranformations to complete the computation.

```
#--> evalf(");

              3.977463261
```

A platform for stepwise refinement

Computer algebra systems have a tremendous potential to support stepwise refinement. For example in Maple we can choose to examine the problem at various levels of detail.

Consider, for instance, the derivation of Simpson's rule for numerical integration (see Figure 1). Ultimately we end up with formulas like

```
#--> simpson(f(x), x=0..2,2);

              1/3 f(0) + 1/3 f(2) + 4/3 f(1)

#--> simpson(f(x), x=0..2,4);

 simpson := proc( F, dx ) local a,b,f,i,h,n,rg,x;

     #    dx is an equation of the form x=a..b.
     #    n is an optional 3rd parameter
     #            for number of steps.
     x := op(1,dx); rg := op(2,dx); # dx is x=a..b
     if nargs > 2 then n := args[3] else n := 4 fi;
     a := op(1,rg); b := op(2,rg); h := (b-a)/n;
     # define a function and evaluate appropriately
     f := readlib(unapply)( F , x );
     h/3* ( f(a) + f(b)
         + 4*sum( f(a + (i*2-1)*h),i=1..n/2 )
         + 2*sum( f(a + (i*2)*h),i=1..(n/2)-1 ) );
 end:
```

Figure 1: Simpson's Rule

```
     1/6 f(0) + 1/6 f(2) + 2/3 f(1/2) + 2/3 f(3/2) + 1/3 f(1)
```

or more generally,

```
#--> simpson(f(x),x=0..2,n);

2/3

                       1/2 n
                       -----
                        \            2 i - 1
    (f(0) + f(2) + 4 (   )   f(2 ----------))
                        /             n
                       -----
                       i = 1

           1/2 n - 1
            -----
             \
     + 2 (    )           f(4 i/n)))
             /
            -----
            i = 1

     1 / n
```

Too often, our testing consists of simply asking the students to recall this formula. But clearly, machines can compute this and the formula can be looked up. Even for our poorer students, the interest must lie in grasping the underlying mathematical model.

We need little more than the notion that a polynomial of degree two can be used to represent an arbitrary function.

We begin with three points. The chosen values of x and the corresponding values for $f(x)$ are shown in the lists below.

```
#--> xvals := [0,1,2];

                    xvals := [0, 1, 2]
```

```
#--> yvals := map(f,xvals);
```

$$yvals := [f(0), f(1), f(2)]$$

Under the right conditions (the ones that usually hold in class) these three points uniquely define a polynomial of degree 2. The polynomial and the function must have these three points in common. The polynomial is given by:

```
#--> interp(xvals,yvals,x);
```

$$\begin{array}{ccc} 2 & 2 & 2 \\ \end{array}$$
```
1/2 f(2) x  - 1/2 f(2) x - f(1) x  + 2 f(1) x + 1/2 f(0) x

   - 3/2 f(0) x + f(0)
```

We complete the derivation of Simpson's rule by integrating the resulting polynomial instead of $f(x)$.

```
#--> Int(",x=0..2);
```

```
                    2                2
Int(1/2 f(2) x  - 1/2 f(2) x - f(1) x  + 2 f(1) x

            2
   + 1/2 f(0) x  - 3/2 f(0) x + f(0),

   x = 0 .. 2)
```

```
#--> expand(");
```

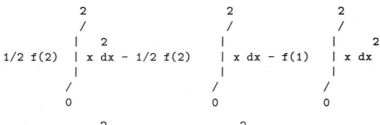

```
#--> Eval(");
```

$$1/3 f(0) + 1/3 f(2) + 4/3 f(1)$$

The Maple commands form a high level description of the solution.

```
xvals := [0,1,2];
yvals := map(f,xvals);
interp(xvals,yvals,x);
```

```
Int(",x=0..2);
expand(");
Eval(");
```

Stepwise refinement is implicit in the question "What really happens when we use `interp()`?".
We can proceed by defining F to be an arbitrary polynomial of degree 2,

```
#--> F := <a*x^ 2 + b*x + c|x>;
                           2
                F := <a x + b x + c|x>
```

We obtain three equations by evaluating both f and F at the three known points and solve the set of equations for the three unkown constants. The result is a quadratic polynomial, say R.

```
#--> eq:={f(0)=F(0), f(1)=F(1), f(2)=F(2)};

   eq := {f(0) = c, f(1) = a + b + c, f(2) = 4 a + 2 b + c}

#--> solve(eq,{a,b,c});

{c = f(0), b = - 1/2 f(2) + 2 f(1) - 3/2 f(0),

   a = 1/2 f(2) - f(1) + 1/2 f(0)}

#--> R := subs(",F(x));

 2
R := (1/2 f(2) - f(1) + 1/2 f(0)) x

   + (- 1/2 f(2) + 2 f(1) - 3/2 f(0)) x + f(0)
```

If we are not comfortable with solving systems of linear equations there is a simple explanation. To solve the system from first principles, choose an unsolved equation, solve it, and update all the other equations with the implied substitution. Repeat this process until done. A possible first step appears below.

```
#--> solve( f(0) = F(0) , {c} );

                        {c = f(0)}

#--> eq := subs(",eq);

eq :=

    {f(0) = f(0), f(1) = a + b + f(0),

       f(2) = 4 a + 2 b + f(0)}
```

Having derived the model, we can now observe how it behaves in specific instances using just a few additional commands.

```
#--> F := makeproc( R , x );

F := proc (x) option operator; (1/2*f(2)-f(1)+1/2*f(0))*x**2
+(-1/2*f(2)+2*f(1)-3/2*f(0))*x+f(0) end

#--> f := < x^ 4 + 3 | x>;

                          4
              f := <x + 3|x>

#--> plot( {f(x),F(x)}, x=0..2);
```

See Figure 2.
Additional cases may also be considered.

```
#--> f := < x + 4 | x >;
#--> plot(f(x),F(x),x=0..2);
#--> f := sin;
#--> plot(f(x),F(x),x=0..2);
```

We have been able to design the model and observe its behaviour all in the same environment.

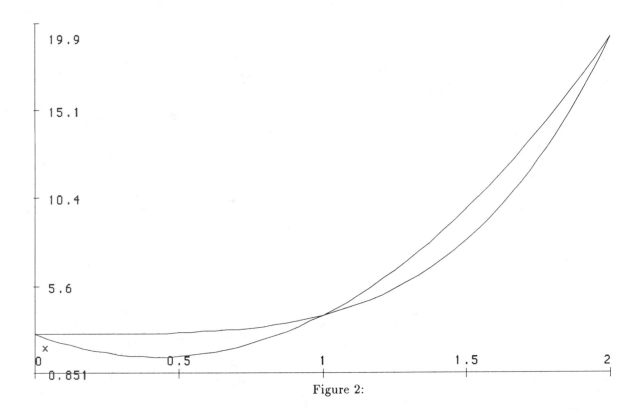

Figure 2:

Accessibility

A note is in order about user interfaces and the Macintosh release of Maple. While much of what has been discussed above is independent of the user interface, depending to a large extent on the functionality of the commands, user interfaces cannot be ignored. The easier the commands are to construct, the more likely it is that we are to try a few "computer experiments" when starting almost any new problem - the more effective our mathematical laboratory will become.

Maple: Symbolic computation for the Macintosh now runs on the Macintosh[R] under Finder[TM] (and MultiFinder[TM])[6] This is a complete version of Maple version 4.2 and performs well on a standard Mac Plus with only one megabyte of memory. It includes plotting and is known to run well under a variety of networking software.

The user interface of Maple on the Macintosh provides you with a *session window* (already open) to capture all the results of typical Maple commands. There are three exceptions to the use of this window for displaying results. The **help()** command displays its results in a read-only text window. Similarly the **plot()**

---

[6] Macintosh is a registered trademark of Apple Computer Inc. Finder and Multifinder are trademarks of Apple Computer Inc. Maple is a trademark of the University of Waterloo while Maple for the Macintosh is published by Brooks/Cole Publishing Co., Pacific Grove, Ca.

command produces its results in a special *plot window*. Finally there is a *status window* that indicates the amount of system resources you have used. See Figure 3.

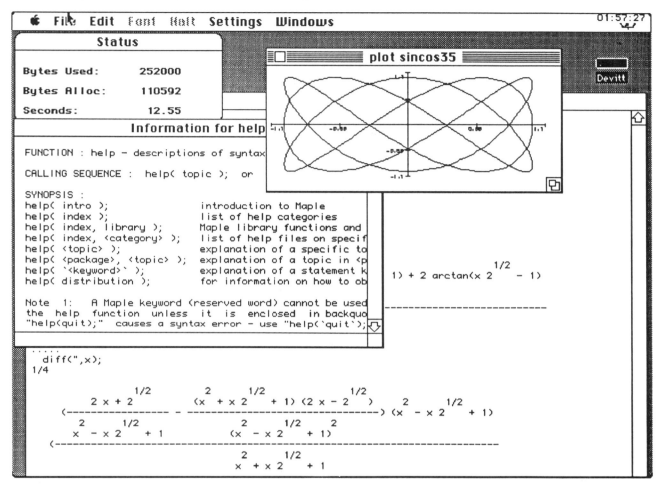

Figure 3: A Sample Screen

If you were never to venture beyond the use of this one session window you could pretend that you were using Maple on a typical large mainframe. You would simply type in your commands using the Macintosh *enter* key at the end of each line, rather than the *return* key[7]. However, a big advantage to the Mac style implementation over others is that you can edit and reuse previous commands all directly from within your Maple session.

The various windows can be thought of as multiple clipboards. Classroom examples are easily prepared and stored on a file server. After solving the problem using Maple, save the Maple instructions together with comments etc. outlining the solution process in file in the class library. By opening such files from within Maple, students have access to these documents. They may try out examples immediately by *selecting* and *entering* the associated commands. The students can easily modify the examples and save their own versions, either on diskettes or on file space provided on the server.

The standard plot command in Maple 4.2 directly supports the plotting of Maple expressions and functions, parametric plots, polar coordinates, and curves generated by lists of points. It uses adaptive techniques and is robust in that singularities, or wildly fluctuating function values are generally handled well. Several output devices are supported including regis, ln03, vt100, tektronics, postscript and ascii character plots. Several

---

[7] These keys have distinct functions on the Macintosh keyboard.

curves may be plotted on one graph by simply passing a set of expressions as the first argument to the plot function. All such plots may be saved as a Maple data structure, and later modified.

Maple for the Macintosh also uses a new output format for plots (`plotdevice := mac;`). By default, plots created by Maple on the Macintosh are displayed in a special *plot window*. This window can be resized, printed, or saved directly. See Figure 4. The graphs can be cut and pasted to the scrapbook or to other applications directly. Figure 5 shows the result of enhancing the graph generated by Maple in Figure 4 through a few simple commands in MacDraw®.[8]

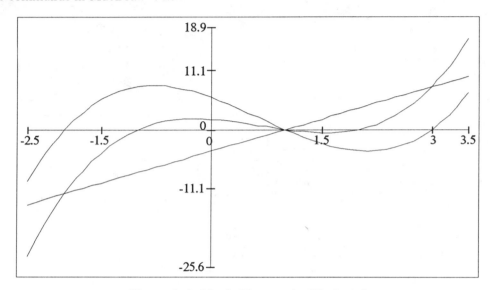

Figure 4: A Maple Plot on the Macintosh

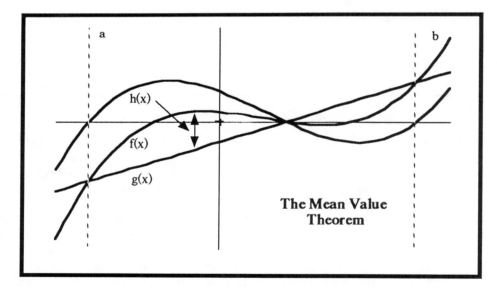

Figure 5: An Enhanced Maple Plot

Maple is small enough that it may be used effectively under MultiFinder. On a two megabyte machine, you can have Maple and at least one other application running simultaneously. Cutting and pasting between applications becomes quick and elegant.

---

[8] This is a registered trademark of Claris Corporation.

While much work remains to be done on the development and design of effective experimental laboratories for mathematics, all the key ingredients are in place. We are starting to address the issue of "how" a computer algebra system should be used for doing experimental mathematics. Maple's student package is a specific attempt to address this issue of style. The Maple Macintosh interface is an important first step and serves as a prototype for emerging interfaces on other work stations. The next decade promises to be an exciting time for mathematics and mathematics education.

References

[1] Char, et. al., *The Maple User's Guide* , WATCOM Publications Ltd., 415 Phillip St., Waterloo, Ontario, 1985

[2] Devitt, J. S., *Teaching First Year Calculus through the Use of Symbolic Algebra* , unpublished classroom notes, The University of Saskatchewan, October, 1987.

[3] Devitt, J. S. *The Full Power of Maple on a Mac Plus?*, the Maple Newsletter, vol 4, 1989, pp.4-9.

[4] *Toward a Lean and Lively Calculus* , MAA Notes, No. 6, edited by R. G. Douglas, The Mathematical Association of America, 1986

[5] *Calculus for a New Century: A pump not a filter*, MAA Notes, No. 8, edited by L. A. Steen, The Mathematical Association of America, 1988

[6] Hearn, Anthony C., *Reduce 2: A system and language for Algebraic Manipulation*, Proceedings of the Second Symposium on Symbolic and Algebraic manipulation, Association for Computing Machinery, 1971, pp. 128-133.

[7] Rand, R. H., *Computer Algebra in Applied Mathematics: An Introduction to MACSYMA*, Pitman Pub. Ltd. London, 1984.

[8] Ore, Oystein. *The Four Color Problem*, Academic Press, New York, 1967.

[9] Zaslove, E. *Problem E2971*, The American Mathematical Monthly, 1982.

[10] Sadowski, J. *Problem E3022*, The American Mathematical Monthly, 1983.

Using Hand-Held Graphing Computers in College Mathematics

Gregory D. Foley

The Ohio State University

The distinction between calculator and computer is no longer clear. In 1984, Brophy and Hannon wrote that, "In mathematics courses, computers offer an advantage over calculators in that they can express results graphically as well as numerically, thus providing a visual dimension to work with variables expressed numerically" (p. 61). This advantage of computers over calculators disappeared in early 1986 when Casio introduced the *fx-7000G*, a programmable scientific calculator with interactive graphics, that is, a hand-held graphing computer. Other similar yet more powerful and sophisticated machines soon followed: Casio's *fx-7500G*, and *fx-8000G*, Hewlett-Packard's *28C* and *28S*, and Sharp's *EL-5200*. In January 1986, only months before the introduction of the *fx-7000G*, at the Tulane conference on calculus reform, Tucker et al. (1986) had considered calculus curriculum revision based on the levels of technology required for various types of computational support. At that time numerical computation could be done on hand calculators, but interactive graphics and symbolic manipulation required micro or mainframe computers. Because of access problems, they shied away from recommending computer-based graphics and symbolic manipulation as a part of mainstream calculus. "The participants at that conference had no idea that a Casio *fx7000-G* [sic] or an *HP-28C* was looming on the horizon" (Tucker, 1987, p. 5).

This paper is an extension of a workshop with the same title presented on October 29, 1988 at the Conference on Technology in Collegiate Mathematics at The Ohio State University in Columbus. Preparation of this paper was supported in part by the *Mathematics Through Technology: Establishing Concepts and Skills of Graphing and Functions in Grades 9 through 12* grant from the National Science Foundation (TPE-8751353). This grant and others from British Petroleum (formerly, Standard of Ohio), the Ohio Board of Regents, and The Ohio State University have provided funding for the Calculator and Computer PreCalculus Project. The opinions and conclusions herein expressed are solely those of the author.

Hand-held graphing computers combine the capabilities of a scientific calculator, a programmable computer, an interactive-graphics computer system, and in the case of the Hewlett-Packard calculators, a limited computer mathematics system that performs symbolic manipulation. These machines are powerful tools for mathematical experimentation and exploration. They are too small to lend themselves to typewriter-style keyboarding, and ultimately this may be the lone distinction that remains between hand-held and micro computers.

What Hand-Held Computers Can Do

The capabilities of these pocket-sized computers have been described in some detail elsewhere (see, e.g., Foley, 1987a, in press; Michel, 1987; Potter, 1987a; Tucker, 1987). Muciño (1988) has even provided a "buyer's guide" for these machines. This section provides a summary of the features of the Casio, Sharp, and Hewlett-Packard hand-held graphing computers as a reference for college mathematics faculty who are planning curriculum and instruction to take advantage of these versatile computational tools.

The hallmarks of these supercalculators are (a) large display screens, (b) interactive graphics, and (c) on-screen programming. When choosing to buy, use, or design a curriculum around one of these machines all three of these factors should be carefully considered. Durability and price are also important; so included below are some notes about the durability and lowest prices as of December 1988 for each model.

The Casio *fx-7000G*, *fx-7500G*, and *fx-8000G* all have eight-line text displays and graphics viewports that are 63 rows by 95 columns of pixels. They can readily produce the graphs of functions, and with some programming, the graphs of polar equations, parametric equations, conics, and even three-dimensional graphs. The viewing rectangle and the scaling units are set using the Range feature. The Trace command allows pixel-to-pixel movement along the most recently drawn function; the Casio displays the x- or y-coordinate associated with each pixel along the way. The automatic zoom feature or the Factor command can be used to zoom-in or zoom-out about a plotted or traced-to point, or as a default, about the center of the current viewing rectangle. Early versions of the *fx-7000G* did not have automatic zoom, but now all models have this important feature. Other capabilities include statistical features and binary, octal, and hexadecimal computation and conversion. Commands and programs can be selectively changed and re-executed. Casio syntax closely parallels standard algebraic syntax. The *fx-7000G* ($50) has only 0.4K memory. The *fx-8000G* ($70) has 1.4K regular memory plus an additional 1.9K for its file editor. It has an input buffer that saves the last prior command (this comes in handy if you accidently hit a wrong key). It can be linked to a printer or to a tape recorder to save programs externally. The *fx-7500G* ($65) has the fastest graphics and the most memory (4K) of the three.

The Sharp *EL-5200* ($75) has four lines of textual display and 32×96 pixels of graphics display. It has an extensive input buffer that saves several prior commands. The graphics are slower than on the Casios, and the user does *not* see the graph being drawn. In addition to all of the features of the Casio, the Sharp permits automatic setting of the *y* viewing-rectangle parameters and has a scrolling screen for tracing along a graph beyond the current viewing rectangle, plus it has built-in equation-solving and matrix capabilities. The programming is awkward compared to the Casio, but the memory capacity is larger (8K) even than the Casio *fx-7500G*. A drawback of the Sharp is that its right-hand keyboard is a touchboard connected to the machine's main circuitry by ribbon cables, which can be damaged by opening the calculator past the flat position.

The Hewlett-Packard *HP-28C* (1.6K) is no longer manufactured and no longer available. It has been replaced by the *HP-28S*, that has 32K of memory. Both machines operate in essentially the same manner. Like the Sharp, the *HP-28* only has four lines of text, but the graphics viewport is a bit wider at 32×137 pixels, and graphs are shown as they are being drawn. The *HP-28* can solve equations and operate on matrices, and in addition, can find derivatives and definite integrals, generate Taylor series, determine antiderivatives of polynomials, and handle complex numbers. The *HP-28* has an operating logic based on Reverse Polish Notation, and its working memory is organized into a stack, or column of entries. The *HP-28* also permits the use of algebraic syntax. The machine is menu driven, and user-developed, stored programs are automatically added to the HP's extensive list of built-in functions. The programming

is flexible and especially nice for experienced programmers, permitting BASIC-, FORTRAN-, and Pascal-like commands. It is the most powerful, and the most expensive ($170), of the machines described here. Customized, stored programs can powerfully personalize the *HP-28* to solve many mathematics problems with just a few keystrokes. Wickes (1988) provides much valuable information for those intending to make substantial use of the *HP-28*.

All of these hand-held graphing computers permit interactive experimentation. The Casios have the largest screen and best graphics. The *HP-28* is the most versatile and powerful. All of them can be used to make mathematics more oriented toward concept development and problem solving and less oriented toward paper-and-pencil computation. The use of hand-held computers can be applied to many areas of undergraduate mathematics, especially precalculus, calculus, and statistics. This paper focuses on the applications of these pocket computers to college algebra, trigonometry, and analytic geometry, especially the interactive graphing of functions and relations and the interpretation and use of the obtained graphs to solve problems.

The Ohio State C^{2}PC Project

The Ohio State University Calculator and Computer Precalculus (C^{2}PC) Project is a three-year field-based project aimed at developing a precalculus course that is rich in problems and takes full advantage of interactive computer graphics technology. The primary objectives of the project are (a) to create instructional materials that make effective use of computer- and calculator-based graphing to strengthen student problem solving skills; (b) to improve student understanding of functions, graphs, and analytic geometry–critical areas of mathematical deficiency in the current college preparatory curriculum; and (c) to increase significantly the number of students adequately prepared to pursue higher education in mathematics, science, and technical fields.

"Graphs of functions and relations are highly valued for their ability to display complex information visually. Yet students come to view graphing as a task to be completed rather than an interpretational aid" (Dick, 1989, p. 13). The interpretation and use of graphs is essential in today's world in which graphs are so widely used to present information. Moreover, these skills are especially important for students who intend to study calculus and to pursue scientific and technical careers. Yet, evidence from the Second International Mathematics Study shows that many 12th-grade precalculus students are weak in coordinate geometry, functions, and graphing, and in particular, they do not make a strong connection between a function and its graph (Chang & Ruzicka, 1985). For instance, most do not realize that the solution for a system of two equations corresponds to the intersection points of their graphs.

The C²PC project materials (Demana and Waits, 1988c; Foley et al., 1988; Osborne and Foley, 1988; Waits and Demana, 1988b) are designed to shore up this sagging section of the curricular fabric. After one and a half years of piloting, these materials were field-tested in some 80 high schools and 40 colleges and universities across the nation during the 1988-89 academic year. Throughout the course, hand-held graphing computers and microcomputer software are used as tools for concept development, problem solving, and exploration. Most field-test teachers participated in summer inservice programs to gain familiarity with the materials and to help them prepare to create instructional environments in the spirit of the project. In order to give the reader a sense of the C²PC instructional environment, some distinguishing characteristics of the project are outlined below. This general outline is followed by a set of illustrative examples.

Characterizing Features. C²PC has three characteristics that together make it unique in establishing the concepts of functions and graphing:

1. *Interactive graphing.* Interactive computer graphing is used to provide a rich array of examples of graphs and functions for students to explore and examine. This gives the student many opportunities to form generalizations and to develop concepts about graphs, functions, and their characteristics.

2. *Problems as means.* Real world problems situations are used as the means to approach and teach concepts and skills instead of merely as exercises after the concept has been taught. Often a problem serves as the stimulus for a discussion of some new mathematics with the new mathematics serving as the conclusion for that discussion.

3. *Calculus topics without calculus.* The mathematics is organized differently from most texts. Many topics that are treated lightly or not at all in other precalculus texts are explored in depth. Limits, asymptotes, extrema, continuity, and other topics that foreshadow calculus are given a thorough treatment.

Classroom Arrangements. We have found that a key factor affecting success in using interactive computer graphing is the arrangement of the classroom. Graphs can and should serve as a stimulus for mathematical discussion, and students must be able to see what is being discussed. We have used a variety of arrangements, operating in one or a combination of the following modes:

1. *Graphing calculator mode* with each student using a hand-held graphing computer.

2. *Demonstration mode* with a single large monitor or with an overhead projection palette tied to a computer.

3. *Laboratory mode* with each student or pair of students stationed at a microcomputer.

Instructional Principles. The C²PC course is organized around five methodological themes. These themes serve as threads that wind their way through the project materials. The themes are:

1. *Active involvement.* Students are actively involved in problem solving.

2. *Verbal interaction.* Students talk about the mathematics they are learning.

3. *Problem revisitation.* Important problem situations are revisited on a regular basis.

4. *Informal language.* The formal language of the mathematical topics is kept to a minimum and is not introduced until there is a need for it.

5. *Generalization.* Student learning is facilitated by encountering many instances from which to make generalizations.

Mathematical Emphases. The use of interactive graphics offers an opportunity to change emphases in the mathematical content of precalculus. Most calculus instructors have based much of their instruction through the years on the assumption that a picture or graph explains all. The assumption is that graphs intuitively provide a great deal of information about associated problem situations and algebraic representations. Our experience indicates that graphs become intuitive only after students have learned how to read and interpret the information they provide. That is, students must be *taught* what is contained in graphs before they can serve as an intuitive base for explanation. In large part, the content of the C^2PC course was selected to extend substantially students' ability to know what they are seeing when they encounter a graph and to establish firmly the connection between an equation and its graph. Following are some of the major content emphases that are exploited to build an understanding of graphs and functions.

1. *Viewing rectangles and scale.* With either paper-and-pencil graphing or a computer graphing utility, one only examines a portion of most graphs. A viewing rectangle specifies the portion of the plane within which a graph is to be examined; that is, the minimal and the maximal values of the x-coordinates and the y-coordinates. Students learn how to pick viewing rectangles to satisfy the purpose of the problem at hand; they learn how to zoom-in, zoom-out, and choose different scales for the two coordinate axes.

2. *Local behavior of functions.* The ability to change viewing rectangles permits students to examine closely the graphical behavior of functions. Students can zoom-in to see at close range such local features as extrema and intercepts.

3. *End behavior of functions.* Alternatively, students can zoom-out to obtain a global view of the graph of a function. The notion of asymptote comes alive for students in a new and richer way.

4. *Graphical solution algorithms.* Equations, inequalities, and systems of equations can be solved graphically through the use of a zoom-in procedure. This graphical approach is powerful. Equations involving any elementary functions can be solved by the zoom-in method; whereas their algebraic solutions require a myriad of paper-and-pencil methods.

5. *Parameters and functions.* Students can explore and discover the effects that equation parameters have on the graph of a function. For example, students can experiment with the equation $f(x) = ax^2 + bx + c$ to determine the effects that $a, b,$ and c have on the graph of a quadratic function.

6. *Mathematical modeling.* Students investigate a wide variety of challenging problem situations. They create algebraic and geometric representations for a given situation and use these representations in the solution of the associated problem.

Examples

Three examples are given below to illustrate the C^2PC instructional approach and the interactive graphing capabilities of a hand-held graphing computer. The Casio is used in the examples because this is the machine used at most of the project field-test sites. The Range feature, which shows all six viewing-rectangle and scale parameters at once, and the fast, large-screen graphics make the Casio a suitable choice for the methods being illustrated. Other pocket graphers or computer graphing software could be used in a similar way to solve the example problems. Figures in this paper are given as they appear on the Casio graphics screen.

The first example is actually a set of examples intended to illustrate several aspects of the C^2PC approach. The example foreshadows calculus in its use of modeling, in its treatment of a function as the object of consideration, and in its inclusion of the extreme value concept. Few, if any, of the subproblems in the example would ever be considered in a traditional precalculus course. Parts (a), (b), and (c) – write an equation that models the problem, draw a complete graph of the equation, and identify the portion of the graph relevant to the problem – represent a typical sequence of steps stressed in the project materials. These steps are all that is asked of students on their first encounter with the full-fledged box problem. Earlier, students encounter the box problem in a restricted setting that leads to a quadratic equation. On subsequent *revisitations* question (d), then (e) is asked. In this way by the time students are faced with the extreme-value aspect of the problem, they can deal with this new and challenging issue in a familiar setting.

Example 1.

An open box is to be made from a rectangular piece of sheet metal 20 cm long and 15 cm wide by cutting and removing equal square pieces from each corner of the rectangular sheet and then bending up the sides.

(a) Write an equation that expresses the volume of the box as a function of the side length of the removed squares.

(b) Draw a complete graph of this equation.

(c) What part of the graph in (b) represents the problem situation?

(d) What size square must be cut and removed to form a box with a volume of 250 cm^3 ?

(e) What are the dimensions of the box with the largest volume? What is the maximum volume?

Solution.

(a) This step is nontrivial for most students when they first encounter a box problem of this sort; this step usually occurs in beginning calculus. In the C^{2}PC materials students face this problem only after they have had numerous experiences in modeling simpler problem situations. An equation for the volume function is $V(x) = x(15 - 2x)(20 - 2x)$.

(b) Graphing $y = x(15 - 2x)(20 - 2x)$ in the Casio default viewing rectangle of $[-4.7, 4.7] \times [-3.1, 3.1]$ yields the graph shown in Figure 1. Notice that the Casio graphics viewport does not include the top row and left-most column of pixels. The point $(0, 0)$ has actually been graphed, but this is revealed only if the Trace command is used.

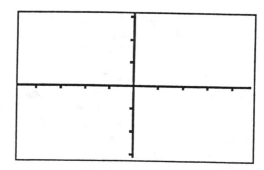

Figure 1. The graph of the volume function $y = x(15 - 2x)(20 - 2x)$ in $[-4.7, 4.7] \times [-3.1, 3.1]$

This graph sheds some light on one of the questions raised by Small, Hosack, and Lane (1986): "Should a student analyze a function to sketch the graph, or just call up a graphing program?" (p. 433). Some thought and experimentation is typically required to obtain a useful computer-generated graph of a given function. Here a *complete graph* is sought; that is, one that displays all key attributes of the function. This is a somewhat subjective and perhaps vague notion, but one that we have found to be pedagogically useful. Goldenberg's (1988) comment about interpreting graphs applies well to the task of obtaining a complete graph: "To interpret graphs correctly, we need mathematical knowledge and expectations, not just perceptual experience" (p.135). In this case a viewing rectangle of $[-5, 15] \times [-500, 1000]$ is suitable (see Figure 2).

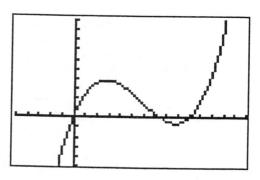

Figure 2. The graph of $y = x(15 - 2x)(20 - 2x)$ in $[-5, 15] \times [-500, 1000]$

(c) The only values of x that make sense in the problem situation are those between 0 and 7.5. The graph shown in Figure 3 extends just beyond these values.

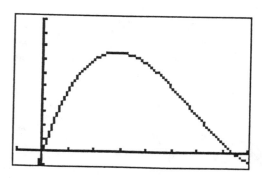

Figure 3. The graph of $y = x(15 - 2x)(20 - 2x)$ in $[-1, 8] \times [-50, 500]$

(d) To determine possible side lengths of removed squares that would yield a 250 cm^3 box, we overlay the graph of $y = 250$ and project down from the points of intersection to the x-axis (see Figure 4). The dashed downward pointing arrows do not appear on the Casio.

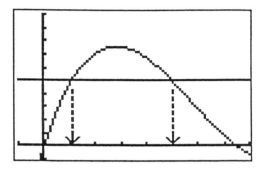

Figure 4. The graphs of $y = x(15 - 2x)(20 - 2x)$ and $y = 250$ in $[-1, 8] \times [-50, 500]$

It appears that an x-value of approximately 1.1 cm or 5.0 cm would yield a 250 cm^3 box. Substitution shows that $x = 5$ exactly satisfies this condition. Using traditional algebraic methods we can determine that the other exact solutions to the equation $x(15 - 2x)(20 - 2x) = 250$ are $x = \frac{25 \pm \sqrt{425}}{4}$, but only $x = \frac{25 - \sqrt{425}}{4}$ is between 0 and 7.5. The approximate value of this second solution to our problem is 1.10 cm.

(e) Using the graph of the volume function shown in Figure 3, we can use the Trace feature to approximate the maximum value at $y = 379.0377662$ (see Figure 5; notice that the Casio viewport is reduced to 55×95 pixels when the Trace function is being used), and then employ the $X \leftrightarrow Y$ command to obtain $x = 2.829787234$. Here x represents the height of the box in cm, and y represents the volume in cm^3. The other dimensions of the box would be 9.34 cm and 14.34 cm. The standard calculus solution is $x = \frac{35 - \sqrt{325}}{6} \approx 2.828707270$

and $y \approx 379.0378082$. This 10-decimal-place accuracy could be achieved by using zoom-in. Notice that our initial approximation for the volume has 7-place accuracy, but our initial approximation for the height has only 3-place accuracy. Why?

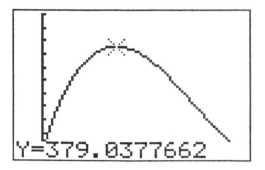

Y=379.0377662

Figure 5. The graph of $y = x(15 - 2x)(20 - 2x)$ together with the readout of the relative maximum value on the interval $0 < x < 7.5$

The next example is a typical trigonometric equation. It illustrates the zoom-in method for equation solving, pointing out that zoom-in can be accomplished in four different ways on the Casio:

1. Key in Range setting parameters by hand.
2. Automatic zooming.
3. Use the Factor feature.
4. Setting the Range within a program.

Zoom-in can also be used to solve inequalities and systems of equations and to locate relative extrema. The equation in Example 2, unlike many equations involving elementary functions, can be solved exactly by traditional methods; so a paper-and-pencil approach is sketched.

Example 2.

Solve $\cos x = \tan x$, $0 \leq x \leq \frac{\pi}{2}$.

Solution.

Using zoom-in. One approach to solving this problem is to graph the functions corresponding to the two sides of the equation on the same coordinate system and then to determine the x-coordinates of any points of intersection that lie between $x = 0$ and

34

$x = \frac{\pi}{2}$. On the Casio this can be done without any programming by entering and executing the dual command Graph Y = cos X : Graph Y = tan X, and then choosing a nested sequence of progressively smaller viewing rectangles each containing the relevant point of interest.

This zooming-in procedure can be accomplished by (a) using the Range feature or (b) using the Trace feature together with automatic zoom or the Factor command. Students unfamiliar with computer graphing can use the Range feature to set viewing rectangle parameters by hand until they gain some facility in making intelligent choices about picking their next view of a given situation. Next automatic zoom (Casio's instant factor function) with its relatively small magnification factor of 2 in each direction can serve as a prelude to the more versatile Factor command.

The following style of program, suggested by a colleague (Shumway, personal communication, 1987), is faster and more flexible than the automatic-zoom approach, and since it is stored in memory, you cannot lose it by hitting one wrong key while trying to zoom-in. It is simple, short, and easily modified.

  "F =" ? → F
  Factor F
  Graph Y = cos X
  Graph Y = tan X

Here's how it works: After entering the program we set the viewing rectangle to the Casio default of $[-4.7, 4.7] \times [-3.1, 3.1]$ with a scaling unit of 1 on each axis, and execute the program. The prompt F = ? will appear on the screen. It is asking for an F value, which will act as a magnification factor. In this case, to preserve our initial choice of viewing rectangle, we enter the value 1, and then continue the execution of the program by pressing EXE. The graph shown in Figure 6 will appear. Tracing along the graph of the tangent function to the pixel that best approximates the unique point of intersection between $x = 0$ and $x = \frac{\pi}{2}$ yields a readout of $x = 0.7$ (see Figure 7). We reexecute the program using $F = 10$, which yields Figure 8. Continuing in this manner, we obtain successive approximations of $x = 0.67$, shown in Figure 9, and then $x = 0.666$, shown in Figure 11. We could continue in this manner to obtain 10-decimal-place accuracy.

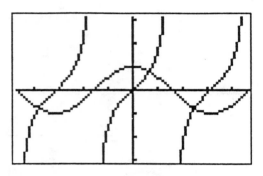

Figure 6. The graphs of $y = \cos x$ and $y = \tan x$ in $[-4.7, 4.7] \times [-3.1, 3.1]$

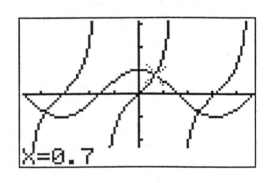

Figure 7. The x-coordinate readout for the pixel nearest the point of intersection in $[-4.7, 4.7] \times [-3.1, 3.1]$

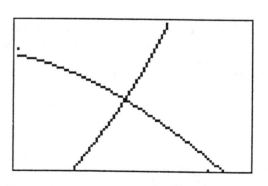

Figure 8. The graphs of $y = \cos x$ and $y = \tan x$ in $[0.23, 1.17] \times [0.49, 1.11]$

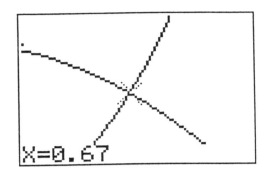

Figure 9. The x-coordinate readout for the pixel nearest the point of intersection in $[0.23, 1.17] \times [0.49, 1.11]$

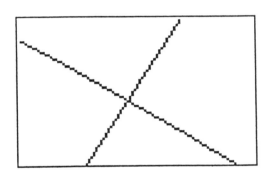

Figure 10. The graphs of $y = \cos x$ and $y = \tan x$ in $[0.623, 0.717] \times [0.759, 0.821]$

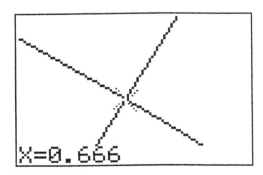

Figure 11. The x-coordinate readout for the pixel nearest the point of intersection in $[0.623, 0.717] \times [0.759, 0.821]$

Alternatively, we could use Vonder Embse's (1988) fancy 19-line zoom-in program. Vonder Embse's program, based upon Stickney's earlier version (1988, of these proceedings), allows the user to zoom-in on a point by tracing to each of two opposite corners of the next viewing rectangle. This emulates a feature of the *Master Grapher* software (Waits and Demana, 1988b) that we have found to be pedagogically advantageous throughout the C^{2}PC project; namely, letting the student see the new viewing rectangle within the old one. Goldenberg (1988) has also found this to be valuable: "When multiple scales are used to represent the same graph, graphing windows should contain internal frames . . . to help students recognize which portion of a distance [*sic*] view is being enlarged in a close-up view" (p. 171).

Traditional approach. The standard traditional method is to seek an exact solution by using trigonometric identities to obtain an equivalent equation that can be readily solved. Such an equation is $\sin^2 x + \sin x - 1 = 0$. Its one solution for $0 \leq x \leq \frac{\pi}{2}$ is $\sin^{-1}\left(\frac{-1+\sqrt{5}}{2}\right)$.

The exact solution gives rise to the following theorem: If one leg of a right triangle is in golden ratio to the hypotenuse, then the second leg is the geometric mean of the first leg and the hypotenuse, and conversely. We probably would have missed this relationship using the zoom-in method. Most traditional precalculus classes would miss it, too.

Moreover, the exact answer does not give the average student any idea of the size of the angle for which the cosine and tangent functions are equal. Many students would not even realize that is what the problem is asking for. To answer the question, "Which method is better?", we must first answer the question, "What is the educational goal of studying this problem?" Or perhaps, "What are the goals of the course, and how does this problem fit into the grand scheme?"

Exact answers tend to please our mathematical souls, yet they are rare. Zoom-in is a very general method that makes the solver of an equation think in terms of the functions involved and their graphical representations. Most students, with sufficient, carefully designed exposure to the zoom-in method, come to realize what it means to solve an equation, and they learn a great deal about functions and graphs in the process.

The final example comes from Larry Thursby, a high school student in the C²PC project. After graphing ordinary rose curves in class one day, he went home and began exploring what I now call *generalized rose curves*. This is an example of what students can and will do on their own if put in the proper learning environment.

Example 3.

Graph $r = 6 \sin 2.5q$.

Solution.

This can be accomplished on the Casio by using the following program:

```
0 → T
Range -9, 9, 1, -6, 6, 1
Lbl 1
6 sin 2.5T → R
Rec (R, T)
Plot I,J
Line
T + 0.1 → T
T ≤ 4π ⇒ Goto 1
```

The portions of the program in **boldface** type may vary from problem to problem. Notice the variable T (used for the angle θ) takes on values from 0 to 4π in 0.1 size increments. The functional value R is computed for each of these values of T . These polar coordinates, (R, T) , are converted to rectangular coordinates; the point is plotted. Then that point is connected by a line segment to its immediate predecessor (except for the first time through the loop when there is no preceding point). The Line command is given in boldface because in some cases you may wish to delete this step to avoid connecting points. The Range parameters are chosen so that the entire graph will appear on the screen and to fit the 3:2 aspect ratio of the Casio screen.

When the program is executed the Casio flashes back and forth between the text window and the graphics window. At the end of execution, the Casio will show the text window, and the graph will be stored in the graphics window. The G ↔ T command will reveal the graph shown in Figure 12.

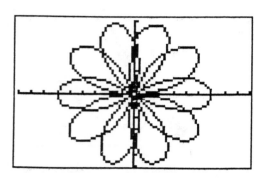

Figure 12. The graph of $r = 6 \sin 2.5\theta$ in $[-9, 9] \times [-6, 6]$

Discussion and Conclusion

Hand-held graphing computers with their interactive graphics capabilities have profound implications for what we can and should teach and how we should teach it. Graphs of functions and relations can be quickly drawn and explored. Interactive graphical methods, such as zoom-in, can be used to develop mathematical connections and to solve realistic problems. The Sharp *EL-5200* goes beyond interactive graphics with its built-in equation-solving and matrix capabilities. For instance, the Sharp's SOLVE key makes short work of Example 2. We can subtract $\tan x$ from each side of the original equation to obtain $f(x) = \cos x - \tan x$, set x -min $= 0$ and x -max $= 1$, and then graph f using the AUTODRAW feature, which picks the y -min and y -max for us automatically. This yields the graph shown in Figure 13. After graphing the function, we need only press the SOLVE key. After a brief wait the solution $x = 0.666239433$ will appear on the screen (see Figure 14; notice that the Sharp viewport is reduced to 24×96 pixels when the SOLVE key is used).

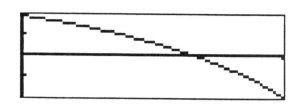

Figure 13. The graph of $y = \cos x - \tan x$ in $[0,1] \times [-1.0711, 1]$

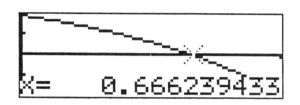

Figure 14. The x-intercept readout obtained by using the SOLVE key

The *HP-28* can solve Example 2 without even drawing a graph. Using the Solve menu, we can store the equation in its original form, provide the HP with an initial guess, say $x = 0.5$, and then in a matter of seconds have the 12-place solution 0.666239432493 with a message that says, "Sign Reversal," indicating an approximate solution.

The C^2PC materials do not include automatic equation solving because that is not in keeping with the project's goal of developing a strong, stable connection between algebraic and geometric representations of functions and relations. When the instructional focus is not on solving the equation or gaining geometric intuition but on using the solution, the automatic solving routines of the Sharp and the *HP-28* are appropriate. This should be the case by the time a student studies calculus, if not before. Hand-held graphing computers, especially the symbol-manipulating *HP-28*, offer many powerful, sometimes controversial new approaches to the teaching and learning of college mathematics. Many more questions can be asked than answered. This article has offered a set of examples that illustrate how these machines can be used in the context of a course in college algebra, trigonometry, and analytic geometry designed to prepare students for calculus. We should all reflect on how

this new breed of calculators can be used to improve the instruction in the courses we teach.

References [1]

Brophy, J., & Hannon, P. (1985). On the future of microcomputers in the classroom. *The Journal of Mathematical Behavior, 4,* 47-67.

Chang, L.-C., & Ruzicka, J. (1985). *Second international mathematics study: United States technical report I.* Champaign, IL: Stipes Publishing.

Demana, F., & Waits, B. K. (1987). Problem solving using microcomputers. *College Mathematics Journal, 18,* 236-241.

Demana, F., & Waits, B. K. (1988a). The Ohio State University Calculator and Computer Precalculus Project: The mathematics of tomorrow today! *The AMATYC Review, 10*(1), 46-55.

Demana, F., & Waits, B. K. (1988b). Pitfalls in graphical computation, or why a single graph isn't enough. *College Mathematics Journal, 19,* 236-241.

Demana, F., Waits, B. K. (Principal Authors), Osborne, A., & Foley, G. D. (1988c). *Precalculus mathematics, a graphing approach* (preliminary ed., 2 vols.). Reading, MA: Addison-Wesley.

Dick, T. P. (1989, January). From ends to means: Using graphing calculators in mathematics instruction [abstract of paper presented at the Graphing Calculators contributed paper session at the meeting of the Mathematical Association of America, Phoenix, AZ]. In *Program: Phoenix Joint Mathematics Meetings* (p. 13).

Dick, T. P., & Shaughnessy, J. M. (1988). The influence of symbolic/graphic calculators on the perceptions of students and teachers toward mathematics. In M. J. Behr, C. B. Lacampagne, & M. M. Wheeler (Eds.), *Proceedings of the Tenth Annual Meeting of the North American Chapter of the International Group for the Psychology of Mathematics Education* (pp. 327-333). DeKalb: Northern Illinois University.

---

[1] In addition to the sources cited in the body of this article, the bibliography includes other works on using hand-held graphing computers and microcomputer graphing to enhance the teaching of mathematics.

Educational Technology Center. (1988). *Making sense of the future: A position paper on the role of technology in science, mathematics, and computing education.* Author, Harvard Graduate School of Education, Cambridge, MA.

Fiske, M. B. (1988, October). Teaching linear regression using the Casio fx-7000G calculator. In this volume.

Foley, G. D. (1987a). Future shock: Hand-held computers. *The AMATYC Review, 9*(1), 53-57.

Foley, G. D. (1987b). Reader reflections: Zoom revisited [response to Montaner, 1987]. *Mathematics Teacher, 80,* 606.

Foley, G. D. (1988a). Let us teach exploration. *The AMATYC Review, 10*(1), 63-66.

Foley, G. D. (1988b). Timeless and timely issues in the teaching of calculus. *The AMATYC Review, 9*(2), 55-60.

Foley, G. D. (in press). Using hand-held computers in calculus and analytic geometry. Essay to appear in the revision of A. Mizrahi & M. Sullivan (Eds.), *Calculus and analytic geometry* (2nd ed.). Belmont, CA: Wadsworth.

Foley, G. D., Edwards, A., Demana, F., Waits, B. K., & Wern, L. A. (1988). *Computer graphing laboratory manual* to accompany *precalculus mathematics, a graphing approach* (preliminary ed.). Reading, MA: Addison-Wesley.

Frantz, M. E. (1986). Interactive graphics for multivariable calculus. *College Mathematics Journal, 17,* 172-181.

Goldenberg, E. P. (1988). Mathematics, metaphors, and human factors: Mathematical, technical, and pedagogical challenges in the educational use of graphical representations of functions. *The Journal of Mathematical Behavior, 7,* 135-173.

Heid, M. K. (1988). Resequencing skills and concepts in applied calculus using the computer as a tool. *Journal for Research in Mathematics Education, 19,* 3-25.

Michel, F. (1987). La HP-28C. *Mathématique et Pédagogie,* No. 62, 21-26.

Montaner, F. R. (1987). Use of the zoom in the analysis of a curve. *Mathematics Teacher, 80,* 19-28.

Muciño, M. (1988). Buyer's guide to graphics calculators. *Mathematics Teacher, 81,* 705, 707-708.

Nievergelt, Y. (1987). The chip with the college education: The HP-28C. *American Mathematical Monthly, 94,* 895-902.

Nievergelt, Y. (1988). The HP-28S brings computations and theory back together in the classroom. *Notices of the American Mathematical Society, 35,* 799-804.

Osborne, A., Foley, G. D. (Principal Authors), Demana, F., & Waits, B. K. (1988). *Instructor's manual* for *Precalculus mathematics, a graphing approach* (preliminary ed.). Reading, MA: Addison-Wesley.

Potter, D. (1987a, August). Seven ounces of pocket graphics: Casio's new fx-8000G and friends. *Pico: The Magazine of Portable Computing,* pp. 22-23.

Potter, D. (1987b, August). Casio's fx-8000G goes to high school. *Pico: The Magazine of Portable Computing,* pp. 23-24.

Shumway, R. J. (1988). Graphics calculators: Skills versus concepts. In J. de Lang & M. Doorman (Eds.), *Senior Secondary Mathematics Education* (pp. 136-140). Utrecht, The Netherlands: University Press.

Small, D., Hosack, J., & Lane, K. (1986). Computer algebra systems in undergraduate instruction. *College Mathematics Journal, 17,* 423-433.

Smith, D. A., Porter, G. J., Leinback, L. C., and Wenger, R. H.. (Eds.). (1988). *Computers and mathematics: The use of computers in undergraduate instruction* (MAA Notes No. 9). Washington, DC: Mathematical Association of America.

Steen, L. A. (1987, October 14). Who still does math with paper and pencil? *The Chronicle of Higher Education,* p. A48.

Stickney, A. (1988, October). "Zooming In" On a Graphing Calculator. In this volume.

Tall, D. O. (1985, March). Understanding the calculus. *Mathematics Teaching,* No. 110, 49-53.

Tucker, T. (1987, January-February). Calculators with a college education? *Focus: The Newsletter of the Mathematical Association of America,* pp. 1, 5.

Tucker, T. (Chair), Goldstein, J., Lax, P., Raphael, L., Rodi, S., van der Vaart, R., Van Valkenburg, M. E., & Zorn, P. (1986). Calculus syllabi: Report of the Content Workshop. In R. G. Douglas (Ed.), *Toward a lean and lively calculus: Report of the conference/workshop to develop curriculum and teaching methods for calculus at the college level* [MAA Notes No. 6] (pp. vii-xiv). Washington, DC: Mathematical Association of America.

Vonder Embse, C. (1988). *Graphing calculators: Programming on the Casio.* Unpublished manuscript, Central Michigan University, Department of Mathematics, Mount Pleasant.

Waits, B. K., & Demana, F. (1987). Solving problems graphically using microcomputers. *UMAP Journal, 8*(1), 1-7.

Waits, B. K., & Demana, F. (1988a). Manipulative algebra–the culprit or the scapegoat? *Mathematics Teacher, 81*, 332- 334.

Waits, B. K., & Demana, F. (1988b). *Master grapher* [Computer software]. Reading, MA: Addison-Wesley.

Waits, B. K., & Demana, F. (1988c). The Ohio State University Calculator and Computer Precalculus (C^2PC) Project. *Mathematics Teacher, 81*, 759, 775.

Waits, B. K., & Demana, F. (1989). Computers and the rational-root theorem, another view. *Mathematics Teacher, 82*, 124-125.

Wickes, W. C. (1988). *HP-28 insights: Principles and programming of the HP-28C/S.* Corvallis, OR: Larken Publications.

Changes in Pedagogy and Testing When Using Technologies in College-Level Mathematics Courses [1]

John G. Harvey

The University of Wisconsin – Madison

One of my favorite cartoons shows a student sitting at a table with four calculators on it. A teacher, while bending over the table and touching one of the calculators, is saying "If you have four calculators and I take one away, how many are left?" I like this cartoon for two reasons. One is that for many years we have been urging teachers to use physically manipulable materials to help students to learn and to do mathematics. When thought of in this way, the teacher in the cartoon is effectively using the calculators as manipulatives to help the student to understand and solve the open number sentence $4 - 1 = ?$. This cartoon also represents to me the belief held by some persons that this is the only way to use calculators and computers in mathematics instruction, while at the same time the cartoon shows how ridiculous and ineffective the use of electronic technologies will be if we permit students to use them only for trivial, sometimes unimportant, tasks.

At present, we do not know a great deal about the effective use of electronic technologies in mathematics. But I opine that as we better define effective ways to use present and emerging electronic technologies in mathematics instruction, we will discover that we must also change the ways in which we teach (i.e., change our pedagogy) and how and what we test. This is probably especially true in mathematics courses taught at the college-level because thus far we have used calculators and computers mostly in relatively trivial ways when compared to the mathematics we are teaching and expect students to learn. In this paper I will discuss some of the ways in which pedagogy and testing change when technologies are used in college-level mathematics courses. However, before I do this I will discuss the issue of *effective use* of calculators and computers in mathematics instruction.

Their Minds Will Turn to Mush!

At present at the school (i.e., pre-college) level, the greatest concern about the use of

electronic technologies in mathematics seems to be about the use of calculators. This may be true because calculators are largely regarded as tools (Taylor, 1980) while computers are regarded either as a technology that engages students in computer-assisted instruction or that students program to solve mathematics problems. The present concern about the use of calculators will undoubtedly broaden to include computers as the use of computers as tools increases.

At the school level, a cry often heard from parents, teachers, and school administrators is that if students use calculators for mathematics, they will become dependent upon them and, as a result, will become even more mathematically illiterate than students are presently; that is, *their* [mathematical] *minds will turn to mush!* Thus, these parents, teachers, and school administrators argue that students should not be permitted to use calculators while learning mathematics. As an aside I would comment that these same persons seem unconcerned that their children and students use calculators in almost every other school subject where they are appropriate including science. But this duality isn't new; when I was a student it was expected that I would use a slide rule in physics and chemistry courses but not in mathematics courses (see Harvey, 1989a).

The argument that student' minds will turn to mush is, at best, a weak one. Hembree and Dessart (1986) identified 79 studies of the effects on students of using calculators in learning mathematics in Grades K – 12. From their meta-analysis of the data from these studies of calculator use, they concluded that:

1. In Grades K – 12 (except Grade 4) students who used calculators in concert with traditional instruction maintained their paper-and-pencil skills without apparent harm.

2. The use of calculators in testing produced much higher basic operation and problem-solving achievement scores than did the use of paper-and-pencil alone. This conclusion held across both grade and ability levels. The overall better performance of calculator using students on problem solving appeared to be a

---

[1] This paper is a revision of a paper with the same title given at the First Annual Ohio State University Conference on Technology in Collegiate Mathematics.

result of improved computation and process selection.

3. Students who used calculators had better attitudes toward mathematics and better self-concepts in mathematics than did students who had not used calculators. This conclusion held across both grade and ability levels. (p. 96)

The 1986 National Assessment of Educational Progress (NAEP) mathematics assessment administered test to students in Grades 3, 7, and 11; at each grade level, one sample of students was permitted to use calculators while another sample was not. At each grade level, the students permitted to use calculators outperformed those who were not ($p < .05$). In addition, data from that assessment revealed that Grade 11 students in the upper quartile of mathematics performance used calculators considerably more in five areas than did students in the lower quartile of mathematics achievement (Dossey, Mullis, Lindquist, & Chambers, 1988, pp. 80-81).

This argument also reveals the view of mathematics and, as a result, the mathematics curriculum that the persons making the argument would seem to have: mathematics is a fixed collection of mechanical skills and techniques. At both the school and college levels it is recognized that mechanical skills and techniques may be important when trying to solve problems but that knowledge of them is only a small component of mathematical knowledge, intuition, maturity, and problem-solving ability (The College Board, 1983a, 1985; Commission on Standards for School Mathematics, 1987; Douglas, 1986; National Council of Teachers of Mathematics, 1980; Steen, 1987a). When, however, the skill-and-technique view of mathematics is adopted, then indeed, calculators should not be used by students because their use takes away the need to learn what is then the substance of mathematics. About the only thing that unites those who see mathematics as a collection of mechanical skills and techniques and those who do not is that both groups are agreed that they want students to learn mathematics.

Human history could be a litany of the tools (i.e., technologies) on which we have become dependent. Physically, these technologies include fire, clothing, wheels, steam and internal combustion engines, locomotives, automobiles, and vacuum cleaners. Mathematically, we rely on the theorems proved and the problems solved throughout history; we regard the mathematics that has been generated throughout the ages not only as knowledge but as a set of tools that we should not avoid but should take and use whenever we need them. Electronic technologies and, in particular, calculators need simply to be regarded as useful tools, and as tools, we need to know when to use them effectively and when not to use them. And like all other tools, we need to know when electronic technologies malfunction, and, when they fail, to have them repaired or replace them.

Finally, even though I have discussed here the arguments advanced by persons interested in the mathematics education of school students, I want to remind college and university mathematics faculty that they may be making the same arguments. I know that I have heard college and university faculty making the argument that graphics calculators or computer algebra systems [2] should not be used in collegiate-level mathematics courses because students will not acquire needed skills and techniques. To some it seems unthinkable that sketching the graph of the function is not the final, mentally consolidating activity associated with computing the first and second derivatives of a function, finding the critical points using the first derivative, and checking those points in the second derivative, or that students do not need to integrate by parts the function f defined by

$$f(x) = x \ \ln(x).$$

I am sympathetic to these arguments since I do not know how much my knowledge of integration, for example, depends upon my skill with integration by parts or partial fractions. Continuing with this example, I do suspect that the skills and my practice of them did little to help me understand the concepts of Riemann sum, partition, and indefinite integration or the Fundamental Theorem of Integral Calculus. My indecision only highlights that we must carefully determine what are effective uses of calculators and computers in both school and college level mathematics.

Effective Uses

Just as some argue that calculators and computers should not be used at all in mathematics, there are others who argue that *any* use of them is appropriate. These advocates believe that no

---

[2] I shall call computer algebra systems and related systems, such as *MATLAB, symbolic mathematics sytems.*

matter how students use calculators or computers to learn mathematics, their mathematics educations will be improved. One consequence of this argument is that we should give students electronic technologies and encourage them to devise their own ways of using them. The result of following this line or reasoning would be, I believe, a proliferation of the poor practices we already see around us. I have, for example, seen people using calculators to find the sum or product of two one-digit numbers. I have also encountered persons who seem to believe that electronic technologies can solve any problem if they "push the buttons" long enough; some of these were my students who, on tests, spent a lot of time making computations and not enough time in analyzing the problem they have been given to solve or in devising a sensible plan for solving the problem.

As I have stated, we know little about the effective uses of electronic technologies in mathematics and especially, in college mathematics courses. We can best judge the effectiveness of uses by specifying what outcomes we expect. I wholeheartedly subscribe to the outcomes sought by the NCTM's Commission on Standards for School Mathematics (1987) and the Tulane Conference (Douglas, 1986); that is, I believe that we should seek uses that promote improved (a) conceptual learning, (b) problem-solving performance, (c) insight into what mathematics is and how it is generated, (d) mathematical intuition, and (e) attitudes and motivation. Effective uses of calculators and computers can help us achieve these outcomes by correctly *focusing student attention* on higher-order learning instead of low-level skills and techniques, and by *reducing or removing instruction* on skills and techniques and replacing it with instruction for those things we seek – concepts, solved problems, insights, intuition, and enthusiasm for mathematics. Here are some examples that indicate to me my faith is well placed.

One example is the conclusions reached by Hembree and Dessart (1986) that were described. There are four additional examples at the college level.

Heid (1988) studied the effects of the use of graphic tools and symbolic mathematics systems on student understanding in an applied calculus course. During the first 12 weeks of instruction the 39 students in the treatment group used these tools to perform routine manipulations; only during the last three weeks was skill development taught. The students in the experimental treatment showed better understanding of the course content and performed almost as well on a final examination of routine skills as did a class of 100 students who had practiced the skills during the entire 15 weeks.

During 1987-88, Kenelly (unpublished) taught the required introductory calculus syllabus to students who used Hewlett Packard *HP-28C* symbol manipulation calculators. Kenelly reported that the entire syllabus could be covered, that students requested they be taught theorems and their proofs so that they might understand the calculator's symbolic manipulations, and that they were enthusiastic and highly motivated.

The effects of using MACSYMA [R] were studied by Palmiter (1986). The experimental group ($n = 40$) studied integration as did her two control groups; the control groups were each comparable in size to the experimental group. The experimental group used MACSYMA while studying the materials; they completed their study of integration in one-half the time required by the control groups. Five weeks and 10 weeks after the end of the instructional treatment, concept and computation tests were given. The test scores of the experimental group was significantly better than those of the control groups on all of the tests. While taking the computation tests, the experimental group used MACSYMA; they completed the test in half the time required by the control groups.

During the Fall Semester, 1988-89, I (Harvey, 1989b) taught a single section of college algebra using *Precalculus Mathematics: A Graphing Approach* (Demana & Waits, 1988) and *Master Grapher* (Waits & Demana, 1986), a graphics tool. The usual syllabus was covered; in addition, considerable time was spent teaching students ways of solving problems using geometric representations generated with the *Master Grapher*. A 25 item algebra test was given as a pre- and posttest to this class of 27 students; there were complete data for 25 of these students. Analysis of those data revealed a statistically significant increase ($p < .001$) in mean achievement from the pretest to the posttest; the mean score on the pretest was 8.76 (s.d. = 2.85) while the mean score on the posttest was 13.31 (s.d. = 3.61).

The Three Ages of Technology

Since we began to experiment with the use of calculators and computers in mathematics education, there have been three technological ages. During the first age (c. 1965-75), the primary use of computers was to teach students a programming language (e.g., BASIC) and to encourage or require them to write computer programs to solve problems. The beginning of the second age coincided with the introduction of both hand-held calculators and microcomputers. During this age, from 1975 to 1985, students may still have been expected to write computer and calculator programs but the primary activities were computer-assisted tutorials and the uninstructed use of calculators.

Because of the ways in which calculators and computers were used during the first two ages, few, if any, changes in mathematics pedagogy were needed. Since students were left to their own devices, expected to write programs to solve the problems outside of class, or to interact passively – again outside of class – to computer assisted tutorials. Together these two ages comprise the passive pedagogic period. During this period mathematics teachers went along doing the same thing they had always done in their classrooms; they used few of these technological innovations on a regular basis in their instruction.

The present age is one in which we are teaching students to use technological tools and in which we are actively using those same tools in classroom instruction on a day-to-day basis. For these reasons, I call this the active pedagogical period. I am not quite certain why mathematics teachers, and especially college mathematics faculty, have moved from a passive to an active stance in their use of technologies.

One reason may be that we have recognized that students may not or cannot discover the appropriate uses of these tools by themselves. My own observation of students using scientific calculators, for example, is that they need to know more about the keys that invoke the built-in mathematical functions and parentheses.

A second reason may be that we have discovered that technological tools can greatly assist us to teach. For example, using *Master Grapher* (Waits & Demana, 1986), I can quickly and easily look at the graph of a function both globally and locally. In addition, the graphs drawn by this graphing tool are better drawn and more accurate than those that I can sketch.

A third, and possibly penultimate, reason may be that they make mathematics more enjoyable while also teaching us new things about mathematics that we thought we had learned and

learned well. For example, the graphing tools (e.g., a Sharp *EL - 5200*) have permitted me to explore geometrically functions that I had never considered before like $f(x) = x \ln(x) - \sin(x)$ and to see graphically that (a) f has a relative minimum (at $x \simeq 0.76$), (b) the first derivative is zero and the second derivative is positive at this point, and (c) that f has an infinite set of inflection points. (The graph of $f''(x) = 1/x + \sin(x)$ shows more clearly that f has that infinite set of inflection points than does the graph of f.) Figure 1 shows that graphs of f, f', and f''.

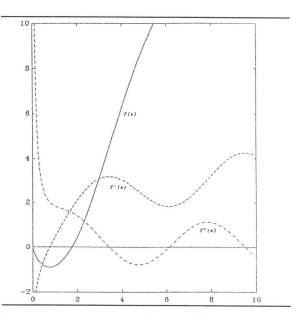

Figure 1. Graphs of f, f', and f'' where $f(x) = x \ln x - \sin x$

Fourth, it may be because these tools have become easier to use and to learn to use and have capabilities that were previously not generally available. One can learn to graph functions using a graphing calculator in less than an hour; at that point, one is ready to begin learning how these calculators can be effectively used in mathematics. Microcomputer symbolic mathematics systems have capabilities only possessed by large, mainframe computers 10 years ago.

What Choices Are There And What Are the Pedagogical Consequences?

At present when considering a choice of technology tools for college-level mathematics courses, there are three choices: programmable and non-programmable scientific calculators, graphing calculators and computer graphing tools, and computer symbolic mathematics systems. The changes in pedagogy that can or should occur when these technologies are used grows as the capability of the tool increases. Thus, the changes in pedagogy that I will discuss when using scientific calculators will also apply to graphing tools and to symbolic mathematics systems.

Scientific Calculators. Among the three tools, scientific calculators are the most widely accepted and the most widely used; a recent survey of 477 college faculty at 183 colleges showed that this is true (Kupin & Whittington, 1988). Table 1 shows the number of respondents who teach each of seven courses and the percent of those who make some use of calculators in those courses.

Scientific calculators have been available longer and are relatively inexpensive (varying in price from about $12 to about $60). When first introduced some college faculty quickly required or permitted their use while other faculty barred them from their mathematics classes. However, scientific calculators–programmable and nonprogrammable–are now familiar equipment in many precalculus and calculus classes. For example, at the University of Wisconsin-Madision we include in our timetable a statement that students taking introductory calculus courses are expected to have a scientific calculator. Some of the uses of scientific calculators in mathematics are these:

1. to facilitiate numerical estimation, approximation, and computation (including numerical differentiation and integration),

2. to make tables so as to search for patterns, to help students sketch graphs, and so forth, and

3. to check results (i.e., answers).

Table 1

Calculator-Use in College Level Courses

Courses	Number of Respondents	Percent of Respondents Indicating "Some Use"
College Algebra	212	77%
Precalculus	274	85%
Statistics		
Non-Calculus	206	94%
Calculus	58	97%
Calculus	426	84%
Discrete Math	140	62%
Linear Algebra	165	81%

Adapted from Kupin & Whittington (1988).

One respondent to the survey conducted by Kupin and Whittington (1988a) may have best summed it up by saying "any topic with calculator-friendly algorithms."

If we actively use scientific calculators, the pedagogic implications of these uses of calculators would seem to be these.

- *We must let students use their calculators as often as they like.* This means, for example, that we can no longer say to students, "You can use your calculators while you solve this set of problems (e.g., you homework) but you cannot use them while you solve that set of problems (e.g., the test problems)."

- *We need to describe and discuss with students the situations in which calculator use is and is not appropriate.* For example, an appropriate use would be the development of a table to graph $f(x) = x \ln(x) - \sin(x)$ in the interval $(0, 5]$ while an inappropriate use would be to use a numerical integration program to find the definite integral of $f(x) = 1/x^2$ in the interval $[1, 5]$. We presently practice this policy with the skills and techniques we teach; thus we simply must extend our policy to include calculators (and computers). If we regularly use a tool in preparing for and teaching our courses, then we will be better able to easily describe appropriate and inappropriate uses.

- *We need explicitly to show students the kind of thinking and planning needed before calculator or computer use begins and after it is concluded just as we do presently with skills and techniques.* By extending this kind of instruction to calculator and computer use we will have additional opportunities to (a) teach problem-solving heuristics and strategies, (b) help students better to understand underlying concepts and principles, (c) give students instruction on estimating and approximating answers, and (d) help students better discern correct from incorrect results.

- *We need consistently to use calculators, both in and outside of the classroom, to show students that calculator and computer use in mathematics is appropriate, important, and acceptable.* I believe that many college students continue to believe that it is acceptable to use a calculator in, say, physics courses but that its use in mathematics courses is unacceptable just as I believed as a student that slide rules were unacceptable. This belief makes students, in a way, calculator and computer phobic and so, afraid that their knowledge of mathematics will not grow appropriately or be adequate if they use these tools while learning mathematics. In other words, they suffer from a sort of self-inflicted "My mind will turn to mush!" argument.

Graphing Tools and Symbolic Mathematics Systems. The use of graphing tools and symbolic mathematics systems is newer, more arguable, and less well explored. In a great part of the undergraduate mathematics curriculum these tools could change what students learn, how they learn it, and how we teach it to them. Many argue that if we fail to include the use of graphing tools and symbolic mathematics systems into the way we teach mathematics the result may be a "Latinized" mathematics curriculum that is studied only by those preparing to be mathematicians (i.e., mathematics scholars); these students would essentially be studying a dead language as far as the rest of the world, including most of the academic world, is concerned (Osborne, of these *Proceedings*; Steen, 1987b) The advantages of using calculators and computers in parts of the college mathematics curriculum have been discussed (Douglas, 1986, pp. vii-xxi; NCTM, 1988; Steen, 1987c; Tucker, 1987; Zorn, 1986; Zorn, 1987). Others have expressed concern or have urged caution in changing the undergraduate mathematics curriculum (Buck, 1987; Gillman, 1987; Stein, 1986). We need to proceed carefully, though expeditiously, in incorporating these tools. Here are some suggestions.

- *We need to analyze carefully the content that we presently teach and that we would like to teach.* That is, we need to carefully reexamine what it is that we want students to learn and to compare that to what we presently teach them and to what we think they *can* learn. Most of the discussions about the inadequacies of present calculus courses (Douglas, 1986; Steen, 1987a) have identified that our present courses teach too many skills and techniques and too few concepts and too little problem solving. My colleagues at the University of Wisconsin-Madison and at the other UW-System institutions acknowledge this is true and that we would like to improve our teaching of concepts and problem solving. However, we also feel that our present calculus courses are very challenging for the majority of our students. The number of students who do not successfully complete

introductory calculus courses on our campuses is about 25% (University of Wisconsin-Madison, Department of Mathematics Calculus Committee, unpublished; Harold Schlais, personal communication); these rates are consistent with other estimates (Douglas, 1986, p. xvii). Only a careful analysis of what it is essential for students to learn coupled with an analysis of (a) what we think students can learn and when they can learn it and (b) which students we want to include and exclude from our mathematics courses will tell us how to improve our undergraduate mathematics curricula. These decisions need to be made locally since they depend upon many factors including the size and admissions requirements of the institution, the qualifications and intended majors of the institution's students, the availability of calculators, computers, and software, the technological literacy of faculty (and teaching assistants), and the willingness and ability of faculty (and teaching assistants), and the willingness and ability of faculty and the institution's administration to change.

- *Once the content of the mathematics curriculum has been examined, we need to determine the ways that particular tools can help us to teach that content.* Undoubtedly, this step is not one that occurs after all of the decisions have been reached about the content of the curriculum, but I have placed it second in my list to emphasize my belief that the choice of appropriate mathematical content is most important (Harvey, 1989a).

- *We must not cling to our present ways of teaching.* There are at least two things that must change; both are related to my earlier observation that to be most effective our instruction will need to include regular use of the tools we select.

First, it seems we will need to give up our roles as expositors, leaders, and "the sources of knowledge" and become instead resource persons to and, on occasion, co-learners with our students. Let me describe a typical session in my algebra class using *Precalculus Mathematics: A Graphing Approach* (Demana & Waits, 1988) and *Master Grapher* (Waits & Demana, 1986). My classroom is equipped with a computer, screen, and color projector. During most class sessions I usually talk for a few minutes about the algebraic techniques that the students are learning and the ways in which they can and should be used; a majority of the class time is spent engaged in problem solving using both algebraic and geometric techniques. During problem solving I basically ask questions, make suggestions, and operate the computer while students supply me with the problems to be solved, the questions to be answered, the procedures to be used, and most of the answers. High school mathematics teachers who are participating in the 1988-89 field test of *Precalculus Mathematics* reported that their classes proceed in about the same way (Bert Waits, personal communication).

Second, we must be willing to rearrange the order in which topic and ideas are presented. Let me relate two experiences here. One comes from my present college algebra course; the other from John Kenelly's experiences in teaching calculus with the Hewlett Packard *HP-28C* calculators.

A typical way of teaching college algebra students to find the real zeros of a polynomial function f of degree three or higher is:

1. Check to see if f has a zero at $0, 1$, or -1 by directly computing the value of f at those points.

2. Estimate the number of real zeros using Descartes' Rule of Signs.

3. Use the Rational Root Theorem to make a list of the possible rational zeros of f.

4. If Step 2 has shown there are no positive or no negative real zeros, delete the appropriate entries from the list developed in Step 3.

5. Check the values remaining in the list by directly computing their images under f to see if they are zero.

6. Use the information gained in the first five steps to factor f as completely as possible, and in most cases, *stop* whether or not the polynomial has been completely factored.

When using a graphing tool polynomials of degree three or higher may not be factored at all. Instead a complete graph (i.e., one that shows the global behavior and as much of the local behavior of the function as possible) is drawn; then, using this graph and successive graphs of local parts of the function, approximations to *all* of the zeroes of the function are obtained. However, since it remains important that students find exact solutions, in

this case rational zeros, to some problems the next procedure also is used. Steps 1, 2, and 3 of the typical method outlined previously are followed. Using the information obtained, a graph of the function is drawn using a graphing tool so that a portion of the domain that contains all of the remaining possible rational zeros is shown. The graph is inspected, and more of the rational zero candidates are eliminated. The remaining candidates are tested by directly computing the value of f at those points. If, at that point, the irrational real zeros of f are desired, then a complete graph is drawn, an the irrational roots are approximated.

This example points out how graphing rearranges the order and the way in which algebraic techniques are used. Graphing has become an integral part of the problem-solving process instead of an end result of it.

John Kenelly (unpublished) reported similar instances in which the content of the typical calculus course was rearranged because of the use of the *HP-28C* in his class; two of the instances are notable. In the first instance he reported that for three decades C students in his classes had been unable to apply the chain rule to three-step problems. When teaching the chain rule to his *HP-28C* class he and his students worked a number of chain rule problems with the sequential differentiation key on the calculator. He reported that on the examinations a number of C students in the experimental section were able to work three and four-step chain rule problems successfully.

Kenelly also reported that graphing was an important tool in his experimental calculus class. "The unit [the *HP-28C*'s] turned around the way that the students looked at graphing and the calculus. The class started out with the graph and used the calculus to understand the process – not the historically opposite direction. This was one of the most successful uses. In fact, the students simply could not fathom that people had to use such long and involved calculations of derivatives just to find the general description of the curve."

● *Using graphing tools and symbolic mathematics systems will enable us to emphasize the liason between geometric, algebraic, and analytical thinking.* In many instances use of these technologies can help establish this liaison because this problem-solving strategy can be applied:

1. find an appropriate algebraic or analytical representation of the problem,

2. use a graphing tool or symbolic mathematics system to develop a geometric representation of the problem or to manipulate the originally derived representation,

3. obtain an (approximation to the) answer from the geometric representation or from the symbolic manipulation, and

4. prove that the answer obtained is correct or is a good one.

- *We should use these technologies to teach students about the global and local behavior of systems and of the interactions between them.* As mathematicians we can predict the behavior of specific functions and relations because we know the class of functions to which they belong and have in our mind's eye a picture of the graphs of the funtions in that class. As a result we usually only need to experiment a little to discover how particular functions differ from our generic images. Students seem not to learn very well that similar functions have similar graphs and, sometimes, vice versa. For example, when you show students the funtion $f(x) = x^2$ and a graph of it and then ask those students what, in general, a graph of $g(x) = 2x^2 - x + 3$ might look like, they often have no idea.

With a graphing tool it is easy to see that the leading term of a polynomial function describes the behavior of the function when $|x|$ is large by showing students that, in an appropriate viewing rectangle as shown in Figure 2, the global behavior of two polynomials of the same degree are quite similar. I have no quantitative data to support this claim, but I firmly believe that the students in my experimental college algebra section have come to know that all quadratic polynomial function are parabolas, that all third-degree polynomial functions have graphs that resemble that of $f(x) = x^3$, and so forth.

At the same time a graphing tool can help students to understand the remaining part of the theorem from theory of equations that I have started to recite; namely, that when $|x|$ is small then terms of lower degree in a polynomial function begin to dictate its behavior. Figure 3 shows another graph of the function $f(x) = x^3 - 4x^2 + 4x$. In the interval from $x = 0$ to $x = 2$, it is clear that the x^3 term no longer describes the behavior of this function. At that point estimating the relative sizes of x^3 and $4x^2$ shows that the latter term is larger as students come to expect. Discoveries about the local behavior of functions lead naturally to a discussion of local maxima and minima and to some knowledge of where they might be found; as a result students should be better able to estimate answers and to determine correctness of their results.

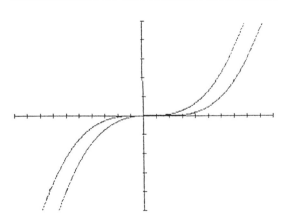

Figure 2. Graphs of $f(x) = x^3$ and $g(x) = x^3 - 4x^2 + 4x$ in the Viewing Rectangle $[-10, 10]$ by $[-500, 500]$

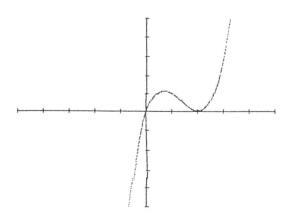

Figure 3. Graph of $g(x) = x^3 - 4x^2 + 4x$ in the Viewing Rectangle $[-5, 5]$ by $[-5, 5]$

48

Study of local behavior in this way can also lead students to understand results that are deep and could not be taught otherwise – at that time or place. From Figures 4, 5, 6, and 7 you see ample evidence of the piecewise linearity of $f(x) = x^3 - 8x + 1$. Figure 4 shows the local behavior of f; in Figure 5, the part of the graph of the function that will be magnified for consideration is shown in the box. Figure 6 shows the result of the magnification; Figure 7 shows the result of a subsequent magnification. After students see this over and over again as they approximate zeros, local maxima and minima, and the points of intersection of curves, they come to understand piecewise linear approximation. [3]

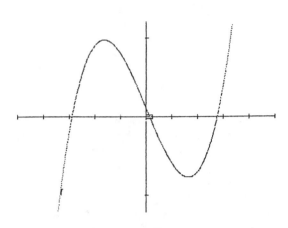

Figure 5. Graph of $f(x) = x^3 - 8x^2 + 1$ in the Viewing Rectangle $[-5, 5]$ by $[-12, 12]$ with the Zoom-in Rectangle Shown

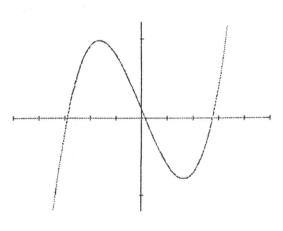

Figure 4. Graph of $f(x) = x^3 - 8x^2 + 1$ in the Viewing Rectangle $[-5, 5]$ by $[-12, 12]$

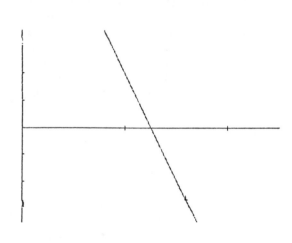

Figure 6. Graph of $f(x) = x^3 - 8x^2 + 1$ in the Viewing Rectangle $[0, 0.25]$ by $[-0.36, 0.36]$

[3] In my classes students came to understand that *Master Grapher* plots points and then connects those points with straight line segments to produce an approximation to the graph of the function being pictured. Thus, when spurious lines and points appear on the scene because an inappropriate viewing rectangle has been used or because the function has a vertical asymptote, students sucessfully sort out the graph of the function from this "garbage."

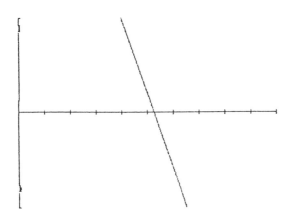

Figure 7. Graph of $f(x) = x^3 - 8x^2 + 1$ in the Viewing Rectangle $[0.12, 0.13]$ by $[-0.01, 0.01]$

Finally, because it is easy to look at related problems, these technologies can also help us to teach students to discover and to generalize. For example, consider the function

$$f(x) = a(x - b)^n + c$$

where n is a positive integer and a, b, and c are real numbers. After beginning with the case of $a = 1$ and $b = c = 0$, it is easy to consider other values of a, b, and c quickly. From this students will generalize the amount and direction that b translates the graph horizontally, the amount and direction that c translates the graph vertically, the way in which a affects the overall shape of the graph, and the rotation in space of the graph about the x-axis when a is negative.

● *Because calculators and computers are discrete machines, both continuous and discrete perspectives must be included.*

● *A part of instruction should be devoted to teaching students how an when to use graphing tools and symbolic mathematics systems.* I realize I have already stated this, but I want to repeat it here because it is especially true for these two technologies because of their added capabilites. The survey by Kupin and Whittington (1988) shows that between 6% and 42% of the respondents that let students use calculators spent "some class time" showing students how to use their calculators and

that 26% of the respondents who permitted computer use taught or assisted their students to use them. The amount of time spent was not quantified. The results of this survey indicates that those presently using calculators and computers in their college-level courses recognize that instruction on the use of these technologies is important and possibly, necessary.

● *We will need to spend time with students talking to and working with them while they use graphing tools and symbolic mathematics systems.* This is an obvious suggestion when the uses of technologies are thought of as important problem-solving strategies. It is also important during this period in time when we know so little about effective uses of technologies and about how and how well students will learn to use them. This kind of information could help us to transform the college mathematics curriculum into one that better teaches concepts, problem solving, and applications.

As already stated, the use of calculators, graphing tools, and symbolic mathematics systems cannot be limited to their use by teachers or to student use in class or while doing homework. Effective uses of these technologies by students will mean that they come to regard them as tools that they use much as they presently use pencils and paper. Thus, tests at all levels will have to assume that students will use these tools, and so, test makers will have to design tests with that assumption in mind.

Testing with Calculators and Computers

Undoubtedly, the college mathematics faculty who presently expect their students to have and to use calculators and computers and who permit the use of those tools on tests already recognize that the tests they administer to their students need to be different. Those faculty have discovered that some of the test questions they previously gave to their students no longer are appropriate in that (a) a correct answer obtained by using a calculator or computer does not reveal if the students understand the underlying concept, algorithm, or technique or (b) the difficulty level of the question has been lowered. In addition, these faculty have also found that the use of calculators and computers in their courses permits them to test understandings not possible when only paper and pencils were used, that test questions may be ones that use realistic data or that have good approximations or estimates as answers, and that they have to be careful in developing their tests to generate questions that test mathematics knowledge and not simply students' abilities to manipulate their calculators or computers. I suspect that if this collective knowledge were analyzed and synthesized, we would have a good, possibly definitive, picture of how both test questions and tests can be validly and reliably constructed to be calculator-based or computer-based. Unfortunately, this is not so; our knowledge of calculator- and computer-based mathematics tests is presently "personal" or "fugitive" knowledge.

In 1975 the National Advisory Committee on Mathematics (NACOME) also recognized that calculator use during testing would "most certainly be invalidated in 'calculator classes'" (NACOME, 1975, p. 42). From about that time to the present the National Council of Teachers of Mathematics (NCTM) has urged that calculators be used in teaching and learning mathematics and in 1986 this organization recomended that "test writers integrate the use of the calculator into their mathematics materials at all grade levels" (National Council of Teachers of Mathematics, 1986).

Until recently recommendations like those made by NACOME and NCTM seem to have had little effect at the national level. Until 1986, there had been little experimentation with calculator- or computer-based mathematics tests at this level. [4] In 1983 and 1984, the College Board's Advanced Placement Program permitted the use of calculators on the Advanced Placement (AP) Calculus Examinations. During the time that calculator use was permitted on the AP Calculus Examinations, the test questions were designed to be "calculator-neutral or calculator immune;" that is, the questions were designed so that calculator use neither enhanced or hindered a student's opportunity to solve the problem successfully. This is contrary to the position I have recommended in that it encouraged students to use calculators while taking their AP calculus courses but denied them an opportunity to use their calculators effectively on the

---

[4] The College Board Advanced Placement Examinations in Physics and in Chemistry permit the use of scientific calculators. The National Society of Actuaries permits calculator use on its actuarial examinations. It seems likely that most of the national examining organizations in the sciences and business permit calculator use on their tests.

test. The reasons the use of calculators on the AP Calculus tests were discontinued were:

> The MSAC [Mathematical Sciences Advisory Committee] recognizes the problems generated by the use of calculators in a calculator-independent situation. The difficulties caused by rapid changes in technology, the lack of equity of access to sophisticated calculators because of expense, as well as administrative and security concerns, are cogent reasons for suspending the use of hand calculators on the AP Calculus Examinations." (Kennelly, 1989b)

The MSAC recognized that there are problems generated when calculators are used in calculator-independent situations. At the same meeting at which they discontinued calculator use on the AP Calculus Examination, the Committee urged that an emphasis be placed on designing tests that allow the use of calculators as aids during testing.

Symposium on Calculators in the Standardized Testing of Mathematics. Because of their common interest in developing mathematics tests requiring the use of calculators the College Board and the MAA jointly sponsored the Symposium on the Use of Calculators in the Standardized Testing of Mathematics in September 1986 (Kennelly, 1989a). Three recommendations from that Symposium are especially relevant here.

1. Mathematics achievement tests should be curriculum based, and no questions should be used on them that measure only calculator skills or techniques.

2. When addressing a particular test item, an important skill is choosing when and when not to use a calculator. Consequently, not all of the questions on a calculator-based mathematics achievement test should require the use of a calculator.

3. There should be no attempt to place an upper limit on the level of sophistication that calculators used on tests should have. Any calculator capable of performing the operations and functions required to solve the problems on a particular examination should be allowed.

While the Symposium participants did not consider or discuss the development of computer-based tests, I believe that these recommendations also extend to the development of those tests as well.

The first recommendation reaffirms and extends to calculators the traditional tenet that the mathematical objectives that are tested should dictate the kind of questions included on the test. It also asserts that items intended solely to test a student's calculator facility should not be included on mathematics tests. The second recommendation has several valid interpretations. One interpretation is that is is not necessary to give tests consisting only of calculator-based (or computer-based) items though on occasion this may be necessary because not enough calculators or computers are available to test all students at the same time. The second recommendation can also be interpreted as meaning that it is important to test whether students know when and when not to use a calculator. The third recommendation may be naive because in 1986 graphing calculators had just been introduced, the Hewlett Packard *HP-28C* had not yet appeared, and graphing tools and symbolic mathematics systems were only beginning to be explored. If tests and test questions will have to be prepared so that, for example, any calculator can be used then it may be that students who have better calculators and who know how to use them will have an advantage on some test items. As examples let me examine the items shown in Figure 8.

Example 1. Determine the number of real solutions of the equation

$$4x^3 - 10x + 17 = 0.$$

Example 2. Find one real solution for the system

$$y = 4 - x^2 + 3x^3$$
$$y = -1 + 2x.$$

© 1988 by Bert K. Waits & Frank D. Demana. Used by permission.

Figure 8. Examples of Mathematics Items On Which Use of Different Calculators Could Produce Inequities

Both examples are very much alike in that a graphing tool makes both of them easier.

When responding to the first item, a student using a scientific calculator without a programmed routine to approximate the solutions could best solve this problem by using Descartes Rule of Signs first to determine that there are two positive real solutions or no positive real solutions and that there is one negative solution. At that point the student would need to sketch the graph of the non-negative domain values to determine how many positive, real solutions there are; a scientific calculator could be used to develop the table of values

for that graph. On the other hand a student who has a graphing calculator would simply develop a complete graph of the function whose zeros are sought and would immediately see that there is only one real, negative solution. A student having a Sharp *EL-5200* calculator could, at that point, press the [**SOLVE**] key and determine that an approximation to the zero is $x \simeq -2.12$; so, changing this question to read "Determine a real solution of ..." would not remove the advantage held by a student having this particular calculator.

The second example asks for an x and a y value that satisfies both equations. Students using scientific calculators would need to follow the steps presently taught: substitute the value of y in the second equation into the first equation, combine like terms, solve $3x^3 - x^2 - 2x + 5 = 0$ for a value of x using the techniques just discussed, and substitute that value into the second equation to find the corresponding value of y. A student having a graphing tool could graph both functions and easily determine the point of intersection using the built-in features of that tool.

Both of these examples show that inequities may occur during testing if students do not have approximately the same technological tools available; the disparities might be even greater if some students have only scientific non-programmable calculators while others have Hewlett Packard *HP-28C*'s. I typically ask students having graphics calculators not to use that facility during tests since other students in my class do not have those calculators. I also give students take home problems that comprise a part of their test so that they each can have the opportunity to work these problems using a graphing tool.

Development of Calculator-Based Tests. Prior to 1986 the Mathematical Association of America's (MAA) Committee on Placement Examinations (COPE) had been discussing the development of calculator-based college-level placement tests; in August 1986, COPE received a grant from Texas Instruments Incorporated to the MAA Calculator-Based Placement Test Program (CBPTP) Project. Development of CBPTP tests began in October 1986. Using the present tests of the MAA Placement Test Program as a starting point, the CBPTP Project test development panels are developing calculator-based placement tests. About 25% of the items on the CBPTP tests will be calculator-active; each CBPTP test will expect students to have a scientific calculator available to them while taking that test. The test panels have been using these definitions of a calculator-active test item and a calculator-based test.

A *calculator-based* [calculator-active] *test item* (a) is an item containing data that can be usefully explored and manipulated using a calculator and (b) has been designed to facilitate active calculator use.

A *calculator-based mathematics test* is one that (a) tests mathematics objectives, (b) has some calculator-based test items on it, and (c) has no items on it that could have been but are not calculator-based except for items that are better solved using non-calculator based techniques. (Harvey, 1989a)

The first two of the six CBPTP tests have been developed and will be published in 1989; they are the Calculator-Based Arithmetic and Skills Test and the Calculator-Based Calculus Readiness Test. Two additional tests are presently being developed and will be published in 1990; they are the Calculator-Based Basic Algebra Test and the Calculator-Based Algebra Test. The remaining two placement tests will be developed during 1989 and 1990 and will be published in 1991; these two tests are the Calculator-Based Advanced Algebra Test and the Calculator-Based Trigonometry Test. As they are published these tests will be included in the MAA's Placement Test Program test packet. An MAA Note describing the development of these tests and their characteristics will be published in 1991.

The College Board has initiated development of a new version of the Mathematics Level II Achievement Test that will be calculator-based (Harvey, 1989c). This test is presently still in the prototype development stages. There are plans to offer this Math II-C version of the Mathematics Level II Achievement annually when it is completed.

Calculator-Active Test Items. An analysis of two of the MAA Placement Test Program tests showed some of the items were not appropriate for calculator-based tests for two reasons: the items no longer tested the objectives they were intended to test or they tested calculator facility instead of mathematical knowledge. An analysis of the items from the Scholastic Aptitude Tests (SAT) (College Board, 1983b) showed that none of those items were not appropriate for calculator-based tests (Harvey, 1989a). The contrast between the two sets of items is that those on placement tests tend

to test the mathematical content (i.e., achievement) of high school mathematics courses and lower-level knowledge while the SAT items measure aptitude instead of achievement. In addition, few of the SAT items were ones requiring much computation and thus, were not calculator-active when judged using a scientific calculator. So, then, what is a calculator-active mathematics test item? Figure 9 gives some examples of calculator-active items drawn from among those developed for the new MAA calculator-based placement tests; the correct answers are marked with asterisks.

Example 1. The approximation of $(1 + 1/6)^4$ correct to 4 decimal places is
(A) 1.0008   (B) 1.1667   (C) 1.8526 *
     (D) 2.1614   (E) 4.6667

Example 2. If $n \times n \times n = 63$, then which of the following is closest to n?
(A) 0.047619   (B) 3.979057 *   (C) 21
     (D) 189   (E) 250,047

Example 3. If a certain buffalo population increases by a factor of 1.1 every year, then in 15 years it increases by a factor of
(A) 1.500   (B) 3.797 *   (C) 4.177
     (D) 16.500   (E) 19.666

Example 4. The sequence of number
$(3/2)^4, (4/3)^6, (5/4)^8, \ldots, ((n+1)/n)^{2n}, \ldots$
approaches
(A) 1   (B) 2.179   (C) 6.192
     (D) 7.389 *   (E) no finite number

Figure 9. Examples of Scientific Calculator-Active Items

The first two examples in Figure 9 were developed for the Calculator-Based Arithmetic and Skills Tests; the third and fourth items will appear on the Calculator-Based Calculus Readiness Test. For the intended student audience each item is one that could not be given if students were not permitted to use scientific calculator. The first two examples test student understanding of exponentiation; Example 1 also tests knowledge of order of operations. Examples 3 and 4 are problem solving items. Example 3 requires an understanding of exponential growth while Example 4 requires students to discern a pattern and choose the best answer. Each incorrect response in each item is based upon the *mathematical errors* that students might make while working problems like these. A description of the errors associated with Example 2 shows what this means.

A. The student reads the problem as $3n = 63$ and solves that problem by dividing 3 by 63.

B. This is the correct answer.

C. The student reads the problem as $3n = 63$ and solves that problem correctly.

D. The student reads the problem as $3n = 63$ and solves that problem by taking the product of 3 and 63.

E. The student solves the problem $n^{1/3} = 63$ correctly.

Figure 10. Description of the Errors Students Might Make While Solving $n \times n \times n = 63$

Development of test questions when calculator or computer use is expected requires considerable skill and a thorough knowledge of the technological tools that students will use. If the suggestions I have made about the ways in which we should use the technologies we choose while teaching are followed, then we will be able to develop reliable, valid tests for our students.

Conclusion

This is an exciting–and disturbing–time for collegiate mathematics. We have an opportunity to restructure our curriculum, to develop new pedagogies, and to test students more accurately if we effectively and appropriately apply present and emerging technologies. On the other hand because of their nature technology tools will probably cause us to reexamine carefully the ways we teach and as a result, to abandon some of those ways. Before the beginning of the 21st century I hope that college mathematics classrooms will be more exciting places in which students are learning mathematics better than ever before–including what each of us remembers as "the good old days" when we were initially learning mathematics!

References

Buck, R.C. (1987). Computers and calculus: The second stage. In L. A. Steen (Ed.), *Calculus for a new century: A pump, not a filter.* (MAA Notes Number 8). Washington, DC: The Mathematical Association of America.

College Board. (1983a). *Academic preparation for college: What students need to know and be able to do.* New York: Author.

College Board. (1983b). *10 SAT's.* New York: Author.

College Board. (1985). *Academic preparation in mathematics: Teaching for transition from high school to college.* New York: Author.

Commission on Standards for School Mathematics. (1987). *Curriculum and evaluation standards for school mathematics.* (Working draft). Reston, VA: National Council of Teachers of Mathematics.

Demana, F. D. & Waits, B. K. (1988). *Precalculus mathematics: A graphing approach.* (Preliminary version). Reading, MA: Addison-Wesley.

Dossey, J.A.; Mullis, I. V. S.; Lindquist, M. M.; & Chambers, D. L. (1988). *The mathematics report card: Are we measuring up?* Princeton, NJ: Educational Testing Service.

Douglas, R. G. (Ed.). (1986). *Toward a lean and lively calculus.* (MAA Notes Number 6). Washington, DC: The Mathematical Association of America.

Gillman, L. (1987). Two proposals for calculus. In L. A. Steen (Ed.), *Calculus for a new century: A pump, not a filter.* (MAA Notes Number 8). Washington, DC: The Mathematical Association of America.

Harvey, J. G. (1989a). Placement test issues in calculator-based mathematics examinations. In J. W. Kenelly (Ed.), *The use of calculators in the standardized testing of mathematics.* (MAA Notes Number 12). Washington, DC: The Mathematical Association of America.

Harvey, J. G. (1989b). Using graphing tools in college algebra. Paper presented at the January 1989 Meeting of the Mathematical Association of America, Phoinix, AZ.

Harvey, J. G. (1989c). Preface. In J. W. Kenelly (Ed.), *The use of calculators in the standardized testing of mathematics.* (MAA Notes Number 12). Washington, DC: The Mathematical Association of America.

Heid, M. K. (1988). Resequencing skills and concepts in applied calculus using the computer as a tool. *Journal for Research in Mathematics Education, 19* (1), 3-25.

Hembree, R. & Dessart, D. J. (1986). Effects of hand-held calculators in precollege mathematics education: A meta-analysis. *Journal for Research in Mathematics Education, 17* (2), 83-99.

Kenelly, J. W. (1989a). A historical perspective on calculator usage on standardized examinations in mathematics. In J. W. Kenelly (Ed.), *The use of calculators in the standardized testing of mathematics.* (MAA Notes Number 12). Washington, DC: The Mathematical Association of America.

Kenelly, J. W. (1989b). *Calculators in the standardized testing of mathematics.* (MAA Notes Number 12). Washington, DC: The Mathematical Association of America.

54

Kenelly, J. W. (unpublished). Calculus classroom experiences with the *HP-28C*.

Kupin, J. L. & Whittington, B. R. (1988). *A survey of the use of hand calculators and microcomputers in college mathematics classes.* (College Board Report No. 88-6 & ETS RR NO. 88-9). New York: College Board.

National Advisory Committee on Mathematics. (1975). *Overview and analysis of school mathematics: Grades K-12.* Reston, VA: National Council of Teachers of Mathematics.

National Council of Teachers of Mathematics. (1980). *An agenda for action: Recommendations for school mathematics of the 1980s.* Reston, VA: Author.

National Council of Teachers of Mathematics. (1986). *Calculators in the mathematics classroom.* Reston, VA: Author.

Palmiter, J. R. (1986). The impact of a computer algebra system on college calculus. Unpublished doctoral dissertation, The Ohio State University, Columbus. (DAI No. DA 8618829).

Steen, L. A. (1987a). *Calculus for a new century: A pump, not a filter.* (MAA Notes Number 8). Washington, DC: The Mathematical Association of America.

Steen, L. A. (1987b). Calculus today. In L. A. Steen (Ed.), *Calculus for a new century: A pump, not a filter.* (MAA Notes Number 8). Washington, DC: The mathematical Association of America.

Steen, L. A. (1987c). Who still does math with paper and pencil? In L. A. Steen (Ed.), *Calculus for a new century: A pump, not a filter.* (MAA Notes Number 8). Washington, DC: The Mathematical Association of America.

Stein, S. (1986). What's all the fuss about? In R. G. Douglas (Ed.), *Toward a lean and lively calculus.* (MAA Notes Number 6). Washigton, DC: The Mathematical Association of America.

Taylor, R. P. (Ed.). (1980). *The computer in the school: Tutor, tool, tutee.* New York: Teachers College Press.

Tucker, T. W. (1987). Calculus tomorrow. In L. A. Steen (Ed.), *Calculus for a new century: A pump, not a filter.* (MAA Notes Number 8). Washhington, DC: The Mathematical Association of America.

University of Wisconsin Department of Mathematics Calculus Committee. (Unpublished). A survey of the grades made by students in introductory calculus courses.

Waits, B. K. & Demana, F. D. (1986). *Master Grapher.* [Software]. Reading, MA: Addison-Wesley.

Zorn, P. (1986). Computer symbolic manipulations in elementary calculus. In R. G. Douglas (Ed.), *Toward a lean and lively calculus.* (MAA Notes Number 6). Washington, DC: The Mathematical Association of America.

Zorn, P. (1987). Computing in undergraduate mathematics. In L. A. Steen (Ed.), *Calculus for a new century: A pump, not a filter.* (MAA Notes Number 8). Washington, DC: The Mathematical Association of America.

Calculus, Technology, NSF and the Future

John W. Kenelly

National Science Foundation and Clemson University

A Perspective

In the rush of technology developments, it is easy to get the impression that changes are arriving faster than you can think. It is not only *easy* – it is *correct*. Before most people had even realized that we had graphing calculators, we had symbol manipulating ones. And before we even started to conceptualize the use of symbol manipulating units in our curriculum, there is over the horizon spread sheet calculators and who knows what else. To a generation of mathematicians that remember square root keys as "major break throughs" (Freiden, 1963) and the total surprise of a "hand held scientific unit" (*HP35*, 1972); the pace and magnitude of the current developments is overwhelming. Yes, things are happening almost at will and the historical chart of dates supports this observation. In Figure 1, we see that it was 2600 years from the abacus to Pascal's mechanical adding machine. Then 260 years from there to the Monroe mechanical calculator. Another 26 years to electrical devices and perhaps 2.6 years to scientific devices. The sequence is not exact, but the pattern of order of magnitude reductions in the number of years is replicated in the approximate 0.26 years between graphing calculators and symbol manipulating units.

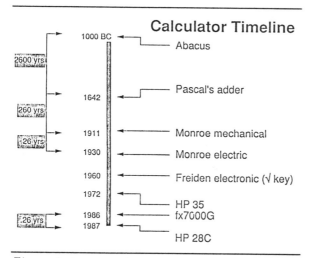

Calculator Timeline

1000 BC	Abacus
1642	Pascal's adder
1911	Monroe mechanical
1930	Monroe electric
1960	Freiden electronic (√ key)
1972	HP 35
1986	fx7000G
1987	HP 28C

2600 yrs / 260 yrs / 26 yrs / .26 yrs / .26 yrs

Figure 1. Important Calculator Dates

The typical mathematics faculty member notes these passing events and even occasionally participates in philosophical discussions on the implications of technology developments. However in their day-to-day practices, their courses – yes, even the ones in statistics – avoid any real integration of hand held calculating devices in the course content. We wonder why NCTM and all the professional organizations that focus on the precollegiate level of mathematics instruction are having such a rough time implementing their calculator recommendations when college level mathematics faculty are turning out to be some of the most conservative practitioners of all.

That will change and there are two observations that support the point. First of all, calculator proficiency is becoming a standard beginning level requirement in engineering schools. In the talks that I have given to mathematics groups, this observation is usually a surprise and quite a shock when they see the level of the expectations. If mathematics instruction is going to hold its on, even in its traditional role of preparing students for science and engineering programs, then major integration of calculators will have to become the norm. To illustrate this, Figure 2 reproduces the calculator proficiency examination that freshmen engineering students at Clemson University must pass two out of three trials with scores of 80% or better in order to remain in the College of Engineering beyond their freshman year.

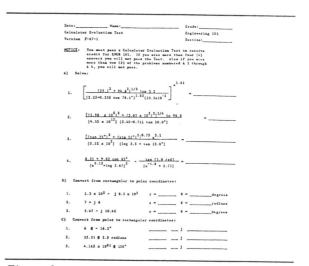

Figure 2. Freshman Engineer Calculator Test at Clemson University

To further support this point, I have very loosely charted the quality level of computer services that faculty and students have received at a typical university over the last two decades in

56

Figure 3. Usually, faculty members have immediate access to a new service when it is introduced and advanced students usually wait about five years before they receive the same advantage on a regular basis. Then it takes another five years for the undergraduates to routinely receive the service. This pattern was consistent throughout the 60's and 70's and it in now collapsing into a uniform level of service to the total university community.

Two significant breaks are worth noting. The initial bend in the undergraduate curve took place when students started receiving improved computer services on a regular basis. This was a result of faculty members realizing that computing was an important part of some undergraduate courses. The second bend is a result of the introduction of micro computers into the secondary schools. Prior to that point, undergraduates arrived on campus and received their computer instruction through variations of several types of batch processes. After all computing was essential to research projects and just tangentially important to undergraduate instruction. All those undergraduates would degrade the interactive systems at most universities and, with the exception of a very few schools like Dartmouth, undergraduates were not given uniform access to the university's interactive systems. When the new wave of high school graduates arrived on campus with screen editing experiences on their Tandy and Apple micro computers and then "graduated" to the big time of batch systems – all hell broke loose. The result was that the undergraduates would not regress to a lower level of service than they were use to in their high school environment.

That same explosion will take place when the new wave of calculator proficient students arrive on campus and they sit in the college classrooms being served mathematics at a lower level than they received in high school. Imagine the reaction of a "Casio" student watching a mathematics instructor spend thirty minutes finding the graph of a function when he has been sitting with it on his calculator screen from the very beginning. Yes, there are important contributions that college mathematics makes to the understanding of a graph, and we had better make them the focus of our curriculum. We need to accept graphing calculator technology and present college material as devices for understanding the graph instead of items that allow us to draw the graph. Otherwise we will have a growing credibility gap with the students and a deterioration of our programs.

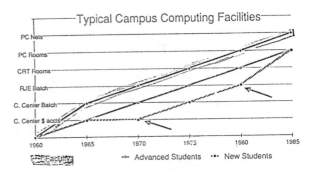

Figure 3. Computer service levels at a typical university

Why Calculus?

The National Science Foundation recognized the need for significant curriculum changes in collegiate mathematics and through its advisory groups established that an initial effort should be made at the calculus level. There was a high level of dissatisfaction with the current calculus course and a great deal of public debate and interest in "calculus reform." A modest conference at Tulane University in January, 1986, had produced a report *Toward a Lean and Lively Calculus* and considerable pubic debate was being generated by the publication. At the national meetings, panel sessions and talks about the problems in calculus were drawing "overflow" audiences. All this interest was focused at a National Colloquium, "Calculus for a New Century" sponsored by the National Research Council in Washington, DC, in October, 1987. Over 600 faculty members attended and a national crusade was launched.

The new NSF calculus curriculum initiative is a major element in the ongoing developments. In the summer of 1988, the initial 25 awards (Figure 4) were announced and automatically the recipient investigators were thrust into the national spotlight. Their programs are directed toward (i) new topic selection and arrangements, (ii) integration of technology, (iii) the integration of research topics and methods into the course and (iv) the establishment of program linkages and cooperative efforts. Their activities are being widely reported in articles in the *AMS Notices*, the *MAA Focus*, and the *SIAM News*. I will not try to detail the individual projects. One NSF grant was awarded to all three of the mathematics organizations to establish a publication, *UME Trends*, that will have as its exclusive focus topics in undergraduate mathematics education.

FY88 Calculus Awards

Multi Year:

Five Colleges Inc, New Mexico State Univ., Rollin s College
Colorado School of Mines, American Math. Society

Planning Grants:

Spelman, Univ. of Illinois-Urbana, Purdue, Iowa State, Boston U., Harvard, Macalester, Duke, Dartmouth, Univ. of NH, Ithaca Col., Rensselaer, Miami Univ., Oregon State, Comm. Col. Of Philadelphia, Penn. State, Univ. of RI, Furman, Washington State

Conferences:

University of Miami

Figure 4. FY 1988 Calculus Curriculum Awards

The NSF Calculus Curriculum program is now well established and the proposal deadline for the next year is February 1, 1989. There are several changes in the progam. As noted in Figure 4, the FY 1988 awards included many planning grants. In FY 1989, the planning grants have been essentially eliminated, and the thrust of the program will be in full scale curriculum projects. The program is seeking both large and small scale efforts. The large scale projects will be expected to have national impact and represent comprehensive approaches. The small scale projects will still be expected to have national impact, but their focus could be on a more restricted set of considerations. You could loosely descibe the difference between the two levels as "mover and shaker grants" and "exploratory pockets." There are two noteworth quotes from the current program guide and they each capture very important expectations in the calculus program.

"The typical calculus course captures little of the spirit and excitement of current developments in mathematics and does little to encourage and initiate the mathematics major." (NSF Calculus Program Guide, p. 1)

"Many of the leaders in the current calculus debate are individuals with international reputations. Their continued advice, counsel and participation in the calculus reform movement are essential." (NSF Calculus Program Guide, p. 3)

Even with all the public debate and discussion, you still encounter the basic question, "Why Calculus?" The selection is best explained with observations about the future need for individuals in the technical work force and the changing demographics of the student population. Everyone recognizes the increasing level of technology sophistication and especially the expansion of the mathematical content of essentially every professional field. Generally this recognition is acknowledged and everyone proceeds to presume that the nation will meet its work force needs with an expansion of its usual methods. The Brookings Institute reports that the change in the entering work force between now and the turn of the century will be only 15% "white male" – yes that's right 15%. When you realize that the typical high school graduate in that group is now beyond the first grade, we see that radical changes must be made immediately. Otherwise, the nation will continue to try to meet the majority of its technical work force needs from the white male population, and the task is impossible. The long time advocates of educational reform that have been motivated by moral and ethical considerations are now joined by the practical considerations of economic survival.

Calculus plays a key role in these considerations. We know that in general terms, from high school to the Ph.D., we lose 50% of the mathematics students each year. However, this loss is not uniform and a massive drop occurs at the beginning year of college and at the calculus course. Calculus is the one course that sits in a choke point position (Figure 5). It is the focus of most of high school mathematics and is the door way into essentially all of the technical careers in college.

The basis of the NSF consideration is that a major curriculum thrust at this important spot in the curriculum sequence might have immediate and consequential results. We'll see!

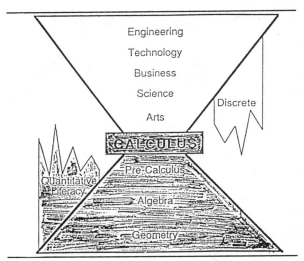

Figure 5. Calculus as a choke point course in the curriculum

Other Important National Science Foundation Programs

Mathematics faculty members should look to a variety of NSF programs to support their curriculum needs (Figure 6). As the community, including many members of this audience, develop the curriculum materials that we need to revitalize the calculus, then we will be left with the challenge of delivering the materials to our students in an effective fashion. In many cases that will involve computers and other technology devices. The proper place for the faculty members to look to at NSF is the Undergraduate Science and Mathematics Education Division (USEME) within the Science and Engineering Education Directorate (SEE). Their support includes the important Instrumentation and Laboratory Improvement Program. This is the ILI program, and it is under subscribed by mathematics departments. I would encourage each of you to familiarize yourself with the program and present a well thought out proposal to the next competition. The deadline is usually in November. The project should be curriculum driven. On that basis, tell the reviewer what you would do in curriculum improvement with equipment if you had the money. If you make a convincing case, then NSF will give you half of the funds that you need. Your institution will be required to provide a matching amount. However the program rules are flexible, and you do not need to have the matching funds "in hand" when you make the proposal. This point should be discussed with an NSF program officer in USEME and in general the matching funds requirement is not a handicap. Inertia on the part

of the mathematics faculty is a far larger problem. We need to realize that NSF responds to what is called "proposal pressure." Unless the mathematics community tells NSF that good mathematics instruction needs equipment, then the NSF funds will continue to be spent in the other disciplines (Figure 7).

Major changes in mathematics instruction will necessitate major retraining of large groups of instructors. Even with the new materials and new equipment, there still remains the major problem of faculty training. Since much of the curriculum innovation talent and many of the leaders in the usage of technology in the mathematics curriculum are part of the audience at this conference, then I would encourage each of you to get familiar with the NSF Faculty Enhancement Program. These funds give support to curriculum training workshops and institutes for faculty members, and it is the program that we should look to for funds to cover our faculty training needs. Mathematics has been an active participant in this program and a number of very successful programs are a credit to our participation. But we should not rest on our laurels. There is still a lot of need for faculty training.

NSF Programs
SEE - USEME

Calculus Curriculum Development
February 1, 1989

Instrumentation & Lab. Improvement
November 21, 1988

Undergraduate Faculty Enhancement
October 14, 1988
March 3, 1989

Figure 6. NSF Programs

	Requests		Awards	
Curriculum Development				
Engineering	193	$70.0	11	$1.90
Calculus	89	$26.0	25	$1.29
Instrumentation & Lab Improvement				
4-yr.	1110	$37.9	323	$9.79
2-yr.	171	$ 6.6	39	$1.10
Ph.D	931	$47.1		
Undergraduate Faculty Enhancement				
	136	$13.5	37	$2.80

(dollars in millions)

Figure 7. FY 1988 NSF Expenditures

The Future

I think that mathematics is on the threshold of very significant changes in its basic service curriculum. The impact of technology will be widespread. The secondary school program will be fundamentally changed by the availablility of simple, inexpensive graphing calculators with matrix capabilities. The college level program will change as a result of a different class of entering students, as well as the availability of symbol manipulating routines in computing devices. Especially the hand held devices.

Clemson University hopes to be on the leading edge of that change. Through grants, we have offered to a wide base of secondary teachers in our "feeder schools" a program that equips the teachers with Sharp *EL-5200* graphing calculators and a full semester course in the proper instructional use of the machines. We feel that graphing technology will have the significant impact that everyone is discussing. However, little debate has been focused on the changes in the curriculum that will result from the ready availablity of matrix operations in the calculators. I feel that capability will be equally significant. Students will be able to routinely solve "just in time inventory" problems and a wide slate of exciting mathematical applications of linear algebra. A serious challenge is now presented to the educational community to equip and train secondary teachers in the new opportunities that technology changes make available.

At the college level, Clemson University has a multiyear FIPSE grant to revise our undergraduate service curriculum in mathematics to include the assumption that each student will have an *HP-28S* calculator. In the pilot developments each student is loaned a *HP-28* and in the final implementation, each entering freshman will be required to purchase the unit to use throughout their undergraduate courses in mathematics. Our classroom and student computer labs will have micro computers for software demonstration and special assignments. We see the desk top and the hand held computers to be units that each have special advantages. In one you have more power, color and a wealth of software. All of these are especially useful in demonstrations and special assignments. In the other hand held units, you have infinite portablity and continual presence. Classes and especially tests will routinely call on the powerful capabilities of the machines. It is an exciting time in mathematics.

Already, the NSF calculus grants are beginning to impact the quality of calculus instruction in the colleges and universities. The presentations are still playing to "standing room only" crowds at professional meetings. The NSF program and the national debate has made it respectable again for mathematics faculty members to express an interest and a concern for the level of calculus instruction in their departments. The funded grants will make major contributions, but their results will be overshadowed by the wealth of improvements that are already taking place through out the departments at large and small institutions.

The basis calculus topics are still fundamental in the background of every educated individual. The presentations and instructional approaches to these topics will change and improve. The manner of how one uses calculus concepts will change, but the quantitative evaluation of change, motion, averaging, and smoothing will remain basic and essential.

TESTING, TEACHING AND TECHNOLOGY [1]

Alan Osborne

The Ohio State University

In higher education mathematics is at an interesting juncture. We are in the midst of a shift in the environment for processing ideas and information. Technology is becoming a pervasive element of the context in which we operate as faculty in colleges and universities. Changes in environment tend to be irreversible. Short of major traumatic events, technology is here to stay as a feature of the environment in which we live and operate.

The challenge is simple: Do we use the technology in teaching mathematics, or do we Latinize the curriculum and our instruction? We can turn mathematics into a dead language. Alternatively, we can examine that environment to see:

- How can we capitalize on the positive characteristics?

- Are there actions we can take to smooth the transition to incorporating constructive use of technology in instruction?

- Are there features that change how we should design curriculum, instruction, and testing?

John Harvey (these proceedings) has made insightful comments concerning testing and teaching. I am electing to elaborate on some of the themes he has established partly to reinforce some of his wisdom and partly to examine some other features of what I perceive as critical in using technology in mathematics instruction. Ten years ago, I could not imagine myself making a presentation of this sort. I resented the computer. I saw it between me and the mathematics I wanted to do or to teach; a clumsy hurdle that I elected not to hurdle. I saw mathematics teachers in high schools teaching courses titled "Computer Algebra II" where the teachers spent a third to a half of the instructional time teaching programming rather than mathematics. I saw that in most instances students were encountering significantly less mathematics and learning considerably fewer ideas and skills.

[1] Preparation of this paper was supported in part by funding by the National Science Foundation Grant *Mathematics Through Technology: Establishing Concepts and Skills of Graphing and Functions in Grades 9 through 12* (Grant #TPE-8751353). Conclusions and findings are those of the author and do not necessarily represent the views of the National Science Foundation.

But the computer software and hardware has changed. Access to mathematics via technology is easy and direct. Menu driven technology and significantly greater power makes the mathematical ideas readily available. Indeed, I can readily shift to instructional methods that help students deal directly with problem solving, modeling, and generalization concerning a variety of mathematical ideas. The computer appeals to my basically conservative orientation to curriculum and instruction: I can feature mainline skills, understandings, problem solving and proof better with technology than without. I project this improving as the technology continues to evolve.

Following, I will identify some critical issues that we need to deal with in teaching and testing mathematics in a technology enhanced environment for mathematics instruction.

Testing

Computers and calculators should be used throughout testing in mathematics. However, we need to enlarge testing and evaluation to encompass one feature of student performance seldom recognized as important in the environment for doing mathematics. We must begin to focus on whether students choose the most appropriate tool for a problem situation in mathematics. Indeed, we need to adjoin this type of behavior to our set of goals for mathematics instruction. Last week I observed a student in precalculus locating the zeros of a function. The function was quadratic and readily factorable. The student in this quiz situation elected to graph the function with a graphing utility. He made some very correct moves indicating he understood the mathematics of functions and graphs. He even dealt with the idea of error correctly. But, he had no decision tree to help him decide whether it was appropriate to work with paper-and-pencil or to use the computer graphing package. He wasted considerable time in changing the viewing rectangle and dealing with error in reading the graph; he should have simply factored the polynomial to obtain the zeros. He did not look at the exercise and make a decision about whether he should use the technology or not.

We must work to teach the judgment of which tool to use to do mathematics. Further, we must design testing alternatives that will examine whether or not students are attaining that judgment. I think that Harvey's comment about the fear of the technology turning students' heads to mush reflects, in part, that we have not recognized that we need to expand what we look for in examining mathematics behavior to encompass the choice of tool to be used for particular mathematics. The National Council of Teachers of Mathematics *Curriculum and Evaluation Standards for School Mathematics* (Commisision on Standards for School Mathematics, 1989) argue that this is an important judgmental skill for every level of the curriculum. Figure 1 indicates the decision tree that the *Recommendations* advocate for every student to understand and employ. I think the decisions are important for students at the university level.

What Do Test Items Measure? An issue of prime concern is whether testing in the paper-and-pencil format does accommodate without significant adaptation to evaluation in a technological environment. We do not always know what students respond to or what understandings trigger correct responses.

Following are four items (Figures 2–5) we used in testing precalculus classes in the Calculator and Computer PreCalculus (C²PC) Project. The items are relatively standard fare for functions and graphing. The items were tested in two formats, one with a graph present and the other without the graph. The data given are the correctness rates on the items in the two formats. In some cases it appears the graph is a significant help; in others, not.

The first item appears on the surface to be of a type that the graph should help the correctness rate. In fact, the data indicate no such power obtains from the presence of the graph. One can argue that the typical student appears to summon the two-point algorithm and, if anything, the graph confuses. In an era when graphing utilities and calculators are readily available, do we know what we are testing?

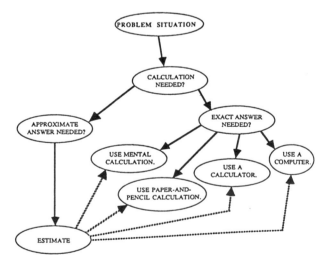

In a Cartesian coordinate system, what is the equation of the straight line passing through point $(0, -5)$ and parallel to the straight line whose equation is $y = 2x + 3$?

A. $x + 2y + 5 = 0$
B. $2x - y - 5 = 0$
C. $2x + 3 = -5$
D. $2x - 5y + 3 = 0$
E. $2x + y + 5 = 0$

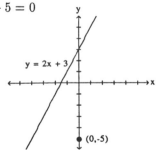

Figure 2. First C²PC Example Item

	Pretest	Posttest
Graph	53	74
No Graph	50	73

Figure 1. Decisions About How to Calculate

The second item appears to be moderately easier without the graph. Can we tell whether students think numerically or examine and interpret the graph? If a student enters this item on a graphing utility, it is the same as having the item given in the graphing format. Apart from the time it takes to enter the function on the utility, do we know what is triggering the students' responses?

If $xy = 1$ and x is greater than 0, which of the following statements is true?

A. When x is greater than $1, y$ is negative.

B. When x is greater than $1, y$ is greater than 1.

C. When x is less than $1, y$ is less than 1.

D. As x increases, y increases.

E. As x increases, y decreases.

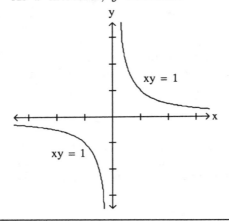

Figure 3. Second C^2 PC Example Item

	Pretest	Posttest
Graph	66	72
No Graph	61	78

The remaining two items have results that are in more of the predicted direction. The graph appears to help in producing correct answers. However, the use of a graphing utility completely rearranges the time demands of the item. Our judgments of how long an item will take go out the window with the use of a graphing utility. And we need to assure some sort of comparability of equipment for our students.

Which of the following, $(x-1), (x-2), (x+2), (x-4)$, are factors of $x^3 - 4x^2 - x + 4$?

A. Only $(x - 1)$
B. Only $(x - 1)$ and $(x + 2)$
C. Only $(x - 2)$ and $(x + 2)$
D. Only $(x + 2)$ and $(x - 4)$
E. Only $(x - 1)$ and $(x + 4)$

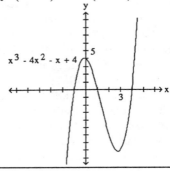

Figure 4. Third C^2 PC Example Item

	Pretest	Posttest
Graph	44	81
No Graph	35	54

The function f, defined by

$$f(x) = \frac{(x - 1)(3x + 1)}{(2x - 1)(x - 2)},$$

is negative for all x such that

A. $-\frac{1}{3} < x < 3$
B. $-\frac{1}{2} < x < 2$
C. $1 < x < 3$
D. $-\frac{1}{2} < x < 2$ or $2 < x < 3$
E. $-\frac{1}{3} < x < 3$ or $1 < x < 2$

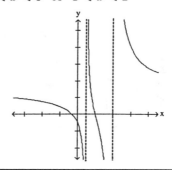

Figure 5. Fourth C^2 PC Example Item

	Pretest	Posttest
Graph	25	64
No Graph	23	39

At issue: What features do we build into items when students have ready access to computers or calculator graphers while taking tests? Do we know what students react to or what they have learned? How does the ready availability of a graph change our information base for judgment concerning what our students know (and how we should assign grades)?

The above item performance was offered without careful analysis nor much comment. In fact, the Ohio State group has been examining more systematically how students acquire and use graphical information. Vonder Embse (1987) explored eye fixation patterns of students in reading graphs of polynomial functions, and Browning (1988) developed a graphing levels test. These studies indicate that the assumption common to mathematics instructors that a graph has intuitive explanatory power simply does not wash for most students. The behaviors and understandings are learned but are significantly more complicated than most mathematicians realize. We must teach students how to use graphs and help them build the intuitions that associate functions and graphs. We need more studies of graphing behaviors in order to tailor instruction to uses of technology. Technology intensifies that old problem of knowing exactly what given items measure. As we move to using computer graphics and symbol manipulators in testing, we must assiduously address this problem recognizing that the new testing environment means that we are operating in a context that is somewhat new and that we will make some mistakes. We are convinced that students should be tested with technology.

Using Technology to Restructure Management of Testing. Third, can we use technology to assess student performance more thoroughly and efficiently? For example, computer-based item pools allow keeping records on a per item basis in order. The power, as well as the effectiveness, of items in assessing given behaviors for particular types of students can be used to improve assessment. A bit of attention to developing such item pools will allow building a better match among courses, curricula and testing. A second element of using technology to restructure assessment can provide a means of decreasing the investment of student time in writing tests. Computer generated testing provides the capability of using the response to a given item i to guide the selection of the next item $(i + 1)$ as indicated in Figure 6 below. The potential of response patterns indicating sorting levels for selection of next

items can shorten the number of items students respond to appreciably. This assumes attention will be given not only to ascertaining the characteristics of items but also what given responses to items mean. Lord (1980) provides a helpful discussion of design, statistical and administrative features of such a tailored testing program.

Students in freshmen and sophomore level courses typically expend five to eight clock hours for exam taking in every mathematics course. How would you use the additional instructional time if you only had to invest one to two hours for testing? The response-dictated item selection process demands a different view of testing than most of us have grown up with and may require that mathematics departments consider thinking about acquiring staff with the appropriate expertise in testing mathematics and using technology to work with the instructional staff.

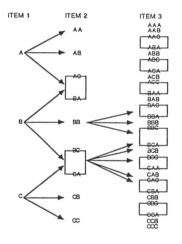

Each item i is multiple choice producing three levels of response. The level of response to item i dictates the selection of the next item $(i + 1)$. Levels of response on three items with perfect branching would "sort" students into up to 27 levels. Assigning items with overlap for the bottom of one level and the top of another as indicated by the boxes would allow students to recoup, in part, poor performance on a given item. Thus, after three items, we would have 19 levels.

Figure 6. Assignment of Levels of Performance through Response Dictated Item Selection

In summary relative to testing, Harvey and I share the strong belief that technology in the form of calculators and computers must be used in testing and evaluation. Evaluation must include how students select which computational tools to use for given mathematical situations. Use of technology means that our insights into what given items measure must be expanded and adjusted for the change in the context of testing. Finally, the use of computers to manage the testing process offers the potential of improving the match between testing and curriculum as well as providing different means of structuring tests. However, part and parcel to restructuring testing means that we must gain more information concerning item and response characteristics.

Teaching

The use of technology will redirect your teaching and your thinking about curricular emphases. Decisions you make concerning content selection and treatment will change. I have visited a number of classrooms during the last two years that are using the technology enhanced curriculum of C^2PC. The following comments concern the redirection of teaching and the decisions that seem obvious to me from my observations.

Exploring, Experimenting and Problem Solving. Use of graphing utilities encourages a more mathematical behavior on the part of students. Given a function, students can try things to see the effect of given parameters. One of our students in working with polar equations accepted the challenge of figuring out the effects of change of parameters in $r = a\sin n\theta$ by experimenting with the choice of n. Most mathematicians have a good sense of what happens with different positive integers. Our student experimented with rational number replacements of n as an extension. (Do you know what happens for $n = \frac{k}{2}$ for different choices of k?)

The good news is that such experimenting and exploring leads to an attitude of conjecture making and testing. We want to encourage such powerful mathematical behaviors. It seems a natural outcome of using computer or calculator graphing utilities if you are experimentally inclined. The bad news is that ready access to graphing utilities does not necessarily lead to experimentation and exploration. In fact, teachers can be as didactically rule-oriented with exploratory tools as they were without. They can damp students' exploratory ventures. Teachers must instruct for the goal of conjecture making and testing; it doesn't happen automatically.

Students who matriculate in primary and secondary schools through mathematics dominated by the traditional three-step instructional process–giving a rule, working an example and directions to go do likewise with 30 similar exercises–need to be carefully introduced to exploration and experimentation as the mode for doing mathematics. Teachers, whether at the school or university level, need help in designing instructional activities to promote the shift from rule-dominated mathematics to a more exploratory mode.

The Information Base for Change. The rigidity of many mathematics department faculty across the country in thinking about the use of technology in lower level mathematics courses is impressive. Harvey discusses the reasons for resistance to change identifying laziness, apprehensions concerning change, the fear of a mush-brain mentality being created within our students and the sheer magnitude of the problem of providing sufficient equipment in many of university settings.

It is hard to quarrel with Harvey's discussion of the reasons for mathematics programs failing to change. I would like to supplement and reinforce his arguments. Many mathematics faculty are not informed concerning effects of using technology in teaching. Research evidence is overwhelming in favor of using calculators. My colleague Marilyn Suydam who ran the National Institutes of Education supported Calculator Information Center for several years carefully collected information about effects of calculators on computational skills for several years. Her compilation of studies for which students were evaluated on paper-and-pencil computation but which compares performance of groups taught with calculator to those taught without summons attention. The results of more than 100 such calculator horse race studies are summarized in Figure 7. A betting person would favor the use of technology; 47 percent of the studies have the groups taught with the calculator winning on the paper-pencil evaluation tasks and only in seven percent of the studies does the paper-and-pencil group win the evaluation race.

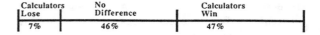

Calculators Lose	No Difference	Calculators Win
7%	46%	47%

Figure 7. The Calculator Horse Race

Most faculty in mathematics departments do not read mathematics education research literature and are not familiar with such results. They argue for or against curricular and instructional reform with no knowledge base often incorporating the specious logic of generalizing from a single personal instance. In an era of profound change in the environment for teaching and learning mathematics, mathematics faculty should read about the attempts to study systematically the effects of using technology in instruction in the literature of mathematics education. Dealing with the ignorance of colleagues while attempting to push through a proposal for a course modification consumes your creative energies that could better be invested elsewhere.

David Cohen made the interesting comment at the recent conference of the Psychology and Mathematical Education convention in DeKalb that we should not necessarily expect the research institutions to lead the way in innovation. Commitments to research and the correlated incentive and resource structures impose a barrier to change of instruction at such institutions. New equipment monies, for example, are invested to support research activities. More rapid implementation may be expected at those institutions more directly committed to investing resources in instruction than at the premiere research institutions until we arrive at a juncture where technological performance capability affects research productivity of faculty and students.

Pressures for Change from Below. A large number of schools and teachers have shown strong interest in the C^2PC precalculus course. Zalman Usiskin has noted comparable inquiry rate for School Mathematics Project materials that make extensive use of technology. Other evidence indicates that many teachers are making extensive use of technology in teaching. Experience suggests that the better high schools, the ones sending a major portion of students to tertiary level education, are the ones who have the resources of money and personnel to explore technology enhanced mathematics and are making pace setting ventures in curriculum and instruction. This is to say; a significant portion of the students to whom we would normally look as the source of good, sound mathematics majors enter the university with experience using technology in doing mathematics.

Query: How will such students behave in university mathematics courses that make no use of technology? Query: If you are required to teach technology-free mathematics, how will you treat such students? Query: Will such students elect to major in mathematics if they encounter freshmen and sophomore level courses that make no use

of technology particularly if other departments do feature technology in course work?

I think that if university mathematics faculties are not careful they will drive many excellent students to fields that have already joined the modern age. Students who come from school mathematics programs that use technology constitute a force for change at the college and university levels. Their expectations need to be honored or they will flee the Latinized curriculum.

An ancillary problem is specific to my field of teacher education in mathematics. Most teachers teach in the manner they were taught. Most of our prospective teachers will teach as they were taught. If our prospective teachers do not encounter use of technology in their mathematics classes, we cannot be surprised if they do not use technology in their teaching at the school level. Prospective teachers need regular, recurring encounters with technology in doing mathematics. We do not have enough time available in methods course work to make it happen without the constructive help of mathematics instructors. They need good models of use of technology in mathematics teaching to serve as a foundation for teaching.

Representations and Generalizations in Mathematics. I am speaking to the converted or as Harvey says, "the convicted." You know using technology changes how you do mathematics. For example, instead of expending hours generating the graphs of a very few parametric equations with pointwise plotting, in a small amount of time you examine several different parametric situations. Learning and using skills has been a labor intensive activity for our students. We have valued having those skills under sufficient control to allow students to operate efficiently; however, using them to extend ideas to new situations is remarkably difficult simply because of the nature of the skills.

Now, those hard won skills are not so important. We can examine many different but related examples easily. Often establishing a generalization has been difficult for a teacher because it takes so long to build the numerous exemplifications of an idea. Indeed, the finding of an instance may require a single computational process that is difficult to apply even though readily understood by a student. Generating the instance may have been so time consuming that the point is lost to the point of interfering with the cognitive processing required to form the generalization.

Our curriculum development efforts with C^2PC and in the seventh and eighth grade level with the Approaching Algebra Numerically (AAN) project convince us that it is easy with the use of technology to build ability to generalize with students. They readily construct many instances. A numerical or graphical problem solving base is accessible to all students. They can extend ideas. In AAN, such a numerical problem solving base generated with the use of scientific calculators is used with good success to establish the idea of variable, a concept domain critical to understanding basic algebra. In C^2PC, examination of many different instances allows students to fix the effects of changes in parameters in their thinking. Students discover and generalize ideas such as phase shift and amplitude change from graphs often before the instructor formally focuses their attention on the key idea. Transformations such as stretching, shrinking, translation and reflection are readily associated with shifts in parameters.

Teaching methodology changes as a result. Instructors find, according to our observations and the reports of teachers, that they focus on different aspects of mathematics. They are able to highlight making generalizations, problem solving and mathematical modeling.

Teachers report an ability to focus on other features of mathematics. We have given mathematics short shrift because of a narrow concentration on computational skills. Our C^2PC teachers report that they are able to focus on communication in mathematics to a much greater extent. Learning to read mathematics has become more important in their classrooms. They think situational aspects of the technology generated graphs produces more talk about problems and representations resulting in sharpened ability to communicate mathematics. As teachers deal with the best ways to use the technology they place a premium on classroom and laboratory arrangements and management processes that extend the conversational, communicative aspects of dealing with the mathematics. We are pleased that teachers are making ventures and experiments to test different methodologies than they have used in previous, technology-free instruction.

Teachers find the ready capability to produce equivalent graphs of a problem situation with the modifications resulting from choices of viewing rectangles and other technology induced variants yields an interest and focus on representations. A natural premium arises in contrasting different algebraic and geometric representations of the same functional relationships. This premium on representations builds mathematical modeling capabilities in the problem solving tool kit of students.

In summary relative to teaching, use of technology appears to force some changes in pedagogical orientation. A common pitfall to the noviate in using technology is to apply blindly techniques of didactic rule-giving that are inconsistent with the experimental, exploratory possibilities inherent in the technology. Our experience indicates, however, that many instructors respond to the capabilities produced by technology to change teaching methodology and to value different, more powerful mathematical behaviors as outcomes of their instruction.

Concluding Statement

We can Latinize mathematics at the university level or we can take advantage of technology. There is powerful impetus to change and, currently, strong pockets of opposition to what I trust is a natural evolution in curriculum and methodology. It is easy to forget that the examples we use and depend on in instruction via textbooks, blackboards and paper-and-pencil activities have evolved over generations. The mathematics community has a collective base of traditions in instruction of techniques and methods that serves to guide what we do. We are bound to make some mistakes in implementing technology due to having little such collective experience to guide problem selection and teaching methodology.

Curricular position statements such as *Toward a Lean and Lively Calculus* (1986) and *Calculus for a New Century* (1987) paint a picture of an undergraduate curriculum that does not fit present realities. Mathematics is more widely used and is applied in a variety of fields seldom represented in current instruction. Present realities include technological advances that should affect content selection and sequencing. This conference demonstrates the wide variety of ways in which the use of technology makes us rethink curriculum and instruction at the undergraduate level.

The NCTM's recommendations *Curriculum and Evaluation Standards for School Mathematics* (1989) prescribe a future for school mathematics consistent with the intents of *Toward a Lean and Lively Calculus* (1986) and *Calculus for a New Century* (1988). Themes of problem solving, communication in mathematics, and higher order cognitive functioning point toward products of the schools that should be quite different than those students who are currently the targets of instruction at the undergraduate level. We must take advantage of the impetus for change in order to be ready for such students.

References

Browning, C. C. (1988). *Characterizing levels of understanding of functions and their graphs.* (Doctoral Dissertation, Ohio State University.)

Commission on Standards for School Mathematics (1988). *Curriculum and evaluation standards for school mathematics.* Reston, VA: National Council of Teachers of Mathematics.

Douglas, R. G. (ed.) (1986). *Toward a Lean and Lively Calculus, MAA Notes #6.* Washington: Mathematical Association of America.

Lord, E. M. (1980). Tailored testing (pp. 150-161) in *Applications of item response theory to practical testing problems.* Hillsdale, NJ: Lawrence Erlbaum.

Steen, L. A. (ed.) (1988). *Calculus for a New Century: A Pump not a Filter, MAA Notes #8.* Washington: Mathematical Association of America.

Vonder Embse, C. (1987). *An eye fixation study of time factors comparing experts and novices when reading and interpreting mathematical graphs.* (Doctoral Dissertation, Ohio State University.)

Using Computer Algebra Systems in Calculus

Jeanette Palmiter

Kenyon College

Kenyon College is in its second year of using Computer Algebra Systems, MACSYMA and Maple, in the entire calculus sequence. After studying reports [1 and 2] on the need for calculus reform, the mathematics department agreed to use MACSYMA, and later, Maple, in the three semester calculus sequence. Our goal was to use the software package to compute limits, derivatives, and integrals that made up the bulk of the calculus curriculum. Now, the bulk of instructional time is devoted to teaching the concepts of calculus. We want our students to understand the meaning of limit, derivative, integral and how to set up and analyze the computations related to these concepts. Students view what instructors say and what instructors spend the majority of instructional time on as important. Therefore, now our students are learning that calculus is not merely a bag of computational tricks.

The syllabus has to change, not so much in the topics that are presented, but in the methods used to present the topics. The majority of lecture time is spent exploring definitions and theorems through graphs, applications, examples, and counter-examples in which the ideas are emphasized and the computer is used as the tool to carry out the computations. In Figure 1 is a list of traditional first year calculus topics that have been omitted, that receive reduced attention, or receive additional attention. Kenyon uses Stein's Calculus and Analytic Geometry textbook. Notice that many of the techniques of integration have been deleted. The time saved by not teaching these computations allows us to delve into certain other topics.

Deleted topics:
   Integration of rational functions
   Integration by partial fractions
   Integration of powers of trigonometric
      functions
   Integration by trigonometric substitutions

Deemphasized topics:
   Computation of limits
   Product, quotient, and chain rules for derivatives
   L'Hopital's rule

Added/expanded topics:
   Bisection method (supported by computer graphics)
   Applied max/min problems
   Newton's method
   Deriving Simpson's method
   Estimating definite integrals

Figure 1. Changes in the Kenyon College Calculus Courses

Currently, the calculus courses are using Maple, Version 4.1, on the time sharing VAX 8600 running VMS. MACSYMA is also available on the system, but using VMS, it can only support a few simultaneous users. Maple is our system of choice for class use. A number of terminals are available around campus and in the dormitories. A terminal with an overhead display device is in each mathematics classroom. Each mathematics faculty office has a terminal.

A lab session in the Computer Center introduces beginning calculus students to Maple at the start of the semester. The Computer Center has 27 terminals so that each individual gets hands-on work with Maple as the program is being demonstrated. The student is also provided a Maple Card, a mini-manual of Maple detailing how to use the program and commands. Students who have never used a computer find Maple easy to learn and use. Programming is not required.

Exams are given in the Computer Center. Maple is used as a super-calculator during these exams. Although it would be foolish to have numerous computations on the exam, some are included to determine how well students use Maple. The bulk of the exams are comprised of conceptual questions. Below are some sample problems taken from final exams and worksheets in the first year calculus course:

1. Let $T(t)$ represent the temperature in degrees at time t (in hours) on a certain day. Explain the behavior of the temperature given the following information:

$$T(4) = 45, \quad T'(4) = 0, \quad T''(4) = 3,$$

$$\frac{\int_0^4 T(x)\, dx}{4} = 30$$

2. Let $f(t) = \int_{-1}^{t} [\cosh(x^2 + 1) - x - 2] \, dx$

Find the value(s) of t on $[-1, 1]$ so as to maximize $f(t)$ and to minimize $f(t)$. (WHAT FOLLOWS (Figure 2) IS A MAPLE SOLUTION)

```
>   y := cosh(x^2 + 1) - x - 2 :
```
                (CONSIDER THE INTEGRAND)
$$y := \cosh(x^2 + 1) - x - 2)$$
```
>   plot (y, -1 .. 1);
```
(PLOT THE INTEGRAND, NOTING THE ZEROS, A AND B)

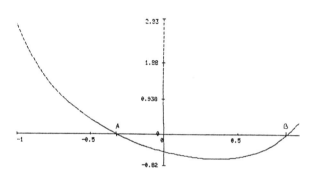

```
>   fsolve (y = 0, x);        (NUMERICALLY
                          COMPUTES A ZERO)
                -.3244352253
>   b := fsolve (y = 0, x, 0.5 .. 1);    (COMPUTE
THE OTHER ZERO BY SPECIFYING AN
        INTERVAL IN WHICH IT OCCURS)
                .8390836331
```
(BY THE GRAPH IT'S OBVIOUS THAT THE MAXIMUM OCCURS AT POINT A. IT IS NOT SO CLEAR IF THE MINIMUM OCCURS AT THE POINT B OR AT THE ENDPOINT $t = -1$.)

```
>   int (y, x = -1 .. b);
```

$$\text{int}(\cosh(x^2 + 1) - x - 2, x = -1 \ldots .8390836331)$$

(MAPLE WAS NOT SUCCESSFUL IN FINDING AN ANTIDERIVATIVE)

```
>   evalf (");
```

(NUMERICALLY EVALUATES THE PREVIOUS DEFINITE INTEGRAL)
                .1953158959

THE MINIMUM OCCURS AT THE ENDPOINT $t = -1$

Figure 2. A Maple Solution for Problem 2

3. Let the sketch in Figure 3 depict the graph of $f(t)$ on $(-\infty, \infty)$

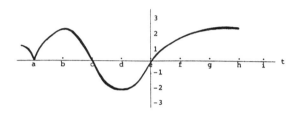

Figure 3. The Graph of $f(k)$

a. Compute $\lim\limits_{t \to a} \dfrac{f(t) - f(a)}{t - a}$.

b. Where does the Mean Value of the Derivative occur on the interval $[c, e]$? EXPLAIN

c. Where does the Mean Value of the Integral occur on the interval $[a, e]$? EXPLAIN

Students are permitted to use Maple on all homework assignments and exams. They soon begin to realize which problems lend themselves better to mental or to pencil-and-paper computations (i.e., differentiating a polynomial). Worksheets, developed by the faculty, supplement the text by extending examples especially suited to the use of Maple.

Several students have had calculus in high school and find the CAS-calculus "different, but not necessarily more difficult." They comment on how much better they now understand calculus.

Although devising more conceptual exams, lectures, and assignments takes some creativity, the payoff is clear from the improvement of understanding demonstrated by the students.

Computer Algebra Systems such as Maple and MACSYMA are becoming more powerful and cheaper each day. Hand-held calculators with computer algebra are already on the market for less than $200. We do our students a disservice by not introducing them to the latest computational devices, especially students in majors where calculus is used as a tool.

References

1. Douglas, R. G. (Ed.). (1986). *Toward a lean and lively calculus.* (MAA Notes Number 6). Washington, DC: The Mathematical Association of America.

2. Steen, L. A. (1987a). *Calculus for a new century: A pump, not a filter.* (MAA Notes No. 8). Washington, DC: The Mathematical Association of America.

Calculus&*Mathematica*™: Background and future

Horacio Porta and J. J. Uhl, Jr.

University of Illinois at Urbana–Champaign

The Nationwide Crisis in Calculus Teaching

That calculus teaching is in a crisis is underscored by two national conferences on calculus teaching within the last three years. The first was a large workshop at Tulane University in 1986. This conference stimulated a second conference sponsored by the National Academy of Engineering, the National Academy of Sciences and the National Research Council in late 1987. Robert M. White, President of the National Academy of Engineering said, "Calculus is a critical waystation for the technical manpower this country needs. It must become a pump instead of a filter in the pipeline... We need to teach the calculus in a way that facilitates complex and sophisticated numerical computation in an age of computers. Somehow or other you have to make calculus exciting to the students."

Lynn Arthur Steen, then president of the Mathematical Association of America, bemoaned the fact that at least 80% of the questions asked in typical American calculus exams can be answered by a command key on a (moderately priced) *HP 28S* and went on to ask whether it makes sense to devote most of our calculus to forcing our students to do by hand rote computations that can be better and faster done by computers. Steen predicted, "Mathematics–speaking machines are about to to sweep the campuses... Template exercises and mimicry mathematics, staples of today's texts, will vanish under the assault of computers. By using machines to expedite calculations, students can experience mathematics as it really is – a tentative exploratory discipline in which failures yield clues to success... Weakness (in high school background) will no longer prevent students from pursuing studies that require college math." Perhaps Steen was thinking of the British philosopher and mathematician Alfred North Whitehead who said, "Civilization advances by extending the number of important operations we can peform without thinking about them."

The problem is acute at Illinois. Of the 900,000 American students currently enrolled in calculus, 3500 of them are taking calculus at Illinois. Yet calculus, the fundamental course that teaches scientific and engineering calculations and computations, is unsupported by computers at Illinois. With this evidence, we can take it as a fact that calculus at Illinois and elsewhere needs a new direction and that direction points squarely at integrating the computer into mainline calculus.

But integrating the computer into today's standard calculus course cannot alone solve the basic problems of calculus teaching. The whole course, from foundations to exercises, must be overhauled. The best way to begin the job of writing the calculus course for a new century is to reflect on what calculus is.

The Evolution of American Calculus Courses

Calculus coalesced under the influence of Newton and Leibniz who transformed formerly disjoint procedures into a powerful instrument for systematic calculation. Their revolutionary contribution shook the scientific community of the late 17th century with its almost magical ability to replace many separate ad hoc procedures by a unified procedure. For the next three hundred years, this magnificent intellectual tool solved concrete scientific and technical problems and furnished the framework for the understanding and development of various branches of science.

Until the 1960's, calculus instruction emphasized problem solving. In America, the pre-1960 era saw the heyday of texts such as Granville, Sherwood and Taylor and a few others. During the 1960's, calculus courses gave way to the "new mathematics" of the sixties which influenced mathematics instruction from elementary school through college. It became the fashion to believe that calculus should be offered as a branch of real analysis and the abstract foundations of real analysis began to work their way into calculus books as foundations of calculus. This prompted a shift to what was believed to be "rigor" which pleased most members of the mathematical community until it was realized that the boom in mathematical rigor was a bust for calculus. We are still on the rebound from this era.

The more recent development of calculus has been influenced greatly by the steady expansion of the list of college majors for whom calculus is required. This has spawned an "applications" boom among calculus book publishers and authors rushed to produce texts replete with problems and examples to illustrate how calculus is used in other fields. These texts do not respect the mathematical integrity of calculus but try to justify calculus in terms of other fields of study. In addition the constraints of available textbook space and the very limited student knowledge of these fields of application often make the applications meaningless and pay only lip service to the underlying mathematics in a frantic dash toward application templates and rote procedures.

The reaction we are experiencing today should have been predictable. The essential content of calculus is lost. Expectations concerning mastery have steadily declined. Rote procedures have taken over. Authors protect their image of erudition by writing defensively and by pointing out limits of the theory instead of emphasizing its positive aspects. Many of the calculations discussed in today's texts appear only in calculus texts and have dubious applicability. The students end up not knowing which is more important: underlying principles, unrealistic problems or familiarity with rote procedures for hand computation.

Henri Lebesgue described it best in an essay about French entry level mathematics of the 1930's:

"The teachers must train their students to answer little fragmentary questions well and they give them model answers that are often veritable masterpieces and that leave no room for criticism. To achieve this, the teachers isolate each question from the whole of mathematics and create for this question alone a perfect language without bothering with its relationship to other questions. Mathematics is no longer a monument but a heap."

Throughout America authors and professors persist in the belief that the foundations of advanced real analysis are the right foundations for calculus. For example, early in the typical calculus text the theorem that guarantees that a continuous function has a maximum and minimum value on each compact interval is stated. The student is told that understanding why this theorem is true is beyond the ability of the student. This message to the student is equivalent to a quick mathematical lobotomy. The student is told that his intuition cannot be trusted. No wonder the student wants to be treated as a machine; we've told the student not to think! Feeling for the positive aspects of

calculus is lost and students willingly offer themselves up as robots waiting to be programmed to produce nice solutions to stock problems.

One trouble professors have with this theorem on continuous functions is that this theorem applies in complete generality even to functions that are everywhere continuous but nowhere monotonic. No one ever tells the student about such functions but these functions lurk in the brains of most professors. Lebesgue also addressed this issue:

"(There) is a real hypocrisy, quite frequent in the teaching of mathematics. The teacher takes verbal precautions which are valued in the sense he gives them but that the sutdents most assuredly will not understand the same way...(many teachers) say with irony 'Fashion dictates precision at one point in the course and all sorts of liberties at another.' The good students have seen enough to be skeptical, too, intead of enthusiastic.... We must attempt an overhaul of the whole structure."

This is the natural result of using foundations that are inappropriately general and abstract. We propose an overhaul of the whole structure from foundations up.

The Right Foundations of Calculus

Here are some key aspects of what we believe are the right foundations and perspectives for our calculus course:

1. The right functions to study in calculus are the piecewise monotone functions. Continuous nowhere monotonic functions are exciting in advanced real analysis. They are useless in calculus. It is clear that any piecewise monotonic function achieves a maximum and minimum value on each compact interval. It is equally clear that a piecewise monotone continuous function has the intermediate value property. Chalk up two victories for the students' intuition.

2. The derivative and integral are used to measure slopes and areas rather than to define slopes and areas. A farmer's field has area even if one border of the field is along a winding river; most farmers can measure this area without the benefit of the abstract definition of the Riemann integral. This view is based on the MAA publication of the notes of Emil Artin's Princeton calculus course of the fifties.

This viewpoint builds confidence and intuition in the mind of the average student. We add that it did not seem to interfere with Hyman Bass's development as a mathematician.

3. Within the context of (1) and (2) above, we proceed with rigor. Within this context the fundamental theorem is easy to prove and understand. Furthermore, formulas for arc length, volumes and the like can be derived in context; within this context, we can expect the students to be responsible for some derivations.

4. Pointwise continuity and pointwise convergence are scrapped in favor of uniform continuity and uniform convergence on compact intervals.

This is also the position taken by Peter Lax and is quite nicely carried off in his Springer UTM calculus course. The reason that we take this view is that uniform convergence and uniform continuity are what the student sees through *Mathematica* plotting. Uniform continuity says simply that for each pixel size, if h is small enough, then the graphs of $f(x+h)$ and $f(x)$ as functions of x are indistinguishable on the screen. Uniform convergence of a sequence to a limits says that for each pixel size the graphs of the late members of the sequence are indistinguishable from the graph of the limit on the screen. What a pleasure it is to see the reaction of students when they watch the difference quotients for $\sin(x)$ converging uniformly to a function whose graph is evidently the graph of $\cos(x)$. Mathematics itself recognizes that uniform convergence is a more elementary idea than pointwise convergence. After all, the uniform limit of a continuous function is continuous, but understanding the possibly pathological behaviour of pointwise limits of continuous functions took the mathematical power of Baire to uncover.

5. The role of approximations is stressed. Knowing that a series converges but not being able to estimate its sum is of dubious value.

Showing that a Taylor series of a function converges at the endpoints without indicating why it converges to the function is a waste of time. In fact the study of series of numbers is not so important as the study of series of functions with estimates on the quality of convergence.

6. Most calculus courses base their study of series of numbers on the axiom that a bounded monotone sequence converges. Operating with this axiom requires a moderate amount of real-variable skill. We replace this axiom with an equivalent axiom:

Given a series, then exactly one of the following two phenomena happens:

(a) The series is convergent.

(b) The partial sums are unbounded or there are numbers a and b with a < b such that infinitely many partial sums are below a and infinitely many partial sums are above b.

This axiom is, of course, equivalent to the axiom that bounded monotone sequences converge. The advantage is that students can understand it and they can work with it. Students at Illinois taught this way in spring 1988 broke the curve on a uniform final in Calculus II.

There is one more point. Students arriving in calculus usually have formed, sometimes in a rather fragile way, a significant foundation of mathematical intuition. Perhaps the most serious mistake made in traditional calculus courses is the destruction of this intuition. The message that students need the intermediate value theorem in order to understand why a cubic polynomial has at least one real root is comforting to the professor and terribly destructive to the students' sets of intuitions. Students who have been working with logarithms for two years and arrive in calculus only to be told that they cannot understand logarithms and exponentials until they see integral calculus emerge confused because their underlying foundation has been damaged. Confidence is lost. Calculus courses and instructors must relinquish the role of acting as curators of dogma and assume the role of builders of mathematical feeling.

Next we give some background on how we plan to integrate *Mathematica* into mainline calculus.

The Appearance of *Mathematica*

Both of us had spent considerable time in the department's curriculum committee on the problems of calculus at the Urbana campus. We both agreed with the nationwide feeling that standard calculus is not successful. On the other hand, we could find no suitable text that did what we and many others thought should be done. As for computers, the only available software was either too primitive or needed exotic expensive computers. Then one day in the winter of 1987-88, we learned of the new *Mathematica* software being developed by Professors Stephen Wolfram, Dan Grayson and Roman Maeder of the UIUC faculty.

Mathematica will do arbitrary precision arithmetic, all manner of symbolic calculations and two and three dimensional plots; it will do all the hand calculations normally associated with calculus (including derivatives, integrals, Newton's method and power series), linear algebra and statistics. It can be used as a calculator or as a programming language and using it is so painless that it blurs the distinction between these two functions. But its real ace in the hole is that it is also a word processor. The word processing capabilities on the Macintosh allow one to create something called "*Mathematica* notebooks." They are live electronic documents which are mixtures of static text and active programs that set up an entirely new vehicle for the teaching of mathematics.

Imagine a calculus book in which every example can be modified on the spot to become infinitely many examples, a calculus book that can plot any calculus function, a calculus book that can differentiate, integrate, find roots, expand in power series – in short a calculus book that can teach the students and act as a slave for them.

This is what we realized one day in the winter of 1987-88. *Mathematica* can be made into an outstanding software package for the teaching of calculus.

One informal session grew into another and some encouraging talks with Stephen Wolfram and Dan Grayson took place. It was not long until we agreed that we would write a calculus course in the form of a *Mathematica* noteboook. We brought in a top notch engineering physics undergraduate student, Don Brown, to help and the three of us have spent the summer and much of this semester on our new course, Calculus&*Mathematica*.

The Electronic Calculus Course, Calculus &*Mathematica*.

The course is a problem course; the text is held to a bare minimum. We believe that the best way for students to involve themselves in calculus is to immerse themselves in a collection of well chosen problems. For this reason, the course proceeds as follows: A snippet of text is followed by a section called Practica of four or five problems. Some of these problems are extensions of the text; others illustrate the utility of the text just studied. This is followed by a collection of problems called Anchors. These problems are chosen to anchor the concepts just learned into the minds of the students.

The whole course takes place on the screen; there is no loss of train of thought in oscillating between the printed page and the computer. The course directs the students to work out concrete exercises and then to embark in their own computational and graphic explorations. This dynamic integration of text and laboratory components of the course is the key feature and it is possible only with *Mathematica*. The computer, the pencil and the text merge into one. Problems are not solved elsewhere but right in their own context. With *Mathematica* handling the rote computations, even ordinary students are able to solve problems far beyond the reach of today's best students. Furthermore, with the computational power of *Mathematica* available at all times, the course can be steered in the direction of underlying concepts and more sophisticated computations than possible in today's courses.

In the course, *Mathematica* is made into a splendid instrument for exposition. For example, students when directed to use *Mathematica* to plot the integrand and the difference quotients for the indefinite integral of the integrand can have little doubt about the meaning of the Fundamental Theorem of Calculus. Students who write a financial planning package should have little doubt about the role that the number e plays in compound interest. Students who have seen the lower Riemann sums fill up the area under the curve of their own choice on a *Mathematica* animation should have little hesitation with the idea of the definite integral as a limit of Riemann sums. Students who have seen the difference quotient converge uniformly to the derivative for the (well-behaved) function of their choice on a *Mathematica* animation should have little doubt about the definition of the derivative. Students who program *Mathematica* animations for the convergence of Taylor and Fourier series should no longer fear these topics the way students of today fear them. Students who can quickly determine that the maximum and minimum values of

$$y = (1 - 2x)^2(x - x^2)(1 - 8x + 8x^2)^2$$

on $[0, 1]$ are 1/64 and 0 respectively should have few problems with generic max–min problems. (This exercise appeared recently in the College Journal of Mathematics as a challenge for professors; its solution using *Mathematica* takes 3 lines and a few seconds.)

In actual problem solutions, C&M routes complex calculations through the computer, thus allowing fundamental principles to appear clearly and unhindered by tedious hand calculation or tedious programming. The fundamental principles are given great emphasis simply because a student cannot use *Mathematica* effectively to do calculus unless the fundamental principles are fixed in his or her mind. And the symbolic power of *Mathematica* allows the C&*M* student to handle a more realistic, sophisticated and complex collection of problems than possible by the rote hand methods that plague today's standard calculus course.

The Calculus&*Mathematica* Scientific Advisory Board

Early on it became clear to us that students of C&*M* will be able to solve much more realistic problems that students in the traditional calculus course only can dream of solving. In an attempt to involve the client disciplines in selecting the right problems for C&*M*, we have formed an advisory board of senior scientists on the Illinois faculty, each of whom has an interest in calculus. At this writing, the following colleagues from the University of Illinois at Urbana-Champaign faculty have agreed to share their expertise with us:

- Roy Asford, Professor of Nuclear Engineering
- Donald Carlson, Professor of Theoretical and Applied Mechanics.
- Judith Liebman, Professor of Operations Research, Chancellor for Research and Dean of the Graduate College
- Arthur Robinson, Professor of Civil Engineering
- Nelson Wax, Professor of Electrical Engineering, emeritus
- Stephen Wolfram, Professor of Computer Science, Mathematics and Physics

The Flexibility of Calculus & *Mathematica*

The electronic course C&*M* is completely flexible on two counts. From the student's point of view, the course is flexible because the student can invade the electronic course at almost any point to work problems, write his or her own notes and then save them for later reference or computation. Imagine a printed text in which at the flick of a switch seven blank pages can be made to appear in the middle of the text for student notes, comments or tips. Imagine a printed text that provides right after each problem statement exactly the right amount of space each student needs for the solution. Imagine a calculus course in which most homework assignments are turned in neatly typed and maybe even well written. All of this is routine in C&*M*. Students will appreciate this flexibility, but from the professor's point of view even more flexibility is possible.

All professors have had the experience of looking at a textbook and deciding that the textbook is very good except that the treatment of topic X in Chapter 4 is totally unacceptable. If the textbook is a traditional printed text, the professor goes on to the next book hoping to find something more acceptable. After all, the professor cannot order a book, physically rip out Chapter 4 and replace it with something to the professor's liking. With the electronic course, C&*M*, changing or replacing any chapter, paragraph or exercise is a routine matter. In the classroom, cuts from C&*M* can be made easily for projection on a screen.

Calculus&*Mathematica* and Minority and Physically Handicapped Students

As we mentioned earlier, Lynn Arthur Steen, past president of the Mathematical Association of America said, "Weakness (in high school background) will no longer prevent students from pursuing studies that require college math" because systems like *Mathematica* can shore up students whose background is not ideal. We are going to put this statement to test by committing about forty percent of the student involvement in the initial evaluation of C&*M* to minority students. We have established contact with Uri Treisman of Berkeley who is nationally renowned for his work with minority students at the Berkeley campus. We also have enlisted the help of Paul McCreary of the University of Illinois who is in the process of setting up a Berkeley–like program at Urbana. We are very happy with the enthusiasm with which Treisman and McCreary have greeted our proposal to involve minority students from the start of the The C&*M* project.

Another possibility which will be developed in the future involves physically handicapped students who have lost use of their hands. It was suggested to us that students who cannot write can have satisfactory interface with a minicomputer like a Macintosh. After all, the British physicist and mathematician Stephen Hawking does all his communication via computer. Since the University of Illinois at Urbana-Champaign is renowned for its programs for physically handicapped students, we were intrigued by this suggestion for C&*M*. As a result we made contact with Professor Janet Floyd of the university's Rehabilitation Center. Floyd has expressed a strong interest in pursuing this possibility. Her view is that many more handicapped students would attempt the C&*M* course than can now attempt the traditional calculus course.

The Calculus&*Mathematica* Educational Advisory Board

The C&*M* Educational Advisory board exists to serve several different purposes. It consists of the following members of the Illinois faculty and staff:

- Peter Braunfeld, Professor of Mathematics and Director of Mathematics Education, Department of Mathematics;
- Sandra K. Dawson, Chair, Department of Mathematics, University High School;
- Janet Floyd, Professor of Rehabilitation;
- George K. Francis, Professor of Mathematics;
- Michael Jeffries, Dean and Director of the Office of Minority Students Affairs;
- Anthony L. Peressini, Professor of Mathematics;
- Kenneth Travers, Professor of Secondary Education.

Professor Floyd deals with the physically handicapped students and Mr. Jeffries and his staff deal with the minority student programs. Professor Persessini is a member of the University of Chicago Secondary Mathematics Writing Project and has a keen interest in precalculus mathematics, high school articulation and testing. Professor Francis is known for his work on computer graphics and has operated the highly successful Apple Laboratory at Illinois for eight years. Professor Travers is well known for his leadership role in the Second International Mathematics Study, and will help in the evaluation of C&*M* for high school students. Ms. Dawson will help in evaluating C&*M* for gifted students. Professor Braunfeld, who is the Director of Mathematics Education in the Department of Mathematics, is our main contact for summer teachers' institutes and other educational programs of the Department.

The National Impact of the Course Calculus&*Mathematica*

Even though most of the course exists currently in a preliminary form and has been under development for only nine months, the course C&M has already attracted a lot of nationwide attention. Some examples are:

1. Only nine universities (Yale, Brown, Dartmouth, Boston, Harvard, Cornell, Drexel, NYU, and Illinois) were invited by Apple Computer to the 1988 MacWorld Exposition to demonstrate what they are doing with Macintoshes at their universities. We were the Illinois representatives. We were the only university demonstration that did not feature hypercard.

2. In August, we were invited to demonstrate our course at the National Bureau of Standards in Washington. Over fifty scientists attended our presentation.

3. The University of California, Berkeley invited us to give a demonstration of our course to the assembled faculties of the mathematics and physics departments in November. More than 250 persons attended the presentation.

4. We were invited to demonstrate our course in a one hour presentation at the Conference on Technology in the Teaching of College Mathematics at Ohio State University in October. More than 350 mathematics professors attended our presentation.

5. We were invited by the computer editor for the publication Notices of the American Mathematical Society to publish an article on our course. The article appeared in the November, 1988 issue.

6. The University of Delaware has invited us to demonstrate our course later this winter.

7. Several universities have asked to be test sites for our course. San Francisco, Oklahoma State University and Knox College are going to try to use some of our preliminary material in some sections of their regular calculus course next fall. Missouri is going to test our course in a special laboratory. Northeastern University and Rochester Institute of Technology have also made contact regarding courseware evaluation next fall.

Evaluation of Calculus&*Mathematica*

We will beginevaluation of the C&*M* software in pilot testing at Illinois during the spring 1989 semester. Revisions will be made over the summer and larger scale evaluation and testing of the course will begin at Illinois in the fall 1989 semester. The scope of this testing and evaluation will depend on the number of computers we can muster. Professor Anthony L. Peressini has agreed to lend his considerable skills at testing to lead in evaluating the software. In addition we will make use of the university's Center for Instructional Research and Curriculum Evaluation.

We regard performance in differential equations classes as the ultimate test of a calculus course. The performance in differential equations of all students who take C&*M* at Illinois will be monitored with an eye toward adjustment of the calculus course.

12. Publication of Calculus&*Mathematica*

The published version of C&*M* will involve three separate related products. First is the software which is the focal point of the course. Supporting the software will be two printed books. The first is a book primarily devoted to theory. It will be written in the style of Artin's Princeton course of the fifties. It will present the theory; it will point out clearly what the important parts of the course are and put the secondary parts in perspective. Finally it will be a guide to the software. The second book will be a problem book for students to work on at the computer or away from the computer.

The publisher will be Addison-Wesley's Advanced Books Program. We chose this publisher because Addison-Wesley's Advanced Books Program does not operate through the usual set of calculus-marketing editors. We will be free of the usual marketing considerations that are in large part responsible for the dreary calculus texts which populate today's calculus book market. Plans call for a preliminary edition to be issued in two years followed by a permanent edition later.

The Effect of Technology on Teaching College Mathematics

Anthony Ralston

SUNY at Buffalo

Both the candidates for President in the recent election spoke about becoming the "Education President." Neither, however, gave the slightest impression that he understands the problems facing education in the U.S. today and the dire consequences if we don't do something about it. If we don't, the best is certainly not yet to come.

There is much compelling evidence that we are failing to educate our youth to be effective citizens in a world increasingly dependent on science and technology. Moreover, the evidence is that things are getting worse. This evidence – which it is not my intention to detail here – consists of statistical data (such as the test scores of American children vis a vis their counterparts in other countries [2,4] and the declining enrollments in scientific and technological disciplines), an increasing body of anecdotal evidence about deficiencies in basic knowledge of American high school and college graduates, the widespread belief of university faculty that students are coming to college ever less prepared to engage in meaningful intellectual activity and mounting indications that American business and industry is, on average, ever less competitive in the world marketplace.

It would be nice to be able to say that technology offers an opportunity to improve the situation just described or even to turn it around. Unfortunately, it's not so. Deep societal and structural problems in the U.S. must be addressed if increasing educational mediocrity is not to result in scientific and technological mediocrity and then economic mediocrity – in short, in second class nationhood. What can be said, however, is that, while not sufficient, it is necessary that technology become widely used in the teaching process in American schools and colleges if we are going to see significant improvements in American education.

Some History

The United States became the world leader in science and technology in the middle of this century without any significant use at all of technology in teaching and virtually none at all in teaching mathematics. So why do we need it now? Very simply, because computers have changed the shape of knowledge and what subject matter it is important to teach in ways that demand the use of technology – computer technology, really – in teaching mathematics.

I had my first experience with using technology in a college course in the autumn of 1950 when I studied Numerical Analysis with Zdenek Kopal at M.I.T. Long hours over a hot Friden – or was it a Marchant? – filling up large ruled sheets with differences and related quantities in the quixotic search for solutions to ordinary differential equations was my introduction to technology used in teaching and to computing technology generally. Probably due to some (temporary?) softening of the brain, I loved it. But not so much that 10 years later when I started teaching numerical analysis, I jumped at the chance to assign students problems to solve on a real computer (an IBM 1620) even though I had to teach them a language (inevitably Fortran – sigh!) so that they could use the computer.

The single most remarkable thing about the use of technology in education is that today, almost 30 years later, essentially nothing has changed in the use of technology in education. An exaggeration? Yes, but not so much of one. Computers are used today in mathematics courses other than just numerical analysis such as linear algebra and differential equations. But they are not widely used even in courses like differential equations where their use should be routine. And they are seldom used in calculus or discrete mathematics courses where they are almost indispensable. And when computers are used at all, it is virtually always in the same old-fashioned way to give students assignments whose results will allow some insight into complex computational problems which cannot be obtained from pure analysis. Rarely – very, very rarely – are computers used in American college *classrooms* as a part – an essential part – of the teaching process. It is the thesis of this paper that much of undergraduate mathematics and almost all of lower division college mathematics cannot be taught properly without the use of computers in classrooms. Which is to say that we shortchange all of our students when we teach calculus or linear algebra or discrete mathematics without the use of computers in the classroom.

Computers in the Classroom

Before I discuss the value of computers in the classroom for teaching mathematics, let me dwell on the reasons why they are so seldom used in this way. I should note that my emphasis here is on computers not calculators although the latter, too, seldom make their way into college classrooms even though the new breed of symbolic calculators have an important role to play in college mathematics.

One reason for the absence of computers in classrooms is that, with few exceptions, American colleges and universities provide inadequate or nonexistent facilities for computer use in the classroom. It is *not* sufficient to provide micros which can be wheeled into classrooms on carts whose screens can then be projected by another portable device on a display screen which can be seen only with difficulty by most of the students in the classroom. What is needed are permanently installed computers, permanently installed projection equipment and overhead monitors or screens which allow good viewing from any seat in the room. Such facilities are not very expensive today [3]. But, nevertheless, most American colleges do not have a single classroom so equipped, and very few have more than one or two classrooms so equipped.

Hardware without software is, of course, of no value. How much software is available which, however user–friendly it may be outside the classroom, also provides instructors in classrooms with facilities which are easy and flexible to use? Not too much yet but there is some for teaching, say, calculus. Although the production of such software is difficult and challenging, it is quite certain that lots more good classroom software is on its way.

The chief barrier to the use of computers in the classroom is, however, neither hardware nor software problems. It is the inability of American college mathematicians to recognize the value of such facilities and their unwillingness to make the effort to use the facilities which are available. Because it is an effort. It is not sufficient to insert an occasional computer demonstration in the same calculus lecture you have been delivering for 20 years (although even this would be worth doing). What is needed is a total rethinking of what those lectures are intended to accomplish and, therefore, the development of, effectively, a new course. However, it may not be lectures that you want to give at all any more, at least not lectures in the classic mold. Instead, you should probably be running more interactive classes, perhaps dividing students into small groups while using the computer as the catalyst to developing understanding and problem–solving abilities. Hardest of all: You will have to replace teaching mechanical skills with teaching mathematics. All of this is *work* – hard work. It is hard to blame college faculty for their unwillingness to undertake such an effort when college administrations are seldom willing to recognize, with rewards or released time, the remaking of an old course in a new light.

There is also, I fear, a still widespread perception among college mathematicians of the computer as an essentially computational device. Despite the fact that computers have always been general symbol manipulators rather than "computers" and despite the development of symbolic mathematical systems (so–called, but wrongly called, computer algebra systems) such as MACSYMA ^{R}, *Maple* and *Mathematica* ^{TM}, the number of mathematics teachers who "think symbolically" when viewing computers is not large. This also needs to change if mathematics instructors are to perceive the opportunities to use computers in the classroom.

I must also mention graphics. Although all mathematics instructors believe that a picture is worth a thousand words and, indeed, regularly draw pictures on chalkboards or overhead projector slides, the potential of dynamic graphics – changing diagrams on a computer screen – has hardly been understood at all. Those who defend calculus as the bedrock of college mathematics usually do so on the basis of the importance of calculus as a mathematics of *change*. (Too often – and incorrectly – calculus is called *the* mathematics of change but no matter.) It is, therefore, ironic that virtually all teaching of calculus uses static tools – chalkboards and overhead projectors – in preference to a tool, the computer, that for the first time offers a dynamic medium for teaching dynamics.

So, if you don't start to teach mathematics with computers in the classroom you will miss major opportunities to enhance student learning and, as well, you will likely perpetuate a mind set that prevents you from making effective use of computers outside the classroom.

I now turn briefly to the use of computers in the classroom for the teaching of the most important courses in lower division mathematics:

Calculus
Discrete Mathematics
Linear Algebra
Differential Equations

Calculus

There is no portion of the current calculus curriculum whose teaching would not be enhanced by a computer in the classroom. The computational, symbolic and graphical facilities of such a system would – one or all – be useful every day in the calculus classroom. Wouldn't a system which allowed you to demonstrate dynamically the progression from secant to tangent aid almost all students in grasping the limit concept? If you still want to teach volumes of revolution, wouldn't it be a big help to see the volume forming and to picture the slices? Wouldn't the computational and graphical facilities of a computer calculus system be just what you need to show the connection between Riemann sums and integrals? And wouldn't a symbolic system designed for the classroom enable a more effective presentation of, say, the chain rule than you are now able to give?

My conclusion is that every period of a calculus class would be enhanced by a computer system designed for the classroom. Such systems are becoming available [1] although as yet they have generally not been designed with the aim of fully integrating the computer into the classroom. And, of course, we need textbooks which look at the computer as an integral part of the learning process not just an add on for those instructors who want some "enhancement" of their courses.

Discrete Mathematics

Ironically, although the movement to teach discrete mathematics to freshmen and sophomores has been motivated by the changes wrought in mathematics by the computer, there has been little use of computers in teaching discrete mathematics. This is particularly surprising because algorithms play an important role in many discrete mathematics courses. Probably the reason for the lack of use of computers in (or out) of the discrete mathematics classroom is the lack of software for teaching discrete mathematics. This is unfortunate but not surprising given that so many of the people most interested in teaching discrete mathematics seem to have been writing books rather than paying attention to other pedagogic matters.

But this should change fairly quickly in the next few years. Two examples of areas where good software for teaching discrete mathematics is obviously needed are:

- *Mathematical induction* which should play a major role in discrete mathematics courses but is often hard for students who have seldom, if ever, been exposed to proofs before. There are lots of ways that computer software could be helpful in teaching induction of which I mention only one. Induction is often taught in what I like to call the "mathematics as magic" mode in which students are presented with the result of a (usually algebraic) problem and then shown how to prove the result correct by induction. A much better approach is to use the *inductive paradigm* – compute, conjecture, prove – to infer the result and then prove it. For the "compute" portion of the paradigm, the computer is invaluable in presenting students with data in which to look for patterns.

- *Graph theory* whose essential visual nature makes it a natural for computer graphics; in addition, the many algorithms associated with graph theory can be derived and explained much better with a computer than without one.

Linear Algebra

If you believe, as I do, that a more algorithmic, computational approach to linear algebra than the classical abstract vector space approach is desirable – but even if you don't – computers have an important role to play in the linear algebra classroom. Systems like MATLAB, although designed more for out-of-classroom than in-classroom use, nevertheless suggest the kinds of things that could be done in a classroom to present a subject such as Gaussian elimination in a more dynamic and understandable fashion than is possible with chalkboard or overhead. But even abstract linear algebra has many topics for which a computer in the classroom would be very useful. For example, linear independence and basis would be easier to grasp using a symbolic–graphical system which allowed bases to be constructed and linear independence to be tested.

Differential Equations

It is my impression that differential equations textbooks of the past decade or two have usually contained a chapter on numerical methods. Such chapters are usually at the end of the book and are, I think, seldom taught. More recently some lovely computer systems for teaching differential equations, notably that developed by John Hubbard and Beverly West at Cornell University, have been developed although intended more for student use outside class than in-class themselves. But differential equations, not just in its computational aspects, generally needs consistent classroom support from a computer. Such bread-and-butter parts of differential equations courses as linear equations with constant coefficients need the support of a computer system to do the algebra and to present examples. Good computer software would also make it possible to show conveniently the close connections between differential equations and difference equations, something which is rarely done in undergraduate mathematics.

I have not even mentioned the impact of the computer – and the symbolic calculator – inside and outside the classroom on *what* you teach in the courses discussed above. Methods of integration is the canonical example of an important topic in calculus courses which needs to be rethought in light of the power of symbolic calculators. More generally, all topics in mathematics whose aim is the attainment of a symbolic or graphical skill can only be justified today if it can be reasonably argued that they aid the understanding necessary to deploy mathematics and to study further mathematics. This is a large topic, beyond my scope here, but one closely related to the use of computers in teaching mathematics.

Technology and Mathematics Pedagogy

In principle, you could use computers in the classroom on a continuous, daily basis without any essential change in your *philosophy* of teaching mathematics. You could still be a professor presenting knowledge to students in a fairly classical manner and giving them exercises and problems to be done outside of class, perhaps also using a computer. But to use the computer in teaching in this way is akin to what was done in the early days of business applications on computers when, essentially, previously used manual systems were emulated on a computer without giving any thought to whether or not the computer suggested new ways of doing business data processing. A computer is a foreign object in a classroom; properly understood, this means it is imperative to rethink the role of a professor teaching students.

I alluded to this earlier when mention was made of the need to consider interactive teaching and small group work. Here I want to discuss generally the impact of computers on the teaching of mathematics. In particular, I want to make a few remarks about *discovery* learning. It is a truism among teachers at all levels from kindergarten through graduate school that what you discover yourself you tend to know and understand better than what you are shown by others – your teachers. Still, it appears that with, perhaps, a very few notable exceptions like George Polya, teachers hardly make any attempt to have their students discover knowledge and, indeed, when they do make such an attempt, it is usually a failure. Do computers – in and out of the classroom – provide opportunities for discovery learning of a qualitatively different kind than any other educational technology – including books – have provided heretofore? I think so although candor requires me to admit that I can adduce no very useful evidence to support my contention. Still, some remarks may be useful in getting others to think more seriously on this matter than I have been able to yet.

Mathematics has been called the "science of patterns" [5]. Any such capsule description of a discipline is bound to be incomplete and even misleading. But it is true that doing mathematics at any level involves the search for and recognition of patterns in symbols, in data, in mathematical representations generally. Much of our teaching consists of defining these patterns for our students and then drilling them in applications of the patterns. How much better it would be if students could be led to recognize the patterns themselves so that the understanding they would achieve thereby would make most of the drill unnecessary and would enhance their ability to apply these patterns in unfamiliar situations.

In this context a computer may be said to be a pattern generator and a pattern manipulator. As in my earlier example of mathematical induction, computers can be used by mathematics instructors to generate and manipulate examples of general patterns in ways which can (may?) lead students to discover the generalizations for themselves. If we can learn to use computers this way, particularly in the classroom, it will surely revolutionize how we teach much – almost all – mathematics and it will surely greatly increase our effectiveness as teachers.

I have neither the space – nor the knowledge – to pursue this further here. But I do believe that all college teachers of mathematics should begin looking at the computer in the classroom as a pedagogic tool dramatically unlike any others which have ever been introduced into the classroom. The potential revolution mentioned above will only come after great effort, and it will probably come so slowly that it will look like evolution. Still, the real importance of symbolic systems and graphical systems is not what they do themselves but what they suggest about how mathematics should be taught generally.

A Final Remark

An immediate contribution that college mathematicians could make to the teaching of mathematics concerns their efforts in teaching prospective precollege teachers of mathematics. For both prospective secondary teachers of mathematics and all prospective elementary teachers there should be a course in the use of technology in the classroom for the teaching of mathematics (and, perhaps, for other subjects as well).

Students need to get used to and to become comfortable with technology in the classroom as early as possible. Not least among the effects of doing this will be that the next generation of college mathematics teachers will embrace technology far, far more rapidly than has the current generation.

References

1. Flanders, H. (1988): *Microcalc*, Mathematics Department, University of Michigan.

2. International Association for the Evaluation of Educational Achievement (IEA) (1988): *Science Achievement in Seventeen Countries*, Oxford: Pergamon Press.

3. Kraines, D. P. and Smith, D. A. (1988): A Computer in the Classroom: The Time Is Right, *The College Mathematics Journal*, Vol. 19, pp. 261-267.

4. McKnight, C. C. et al (1987): *The Underachieving Curriculum: Assessing U.S. School Mathematics from an International Perspective*, Champaign, IL: Stipes Publishing Company.

5. Steen, L. A. (1988): The Science of Patterns, *Science*, Vol. 240, pp. 611-616.

The Twilight of Paper-and-Pencil: Undergraduate Mathematics at the End of the Century

Thomas W. Tucker

Colgate University

In the final decade of this century, paper and pencil will take its last stand in mathematics. The signs are everywhere. When was the last time you multiplied together two 3-digit numbers? Or performed long division? Or computed a square root to two decimal places by hand? Or used a table of logs or cosines? We do it on a calculator or in our heads; we don't do it on paper. In the same way that simple calculators have washed away paper-and-pencil arithmetic (at least for adults), more sophicated calculators and computers will soon wash away pencil-and-paper algebra. The tide is rising, and, like any tide, there will be no way to stop it.

And yet, grade schoolers still are drilled endlessly in long multiplication and division. Some probably are still being taught to take square roots by hand. My tenth grade daughter, Emily, still has to do twenty problems a night "simplifying" $\sqrt{128/3}$ to $(8/3)\sqrt{6}$ (it makes you wonder what's left for punishment when the class gets unruly, as it probably should). High school students are still taught how to use tables of logs. Every calculus book still has tables of trigonometric functions. And perhaps most tellingly, no national mathematics test taken by student under the age of 18 allows calculators at the present time.

Other fields seem perfectly willing to embrace technology. I don't hear my colleagues in physics wailing because calculators now contain all important constants and conversion factors so their students don't know five digits of Avogadro's number anymore. Home economics classes use calculators or mental arithmetic; Emily says she was taught to compute the tip at a restaurant just by doubling the 7 1/2% state tax already on the bill. No English professors have banned word processors on their paper assignments. Calculators are allowed on, for example, the College Board's Advanced Placement Chemistry examination.

Why are mathematicians so slow to adopt technology? There are some reasonable excuses. It is difficult to keep up with developments. At the very time the Content Workshop at the 1986 Tulane calculus conference was recommending the use of calculators with integration and root-finding buttons, Casio was releasing the first graphics calculator and Hewlett-Packard was waiting in the wings with graphing and symbolic manipulation. In 1982, Herb Wilf [3] warned the mathematics community about "The disk with a college education," which gave a micro the symbolic power of a mainframe. Five years later I was warning [2] about "The calculator with a college education," which had much of the power of Wilf's micro. Do we even dare to guess what the next five years will bring?

The logistics of using calculators and computers is also daunting to mathematicians. After all, we became mathematicians because we didn't like balky laboratory equipment. Chalkdust runs in our blood. We deliberately sought the asceticism of paper and pencil. Ask any waiter: our napkins are covered with writing, not food. Even if we do put up with an occasional demonstration in class or a few problems in the textbook marked with a calculator icon, we have no intention of allowing them in the examination room. The logistics are just too complicated.

Paul Zorn [4] has said that maybe the problem is that computers and calculators strike much closer to home for mathematicians. The numerical computations we do in chemistry were never really the point of the lesson; so it's fine if a computer does it for us. But numbers and symbols are the point in mathematics. Or are they? Aren't mathematicians always trying to live down their supposed computing ability? "So you're a mathematician. It must be awful adding up all those numbers." The public believes that computers have made mathematicians obsolete. Isn't it ironic that we reinforce that belief by banning technology from our classrooms? Of course, the more sophisticated public knows mathematicians don't just do arithmetic, they do algebra instead, factoring bigger and bigger polynomials I suppose. Does that mean our obsolescence was only postponed until computers learned to do algebra? I'll be honest. My students are frequently better at algebra than me. In all my years in the AP Calculus program, the hardest problem for me was always that one precalculus problem on the AB exam where you had to factor a cubic—of course the students did great on that one.

I refuse to believe that we became mathematicians because we enjoyed mechanical manipulation so much. Do computers and calculators strike too close to home for mathematicians? I certainly hope not.

No, I suspect the reluctance of mathematicians to use technology is based on a much deeper fear, a fear that something beautiful, pure, and eternal will be lost. Tying oneself to technology is tying oneself to the messy, the experimental, and the present. Somehow, paper and pencil are the tools of civilization, while computers are the tools of barbarians. I think the following benchmark problem does a reasonable job of measuring that fear. It is a multiple choice question. You are, in the best ETS tradition, to circle the "best" answer.

$$\int_0^1 \frac{1}{1 + x^2}\, dx =$$

a) .785398163398   b) .785351335767
c) arctan 1 - arctan 0   d) ln2 - ln1
e) 1 - 1/3 +1/5 - 1/7 ... f) $\pi/2$
g) $\pi/4$ h) $'\pi/4'$

Here is an explanation of the answers (the "distractor analysis" as it is known in the trade). Option (a) is the correct answer to 12 significant digits. Option (b) is what my HP28S gives me when I push the integral key with the accuracy set at 0.01. Option (c) is a favorite of calculus students, which will turn into (a) if you allow calculators on the exam. Option (d) is another favorite. When you try to civilize the barbarians, they can still be pretty barbaric. Also, when you live by the mechanical manipulation sword, you die by it (and the death can be gory). Option (e) is an answer Euler might like. Some mathematicians might like it the best. I like it too. But numerically, it's useless. Why, it even needs 4 terms to do better than option (d)! Option (f) is an error that both I and Paul Zorn made in rough drafts of [2] and [4]. I suspect many mathematicians would prefer this answer to (a).

Finally, we come to options (g) and (h). Option (g) is, I presume, the one most mathematicians prefer, I might even say revere. There is mystery and beauty here. There is a long story, a thread leading from circles and geometry to trigonometric identities to the fundamental theorem of calculus. That the area under the curve $y = 1/(1+x^2)$ has something to do with π should provoke awe. The student who answers (g) will have experienced that wonder. Maybe. Probably

not. It's just another mechanical manipulation. The answers always come out nice so where is the surprise?

There is still option (h). It is the answer my HP28S gives when I apply the PI? button on my USER menu to the number given in option (b). I suspect that option (h) leads to more curiosity, more questions, more respect for mathematics than option (g). Is the answer really $\pi/4$? Let's try more accuracy. Why is it $\pi/4$? How does that PI? button work anyway? By the way, to answer the last question, the PI? button divides the tested number by π and then uses the Euclidean algorithm to estimate the greatest common measure between the given number and the integer 1. More history, more questions! One could also use continued fractions to uncover the fraction behind the 12 digit decimal.

It has been said that exact answers are so precious because they are so infrequent. Yet they are the only answers our students see. Peter Lax complains [1] that calculus classes spend too much time "pulling exact integrals like rabbits out of a hat, and, what is worse, in drilling students how to perform this parlor trick." Another $\pi/4$ pops out of another integral and the audience yawns. I'd be willing to bet, however, that at least one student in a HP28S calculus class first learning about Riemann sums will gasp when the PI? button produces option (h).

I would argue that none of these excuses are valid, even the deeper fear of cultural loss. I'm afraid, however, that many will still see technology as just too much trouble. To show you that technology can be just as easy, just as flexible, just as portable, and just as instructive as pencil and paper, I'd like to give a brief demonstration with an HP28S calculator. It is a lesson about numerical integration, and it is live—no programs written beforehand, just the bare calculator and me.

First, a subdirectory for numerical integration is created on the USER menu:

'NM.INT' CRDIR

Then we write a program to store an integrand F and limits of integration A and B:

<< 'B' STO 'A' STO 'F' STO >> 'FABSTO' STO

We're going to study how well $\int_1^2 (1/x)\, dx$ is approximated by various rules: Left rectangle, trapezoid, midpoint, Simpson. We enter ' $1/x$ ', 1,2 and press FABSTO. The NM.INT menu now includes the names F, A, B as well as FABSTO. We are going to try different values for N, the number of subdivisions. Here is program that stores a value of N and also stores the subdivision size $(B - A)/N$ in the variable H:

<< 'N' STO B A - N / 'H' STO >> 'NSTO' STO

Finally, we have the following "workhorse" program, which sums up N values of F at step size H beginning at a given input value, and multiplies the sum by H:

<< 'X' STO 0 1 N START F EVAL +

H 'X' STO+ NEXT H * >> 'SUM'STO

I won't try to explain this simple program in detail, except to say the the calculator works in reverse Polish taking arguments off the stack. Thus 1 and N before START are the index bounds for the START-NEXT loop. To figure out how the program works, envision the stack before and after each command.

The basic rectangle rules - left, right, and midpoint - are now given by these programs:

<< A SUM >> 'LRECT' STO

<< A H + SUM >> 'RRECT' STO

<< A H 2 / + SUM >> 'MID' STO

Let's compare LRECT and RRECT for our given integral. First write a program to compute the difference between a computed value and the correct answer, ln 2:

<< 2 LN - >> 'ERR' STO

Trying $N = 2, 10, 50$ we get the following errors:

	$N = 2$	$N = 10$	$N = 50$
LRECT	.1401...	.0256...	.0050...
RRECT	−.1098...	−.0243...	−.0049...

Clearly, the two rules have errors of approximately equal magnitude but opposite sign. Not quite so obviously, the errors go down by a factor of 5 as N goes up by a factor of 5. Both of these properties of the errors are easily explained with a picture. Note that the linear dependence on H means that each extra decimal digit of accuracy requires an increase of N by a factor of 10. Since $N = 50$ gets two digits of accuracy in 5 seconds, 12 digits of accuracy will take 5×10^{10} seconds or about 1500 years (that's maybe 100,000 sets of batteries).

An intelligent thing to do at this point is to average LRECT and RRECT since their errors apparently cancel. The result is the trapezoid rule:

<< LRECT RRECT + 2 / >> 'TRAP' STO

Let's compare TRAP and MID errors:

	$N = 2$	$N = 10$	$N = 50$
MID	−.007438...	−.003118...	−.0000124...
TRAP	.015861...	.006242...	.0000249...

The error pattern is clear again. The error of MID is half that of TRAP and has the opposite sign. Also, the errors go down by a factor of around 25 as N increases by a factor of 5.

The next improvement should be obvious: a weighted average of MID and TRAP which cancels errors. The result is Simpson's rule:

<< MID 2 * TRAP + 3 / >> 'SIMP' STO

The errors for SIMP now go down by a factor of around 625 as N increases by a factor of 5:

	$N = 2$	$N = 10$	$N = 50$
SIMP	.000106...	.000000194...	.000000000307

Since $N = 50$ gets better than 9 digits correct in about 10 seconds and increasing N by a factor of 10 decreases the error by 10^4, we can now get 12 digits of accuracy in a minute with SIMP, a noticeable improvement over 1500 years.

I haven't tried to explain why the midpoint rectangles do so well or how the error depends on the derivatives of the integrand. There are limits to what one can do in 20 minutes, but we have written the programs, tried an example for different rules and different numbers of subdivisions, and uncovered the basic behavior of the errors. This demonstration explains the relationships between the rules and their errors better, I dare say, than all the formulas given in most calculus books. In fact, I suspect few mathematicians would guess that even a CRAY would take an hour or more to compute 12 digits of ln 2 using the left rectangle rule but only a few milliseconds using the midpoint rectangle rule.

It is not by accident that this demonstration did not use the symbolic manipulative powers of this calculator. The truth of the matter is that the HP28S is not a great symbolic algebra system. There will be far more impressive demonstrations of symbolic manipulation later at this conference. *Mathematica*™ could give Euler or Ramanujan a run for the money. But even this little *HP28S* would impress Gauss for its numerical computation, and that is the point. Human beings were not built to do repetitive numerical calculations. That's one reason why we resort to algebra. That's what the Fundamental Theorem of Calculus is for. That's also why we teach so much algebraic manipulation in our calculus classes and so few numerical methods, even though the latter apply effectively and universally and the former do not.

I hope the twilight of paper and pencil means the end of mechanical algebraic manipulation in undergraduate mathematics. Ed Dubinsky at a recent conference on calculus asked some representatives of client disciplines how they would feel if students coming out of mathematics courses could do the simple algebraic manipulation reliably, but ran to a computer whenever the algebra got messy. I claim we should never have let the algebra get that messy in the first place. The only reason we asked students to differentiate $(2+\cos x)^x$ was because it could be done by hand and we could teach students to do it. All the while, we did not teach more important numerical and graphical methods because they could not be done by hand. If our future students are running to the computer with messy algebra, it means we are teaching the wrong things. It means we are asking computers to do things human beings used to be able to do. We should be asking computers to do the things human beings never were able to do.

References

[1] Lax, P. (1986). On the teaching of calculus. In R. Douglas (Ed.), *Toward a Lean and Lively Calculus.* (MAA Notes Number 6). Washington, DC: the Mathematical Association of America.

[2] Tucker, T. (1987, January). Calculators with a college education?, *Focus, 7* No.1 (Jan. 1987).

[3] H.S. Wilf, The disk with the college education, *American Mathematical Monthly, 89*(Jan. 1982), 4-8.

[4] Zorn, P. (1987). Computing in undergraduate mathematics. In L. A. Steen (Ed.), *Calculus for a New Century: A Pump, not a Filter.* (MAA Notes Number 8). Washington, DC: The Mathematical Association of America.

An Introduction to *Mathematica*™

Stephen Wolfram

The University of Illinois at Urbana-Champaign

What Is *Mathematica*™?

Mathematica is a general system for doing mathematical computation and can be used in many different ways. This section gives a brief survey of some of them.

One way to use *Mathematica* is as a "calculator". You type in a calculation, and *Mathematica* immediately tries to do it. *Mathematica* does far more than a traditional electronic calculator would: as well as numerical operations, *Mathematica* can do symbolic and algebraic operations, and can generate graphics.

When you use *Mathematica* like a calculator, you are drawing on its built-in mathematical capabilities. But *Mathamatica* is also a language, in which you can make your own definitions. You can write programs in *Mathematica*, working not only with numbers, but also with symbolic expressions and graphical objects.

You can use *Mathematica* as a language for representing mathematical knowledge. You can take mathematical relations from handbooks and textbooks and enter them almost directly into *Mathematica*. The basic approach is to give a sequence of "transformation rules" which specify how *Mathematica* would treat expressions with different forms.

As well as being a language, versions of *Mathematica* on many computers also serve as complete environments for computation. You can, for example, create "notebooks", which consist of ordinary text, mixed with graphics and "live" *Mathematica* input.

Mathematica is set up to fit in with other standard programs. You can use *Mathematica* to prepare input, or analyse output, from external programs. What makes this possible is that *Mathematica* supports many standards, such as UNIX pipes and POSTSCRIPT, which are common to many modern programs.

Mathematica Is a System for Doing Calculations

This section gives examples of the three main types of calculations that *Mathematica* can do: numerical, symbolic, and graphical. Each example consists of a short "dialog" with *Mathematica*. The lines labelled IN[n] are what you would type in; the ones labelled Out[n] are what *Mathematica* would type back.

Numerical Calculations

At the simplest level, *Mathematica* will do numerical calculations, just like a standard electronic calculator (Figure 1). *Mathematical* can, however, go far beyond a standard calculator. It can, for example, calculate with numbers of arbitrary precision. It can evaluate a wide range of mathematical functions, including all standard special functions of mathematical physics.

Mathematica works not only with single numbers, but also with more complicated structures. You can use *Mathematica*, for example, to do operations on matrices. The standard operations of numerical linear algebra are built into *Mathematica*. You can also use *Mathematica* to find Fourier transforms, least-squares fits, and so on. *Mathematica* can do numerical operations on functions, such as numerical integration and numerical minimization.

Example: Find the numerical value of $\log(4\pi)$.

Log[4 Pi] is the *Mathematica* version of $\log(4\pi)$. The N tells *Mathematica* that you want a numerical result.

```
In[1]:= N[ Log[ 4 Pi ] ]
Out[1]= 2.53102
```

Unlike a standard electronic calculator, *Mathematica* can give you answers to any number of decimal places. Here is $\log(4\pi)$ to 50 decimal places.

```
In[2]:= N[ Log[ 4 Pi ], 50 ]
Out[2]= 2.5310242469692907929778915942694118477982950816
```

Figure 1. Numerical Value of $\log(4\pi)$

Symbolic Calculations

The ability to deal with symbolic formulae, as well as with numbers, is one of the most powerful features of *Mathematica* (Figure 2). This is what makes it possible to do algebra and calculus with *Mathematica*.

Example: Find a formula for the integral $\int x^4/(x^2 - 1)\,dx$.

Here is the expression $x^4/(x^2 - 1)$ in *Mathematica*.

```
In[3]:= x^4 / (x^2 - 1)
             4
            x
Out[3]= --------
             2
        -1 + x
```

This tells *Mathematica* to integrate the previous expression. *Mathematica* finds an explicit formula for the integral.

```
In[4]:= Integrate[%, x]
             3
            x    Log[-1 + x]   Log[1 + x]
Out[4]= x + -- + ----------- - ----------
            3         2             2
```

Figure 2. A formula for $\int x^4/(x^2 - 1)\,dx$

Mathematica does many kinds of algebraic computations. It can expand, factor and simplify polynomials and rational expressions. It can solve polynomial equations, or systems of such equations. It can get algebraic results for many kinds of matrix operations.

Mathematica can also do calculus. It can evaluate derivatives and integrals symbolically. It can derive power series approximations.

Graphics

Mathematica does both two and three-dimensional graphics (Figure 3). You can plot functions or lists of data. The three-dimensinal graphics that *Mathematica* produces can be quite realistic: they can, for example, include shading, color, and lighting effects.

You can use *Mathematica* to make two- and three-dimensional pictures. You supply a symbolic representation of the objects, say polygons, in the picture, and *Mathematica* will produce a graphical rendering of them. *Mathematica* also has powerful animation capabilities.

Example: Plot the function $\sin(xy)$ for x and y between 0 and π.

This generates a three-dimensional plot of $\sin(xy)$ as a function of x and y. There are many options for controlling graphics in *Mathematica*.

```
In[5]:= Plot3D[ Sin[x y], {x, 0, Pi}, {y, 0, Pi} ]
```

Figure 3. The Graph of $\sin(xy)$ for $0 \leq x, y \leq \pi$

Mathematica is a Programming Language

You can write programs in *Mathematica*, must as you would in a language like C (Figure 4). *Mathematica* is an interpreter: you can run your programs as soon as you have typed them in.

This defines a function t which makes a table of the first n prime numbers.

```
In[6]:= t[n_] := Table[Prime[i] {i, n}]
```

You can use the definition of t immediately. Here is a table of the first 10 prime numbers.

```
In[7]:= t[10]
Out[7]= {2, 3, 5, 7, 11, 13, 17, 19, 23, 29}
```

Figure 4. A Program in *Mathematica*

Mathematica Is a System for Representing Mathematical Knowledge

Mathematica gives you a way to represent, and use, the kind of information that appears in tables of mathematical formulae.

Fundamental to much of *Mathematica* is the notion of "transformation rules", which specify how expressions of one form whould be transformed into expressions of another form. Transformation rules are a very natural way to represent many kinds of mathematical relations.

Some relations that you could use to define your own logarithm function in *Mathematica* are given in Figure 5.

Mathematical form	Mathematica form
$\log(1) = 0$	`log[1] = 0`
$\log(e) = 1$	`log[E] = 1`
$\log(xy) = \log(x) + \log(y)$	`log[x_ y_] := log[x] + log[y]`
$\log(x^n) = n\log(x)$	`log[x_^n_] := n log[x]`

Figure 5. Relations to Define the Logarithm Function in *Mathematica*

Mathematica Is a Computing Environment

Mathematica gives you an environment in which to set up, run and document your calculations and programs.

There are two pieces to *Mathematica* on most computers: the "kernel", which does computations, and the "front end", which deals with user interaction. (These pieces do not necessarily have to be running on the same computer – you can use *Mathematica* over a network.)

The kernel works in the same way on all computers that run *Mathematica*. The front end, however is set up to take advantage of the different capabilities of different kinds of computers.

On the Macintosh, for example, the *Mathematica* front end lets you use graphical tools to manipulate your input and output, and to insert extra text.

The *Mathematica* front end on the Macintosh takes advantage of the Macintosh's graphical capabilities. Many *Mathematica* front ends support "notebooks", which contain a mixture of text, graphics and *Mathematica* definitions (Figure 6).

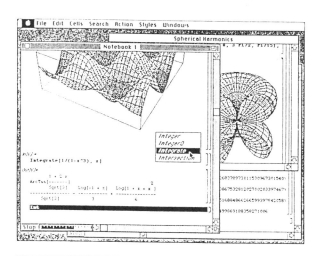

Figure 6. A *Mathematica* Front End Example

There is a growing library of *Mathematica* notebooks which serve as "live textbooks" on a variety of different topics. You can read the text in a notebook to learn about a topic, and then use the *Mathematica* definitions in the notebook to do calculations.

A part of a *Mathematica* notebook about orthogonal polynomials is given in Figure 7. The material in the notebook is arranged in a hierarchical fashion, so you do not need to see details unless you want to.

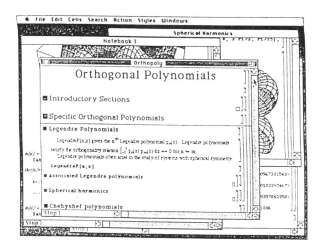

Figure 7. Part of a *Mathematica* Notebook about Orthogonal Polynomials

Mathematica Is a Tool in the Standard Computing Environment

Mathematica interfaces to many elements of standard computing enviornoments. Under UNIX, *Mathematica* can communicate with external programs through pipes.

You can use output from *Mathematica* as input to many kinds of programs. *Mathematica* can write out expressions to use as input for C or FORTRAN programs. It can also produce input for T_EX, which you can use to typeset papers and presentations.

Mathematica produces graphics using POSTSCRIPT. You can render the POSTSCRIPT on many different kinds of devices. You can also read it into other programs.

The *Mathematica* Book

The main documentation for *Mathematica* is the book: "*Mathematica*^{TM}: A System for Doing Mathematics by Computer", by Stephen Wolfram, published by Addison-Wesley Publishing Company (June 1988). The book is available at most bookstores.

- Softcover version (ISBN 0-=201-19330-2): $ 29.95
- Hardcover version (ISBN 0-201-19334-5): $ 44.25

To order the book directly, call Addison-Wesley at 1-800-447-2226.

Buying *Mathematica*

Mathematica for the Macintosh. There are two versions of *Mathematica* for the Macintosh:
- **Standard Version** for Macintosh Plus, SE and II: $ 495
- **Macintosh II Version** for Macintosh II only: $ 795

The only difference between the two versions is speed. The Standard version will run on a Macintosh II, and supports color, but does not make use of the Macintosh II 68881 numeric coprocessor, and is consequently several times slower than the Macintosh II version for operations such as graphic rendering.

Note: Both versions of *Mathematica* for the Macintosh require 2.5 megabytes of RAM. We recommend 4 megabytes or more.

To buy *Mathematica* for the Macintosh, visit your local software dealer, or call Wolfram Research at 217-398-0700 (for orders only: 1-800-441-MATH).

Mathematica **Front Ends.** *Mathematica* consists of two pieces: a kernel that does computations, and a front end, which handles interaction with the user. You can run these two pieces on separate computers. Thus, for example, you can have the front end on a Macintosh, with the kernel running on a remote computer server.

The Macintosh from end for *Mathematica* will be made available as a separate product from Wolfram Research. It will run on Macintoshes with only one megabyte of memory. Call Wolfram Research at 217-398-0700 for information.

University Arrangements. Special arrangements are available for universities which distribute a copy of *Mathematica* with every Macintosh II (or every Macintosh) that they resell. University administrators interested in this program should contact Wolfram Research at 217-398-0700 for further information.

Mathematica **for 386-Based MS-DOS Systems.** There are three versions of *Mathematica* for 386-based MS-DOS systems:

- **386 Version** (no numeric coprocessor required): $ 695
- **386/7 Version** (287 or 387 numeric coprocessor required): $ 995
- **386/Weitek Version** (Weitek numeric coprocessor required): $ 1295

MS-DOS 386 *Mathematica* requires 640k of RAM, and at least 1 megabyte of extended memory. It also requires 5 megabytes of hard disk space.

MS-DOS 386 *Mathematica* supports CGA, EGA, VGA, Hercules, 8514 and other graphics standards. It supports PostScript, LaserJet, Epson FX and Toshiba P3 compatible printers.

MS-DOS 386 *Mathematica* includes DOS-style editing capabilities, but does not support *Mathematica* Notebooks.

To buy MS-DOS 386 *Mathematica*, visit your local software dealer, or call Wolfram Research at 217-398-0700 (for orders only: 1-800-441-MATH).

Other Systems. Versions of *Mathematica* are available for the following systems:

- **Ardent** - for Tital systems
- **IBM** - for RT/AIX systems
- **MIPS** - for M/120 systems
- **NeXT** - bundled as part of standard software on all NeXT Computers
- **Silicon Graphics** - for Iris 4D systems

- **Sony** - for NEWS systems
- **Stellar** - for GS-1000 systems
- **Sun** - for Sun-3, Sun-4 and Sun-386i systems

These versions are available directly from the hardware manufacturers. Contact your local sales representative for information.

For information on possible other versions of *Mathematica*, contact Wolfram Research at 217-398-0700.

Algebraic, Graphical, and Numerical Computing in Elementary Calculus: Report of a Project at St. Olaf College[1]

Paul Zorn

St. Olaf College

Introduction

Computer algebra systems (hereafter, CAS's) offer important new possibilities for mathematical teaching and learning. CAS's provide, for the first time, a powerful and varied range of mathematical computing tools. Using these tools, students can study and represent mathematical ideas more effectively, more efficiently, and more flexibly than before. Just as important, these tools come in a single, easy-to-use package, at an acceptable cost in time and distraction from mathematics itself.

In this paper I describe a CAS-oriented project at St. Olaf College. The project aims to bring algebraic, numerical and graphical computing power to bear on the teaching and learning of elementary calculus.

Educational computing in general; CAS's in particular

Much has been said and written recently (see, e.g., [1], [2], [4], [5], [8]) about how computing can help improve undergraduate mathematics. Among the candidates for improvement, elementary calculus courses are clear contenders. Among the proposed improvers, CAS's are frequently singled out, for two main reasons. First, CAS's do more than other programs; second, they do what they do more easily. (For more information on CAS's, see, e.g., [3], [6], [7], and parts of [2] and [5].)

A few background observations on CAS's will put things in context.

- The term "CAS", although standard, is misleading; CAS's do much more than algebra. They are better seen as mathematical toolkits, with graphical and numerical, as well as symbol-manipulating capabilities.

- CAS's can operate *without programming*, in "calculator mode"; a one-line input command is usually enough. (On the other hand, a CAS *can* be regarded as a high-level mathematical programming language, in which mathematical algorithms are easily realized. This raises its own pedagogical possibilities, but I do not pursue them here.)

- Elementary calculus students can easily learn to use a CAS effectively at an appropriate level; no great mathematical sophistication is needed. (Obviously using a powerful CAS to its *full* mathematical potential which goes far beyond elementary calculus, requires commensurate mathematical knowledge.)

Calculus students at St. Olaf College use SMP, typical of powerful computer algebra systems. A few local features were added, such as a simple help system and several comands for numerical methods. For example, the Midpoint command used below was created locally. It is easy to create such local variations for special pedagogical purposes.

Figure 1 gives a brief SMP session. Commands are chosen to illustrate the combination of algebraic, graphical, and numerical capabilities, and to show how easily the power of the CAS is tapped. Input lines (the ones typed by the user) begin with # I[ ]::. Outputs, marked with #O[ ]:, are returned by the machine. The meaning of most of the lines would be clear from context: D denotes *derivative*; Int means *integral*; Midpoint refers to the midpoint approximation to a definite integral; input command 6 forces a decimal version of the previous line.

[1] Supported by grants from the National Science Foundation (College Science Instrumentation Program), and from the Fund for Improvement of Postsecondary Education.

```
# I[1]:: D[Sin[x],x]
# O[1]: Cos[x]

# I[2]:: Graph [Sin[x], x, 0, 1]
# O[2]:   < a graph appears; units are shown on
            both axes >
# I[3]:: Int[Sin[x],x]
# O[3]: − Cos[x]

# I[4]:: Int[Sin[x], { x,0,Pi } ]
# O[4]: 2

# I[5]:: Midpoint[Sin x,x,0,Pi,6]
# O[5]:
```

$$\frac{Pi(Sqrt[2] + Sin\frac{Pi}{12} + Sin\frac{5Pi}{12} + Sin\frac{7Pi}{12} Sin\frac{11Pi}{12})}{6}$$

```
# I[6]:: N[%]
# O[6]: 2.02303
```

Figure 1. SMP Session

Specifics of the St. Olaf College CAS project

The St. Olaf College mathematics department is carrying out a three-year project to incorporate symbolic, numerical, and graphical computing into elementary calculus courses. Support has been provided by the National Science Foundation and the Fund for Improvement of Postsecondary Education (FIPSE).

Four special sections of elementary calculus, enrolling about 120 students, are taught each semester. Students in these sections use the computer algebra system SMP, to complete specially-prepared homework assignments and, sometimes, in taking examinations. A network of ten powerful, graphics-capable SUN workstations, together with additional remote terminals, is provided for project use.

The syllabus for project sections includes (at least) the nucleus of the usual course. A few non-standard topics, such as numerical methods for integration and root-finding, are covered. Perhaps the most distinctive properties of the special sections, though, are their general viewpoints: to emphasize ideas over mechanics (though not to neglect the latter), and, to draw freely on geometric, numerical, and algebraic methods for learning, illustrating, and applying ideas. Thus, for example, while traditional courses often emphasize graph-*sketching*, project sections freely use computer graphics to illustrate and emphasize connections between analytic properties of a function and geometric properties of its graph.

Students' special experience in project sections stems mainly from working special homework exercises and tests, designed to exploit SMP's various computational powers. Writing and improving these course materials is a major faculty effort. The purpose of the materials is at least twofold. First, they help students use calculus ideas and techniques more powerfully than usual (as, e.g., in combining exact and approximate techniques for integration). Second, and just as important, they can be used to illustrate and teach calculus ideas more clearly in the first place.

Philosophy of the St. Olaf project

It is to be emphasized that the St. Olaf project does *not* aim to "automate" a standard elementary calculus course. Perhaps, some routine calculus manipulations *should* be mechanized, but doing *only* that would be worse than useless. The principal goals of the St. Olaf College project are conceptual: to help students understand calculus

ideas more deeply and to apply them more effectively.

The special opportunity CAS's offer is to supplement the usual narrowly algebraic approach to the calculus with numerical and graphical viewpoints. In standard calculus courses, key concepts are represented and manipulated almost always in *algebraic* form. Thus, for example, students usually treat limits, derivatives, and integrals–all *analytic* objects–only as *algebraic* operations on *algebraic* functions. When such closed-form methods fail, students have little recourse. Such a circumscribed view of important ideas leads to a rigid, shallow, mechanical calculus course, ill-suited to convincing applications.

CAS's allow a variety of "representations" of mathematical ideas. Numerical and geometric viewpoints that traditionally require forbiddingly tedious and repetitive operations are possible with a powerful CAS.

A calculus example using SMP

Suppose $g(t) = \sin(t^2)$ and $G(x) = \int_0^x g(t)dt$.

a. Consider the problem of determining the value of $G(1)$. Notice the fundamental theorem does not help, because g has no elementary antiderivative. Estimate $G(1)$ in two ways: first geometrically, as area under a graph. How accurate can your estimate be? Next, approximate $G(1)$ using a midpoint sum with 10 subdivisions.

b. In order to obtain a useful graph of $G(x)$ on $[0,3]$ we need many (approximate!) values of G. Let SMP do the work, as follows. Give the SMP commands:

    H[$x]:: Midpoint [Sin[t^2],t,0,$x,10]

    Graph [H[x],x,0,3]

(The first command defines a function H that numerically approximates G. The $x notation is simply SMP syntax for function definition. The second command sketches a graph of H.) Discuss the relationship between f and G, by examining graphs of both functions.

SMP commands for the example

Several SMP input and output lines related to the given problem are given in Figure 2.

```
#I [1]:: Graph [Sin[t^2], { t,0,1 } ]
#O [1]:   < a graph appears; I omit it. >


# I[2]:: NMidpoint [Sin[t^2], { t,0,1 } ,10]
# O[2]:  0.309816


# I[3]:: NMidpoint [Sin[t^2], { t,0,2 } ,10]
# O[3]:  0.809254


# I[4]:: NMidpoint [Sin[t^2], { t,0,3 } ,10]
# O[4]:  0.795971


# I[5]:: H[$x]::NMidpoint[Sin[t^2], { t,0,$x } ,10]
# O[5]:  'NMidpoint [Sin[t^2], { t,0,$x } ,10]


# I[6]:: Graph[Sin[x^2],H[x], { x,0,3 } ]
# O[6]:   < a graph appears; see Figure 3. >
```

Figure 2. Related SMP lines

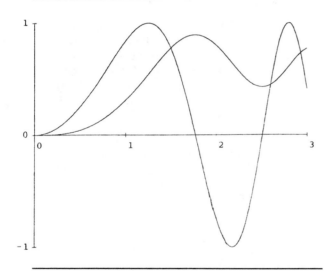

Figure 3. Graphs of $\sin(x^2)$ and H

Remarks on the example

Although the problem uses modern technology, it does so for an entirely traditional purpose: to illustrate and reinforce an important theoretical idea–the connection between derivative and integral. The effect is not to reduce the calculus course to thoughtless button-pushing, but, rather, to shift attention from mechanics to central ideas.

The idea of a function defined by an integral–as opposed to antidifferentiation–is crucial to understanding the fundamental theorem of calculus.

Many students miss this important distinction, and therefore see the fundamental theorem as a tautology: the derivative of an antiderivative is the original function. Here the choice of f forces the crucial distinction to be made.

Problems like this require too many calculations for "manual" methods. With a CAS, the full gamut of algebraic, numerical, and graphical capabilities are used. On the other hand, all calculations carried out by the machine are simple and straightforward. Thus, the machine operates "transparently", rather than as a "black box."

Results of the St. Olaf College project to date

Although the project is not completed, some preliminary results and observations can be mentioned.

- Students seem to perceive the CAS-aided calculus courses as more difficult, but also more valuable, than standard courses.

- Learning SMP (its syntax, etc.) is not a serious obstacle for the average calculus student. On the other hand, not surprisingly, mathematically stronger students also make better use of CAS.

- Many beginning calculus students *expect* routine calculations, not concepts, to be the main focus of their courses. More conceptual courses must contend with such expectations.

- Gauging the success of CAS-aided calculus courses relative to standard courses is important, but also difficult. Common examinations, for example, are impractical when course syllabi differ significantly. Nonetheless, early indications (from St. Olaf College and elsewhere) suggest that CAS's offer significant advantages for mastering conceptual topics and that hand manipulation skills are not sacrificed.

Conclusion

CAS's have existed for many years, but they are just now becoming widely available. Soon even handheld machines (like the *HP-28S*) will be able to do much of what we traditionally teach (and test) in elementary calculus courses. Whether we like it or not, courses must change.

Although powerful CAS's are sometimes seen as threats to traditional mathematical education, they should more properly be seen as offering new opportunities to deepen and revitalize mathematics courses, focusing them more sharply on ideas rather than mechanics.

References

1. R. S. Cunningham, D. A. Smith, "A Mathematics Software Database Update." *The College Mathematics Journal,* 18:3(May, 1987) pp. 242-247.

2. R. G. Douglas, Ed., *Toward a Lean and Lively Calculus: Report of the Conference/Workshop To Develop Curriculum and Teaching Materials for Calculus at the College Level,* Mathematical Association of America, 1986.

3. J. M. Hosack, "A Guide to Computer Algebra Systems," *College Math. J.,* 17(Nov., 1986) pp. 434-441.

4. *The Influence of Computers and Informatics on Mathematics and its Teaching–Proceedings of a 1985 ICMI Symposium,* Cambridge U. Press, 1986.

5. L. A. Steen, Ed, *Calculus for a New Century,* MAA Notes, Number 8, Mathematical Association of America, 1988.

6. T. Tucker, "Calculators with a College Education?" *FOCUS,* 7:1 (January-February, 1987) p. 1.

7. H. E. Wilf, "The Disk with a College Education." *The American Mathematical Monthly,* 89:1 (January, 1982) pp. 4-8.

8. P. Zorn, Computing in Undergraduate Mathematics, *Notices of the American Mathematical Society,* 34(Oct. 1987) pp. 917-923.

LIVE INTERACTIVE TELECLASSES: TUNING IN ON REMOTE LEARNERS
by Hollis Adams, Portland Community College

Community colleges can no longer afford the luxury of waiting for students to walk through the front door. Classes must be made accessible to the community and to business and industry by offering convenient times and convenient locations. Yet the quality and rigor of classes must be maintained at high collegiate standards. Fortunately, these two considerations are not mutually exclusive; Portland Community College is resolving them through the use of live interactive teleclasses. The mathematics department at PCC was quick to take advantage of this opportunity and has begun to reap the benefits.

THE SYSTEM

PCC chose to operate the live interactive courses using one-way video and two-way audio. Classes are narrowcast live from a standard classroom outfitted with television equipment. Students enrolled in the television production program operate the equipment, so one teleclass produces opportunities for learning on several different levels. One TV student, dubbed the "wizard", operates all the studio equipment for a single class. The wizard controls three cameras remotely, moniters sound levels, composes the video-out image, and is responsible for beginning and ending computer graphics. From the studio/classroom, the signal is transmitted to a local TV tower which in turn rebroadcasts it over the greater Portland area. The signal's frequency is much higher than that of standard TV signals, and therefore special receiving devices are required at the remote sites in order to downlink.

The remote students have classrooms equipped with microphones, a special telephone, and a standard television which receives the audio-video signal from the studio/classroom. The remote students talk to the instructor through microphones which are connected to the studio/classroom using dedicated telephone lines. When a remote student from one site is talking, the voice comes through a loudspeaker into the studio and is then transmitted over the system so that all the sites can hear comments regardless of their point of origin. The students from one site never see the others in class from the other sites, yet they quickly adapt to pure audio communication. By the end of the term the remote students' personalities have been firmly established to everyone in the class and we are only left wondering what everyone looks like.

PROBLEMS

Clearly, teaching a group of students scattered throughout the city requires extra instructional support and a high degree of organization. Some of the problems encountered were predictable and easily resolved; others proved more formidable. The college had to create several new lines of governance in order to accommodate the special needs of the teleclasses.

A major problem encountered early on was the lack of a smooth flow of paper to and from the instructor. This has been resolved by having each site designate a person responsible for mailing assignments turned in by remote students to the instructor. The same person receives mail from the instructor and delivers it to the teleclassroom in a timely manner. The mail moves from site to site using a combination of campus and U.S. mail. Most instructors report a minimum of one week turn around time, which is sometimes a source of frustration to students and instructors alike.

The next major problem was the lack of a testing procedure. After a series of false starts, including the use of existing placement testing centers, the procedure currently in place utilizes testing proctors at each remote site. Tests are administered to the remote students in their regular teleclassrooms during class time, just as in a conventional class, then mailed to the instructor.

Given the physical constraints of students in as many as seven different locations, the instructors are forced to be more organized than would be necessary for a conventional class. Lead time for handouts is increased, the timing of tests must be carefully planned, and a fair amount of telephoning on the part of the instructor must take place to insure that tests have arrived on time, etc. The math department compensates first time teleclass instructors with release time from one class of their regular load.

The administrative organization of the college did not accomodate one class and one instructor simultaneously occurring in several locations, and a special policy had to be created. The current arrangement is that each location/campus is responsible for the students registered at that site. This includes having input as to which courses will be offered as teleclasses, provision of the mail person and test proctor, and the provision of administrative support to students who encounter problems. Each location receives the FTE for the students registered at that site.

Many students are apprehensive about taking a teleclass, due to the apparent disadvantages over a conventional class. Students at the remote sites need to be somewhat more independent and assertive in order to use the system to their full advantage. It does require more self-discipline to attend a class where the instructor cannot physically see the students. It is also harder for students to use the telephone to get help from the instructor instead of visiting during office hours. The most significant drawback occurs when there happens to be only one student at a particular site. In such cases, it is virtually impossible to overcome the feeling of isolation. Some students do drop teleclasses for these reasons, but the college has found that

most students rise to the occasion. By the end of a term, most
students feel that the benefits outweigh the disadvantages.

BENEFITS

While it is possible to teach mathematics teleclasses in
virtually the same way as conventional classes, most instructors
have found ways to exploit the system to their students'
educational advantage. Many of these were anticipated, but there
were also a few pleasant surprises.

The studio/classroom was designed to accommodate multiple modes
of presentation. It is equipped with an extensive array of
electronic equipment including video and audio tape players, a
lazer disk player, a slide projecter, and an IBM PC clone. All
instructors are encouraged to use the equipment in any way they
see fit. The studio is equipped with a conventional dry
whiteboard, but most instructors write on plain 8 1/2 X 11 inch
paper located on a desk console in the front of the class. A
camera mounted in the ceiling over the desk allows the students
to see what is written . The camera is capable of zooming in,
making it possible to share illustrations from other sources or
read the display of a calculator.

Using the paper instead of the board has several advantages. The
instructor has complete control over what is on screen at any
given moment instead of depending on a camera operator with no
math background to be able to find "the parametric equations in
the previous example." Further, the readability is greatly
improved using a camera located 5 feet away from the desk rather
than shooting the entire board with a camera across the room. But
the most valuable aspect of using paper rather than the board is
that it produces a written copy of the lecture. Instructors can
refer back to problems worked earlier, which in a conventional
class have invariably just been erased. Students often ask to see
a problem again after class has ended, and no one can argue about
the way a particular definition or theorem was phrased because
there is a written record.

The school was surprised to discover that having a class broken
into small remote groups can actually be advantageous to the
students. School observers and instructors have found that the
students within a site bond to each other more closely than is
typical in a conventional class. They tend to study together and
tutor each other. During class time they often interact with each
other regarding what the instructor has said. Consonant with the
dynamics of small group interaction, one student generally
evolves as spokesperson for the site's concerns. Some instructors
actively encourage this type of interaction by asking each remote
site to venture an opinion on some issue in the lecture. More
often than not, the students confer among themselves and then
give a single answer to the class. We have just scratched the
surface of possibilities in this unusual class arrangement.

FUTURE PLANS

This is the third year of operation for the PCC Television Network and this term there are five mathematics classes (three calculus and two pre-calculus) on the system. Both the number of remote learners and the number of remote learning sites are growing steadily. This term, in addition to students at all four campuses of PCC, the math teleclasses also have students who take classes at their place of employment. Tektronics, Mentor Graphics, Intel, and Portland General Electric's Trojan Nuclear Plant are presently participating in the system, and PCC is actively working on future expansion with other major companies in the Portland area. We are offering classes on TV which otherwise would only be available at the main campus. For the most part, the math classes are offered in the early morning and late afternoon to accomodate part time students who put in a full day at work either before or after class. The response has been so positive that PCC now faces the problem of not having enough time slots available to handle the demand for classes taught in this manner. We hope to open a second channel by Spring term of the 88-89 school year. Offering live interactive teleclasses requires a tremendous amount of work by many individuals at many different locations. But PCC firmly believes that it is educationally sound, and we are committed to this and other innovations that will enhance our students' educational opportunities.

Hollis Adams, Department of Mathematics
Portland Community College
12000 SW 49th Ave., Portland, OR 97219

The Pilot "Calculus with Computers" Course at Appalachian State University. [*]

Wm. C. Bauldry

Dept. of Mathematical Sciences
Appalachian State University
Boone, NC. 28608

1. Introduction

1.1. The Rationale.

Currently there is a national focus on the calculus curriculum. While technological advances have come extremely rapidly, changes in curriculum have moved very slowly. For lack of resources, we are teaching "cookbook" calculus – this approach is no longer relevant, we need to update our curriculum to properly serve the present audience. Just as logarithms have shifted from a computational necessity to a conceptual tool, we need to shift our students' view of the computer as a calculating device to a view of the computer as a manipulator of concepts. We must alter our pedagogical approach to calculus and remove the limiting components by relegating them to microcomputers. Further, we must teach our students to make full use of the analytical properties of the machines. The challenge lies in making use of microcomputers without forcing the students to become Computer Scientists. They must be shown how to use the computer as a tool.

Because of page limitations for the manuscript, we will be extremely terse in our descriptions; we do invite and encourage any interested correspondence.

1.2. The Project

With the arrival of new computer systems with graphically oriented interfaces and software packages that allow symbolic manipulation, we have the tools necessary to build a new approach to the calculus. We envision a classroom where the instructor has the materials available to utilize the computer both as an oracle to provide instant nontrivial examples and develop motivation, and also as an assistant to work quickly

[*] This paper is based on a project supported by the National Science Foundation through the ILI program under grant no. USE–88–53085.

through the many possibilities of approaching a concept, i.e. an assistant that handles "what if ?". We also wish to introduce the student to the computer as a conceptual tool, not merely to regard the machine as an incredible calculator that also plays games. In this vein, we used an existing microcomputer lab and portable demonstration units to augment our calculus course; we also employed a supplemental text for the computer related material. The students were led through introductory material concerning the computer but were not taught programming. We feel it is counter productive to teach simple programming to calculus students when very powerful sophisticated packages are available.

2.   The Pilot Course

2.1.   Development

During the summer of 1987, texts were reviewed for use in the project. At this time we decided to use the format of four normal one hour lectures (as in our traditional course) with the instructor having access to a demonstration unit and an additional one hour per week of laboratory time where machines would be available to the students. In order to accommodate the extra lab hour, a "special topics" course number was used with a notation on the transcript regarding the equivalence to Calculus I; to account for the extra time, the student earned five, instead of four, semester hours.

2.2.   Materials Used

We looked at many texts and chose to adopt Oberle's <u>Calculus and the Computer</u> to supplement Swokowski's <u>Calculus with Analytic Geometry</u>.

We used a microcomputer laboratory which had seven IBM Personal Computers, six Sperry IBM PC compatible machines, and fourteen Apple][e microcomputers, and, for classroom demonstrations, an Apple][e or the author's Macintosh. The Apple][e's were consistently chosen by the naive computer users. The software used included the Waits–Demana Grapher, ArbPlot, Basic with the programs written in Oberle or by the instructor, along with True Basic's Calculus module for classroom demonstrations.

2.3.   Strengths and Weaknesses

The main strength was that the students' graphical perception was greatly enhanced and the geometric intuition that was developed as a result was markedly superior to that of students not using the microcomputer. The students demonstrated a willingness to experiment with unusual functions and began to exhibit the ability to derive conclusions from the results — not just the normal application of formulæ to expression that is the routine of present "cookbook" calculus classes.

The main weaknesses were the limited equipment and software available. The existing microcomputer lab was used for the pilot project, but was not suitable for permanent use for various reasons. Foremost was the lack of an instructor workstation for teaching purposes; also the lack of uniformity in the equipment rendered the laboratory inappropriate. These problems made it very difficult to conduct a reasonable size class without getting lost in the intricacies of the different machines. In addition, this room also experienced a very heavy demand from the computer science courses and there was much competition for time on the computers. The laboratory could not carry the increased load of additional students with any reduction in time necessitated by use as a classroom.

2.4. Students

The class consisted of majors in Mathematics Education (40%), Computer Science (30%), Mathematics (20%), and others (10%). Of the Math. Ed. students, almost all were recipients of the North Carolina Teaching Fellows award. This scholarship program seeks to enhance the quality of teachers in North Carolina by attracting and supporting the brightest students. Sixty percent of the students had never used a computer prior to the course. Three students had completed previous computer science courses.

3. Laboratory Under Development

3.1. Machines and Software

We have chosen the Apple Macintosh SE microcomputer with an internal hard disk drive HD20 as the student workstation. The Macintosh has a graphically oriented interface to its operating system that is well suited to the novice — we don't have to "teach the machine" to the extent necessary with IBM compatible equipment. For the instructor we have a Macintosh II microcomputer with a color monitor and an interface to an overhead projector. The Macintosh II's capabilities and expansion possibilities make this a very attractive choice for many different users. We have already been approached by the Math. Ed., Computer Science, and Statistics faculties who are interested in using the lab.

Our major software is University of Waterloo's Maple. This is a symbolic algebraic manipulation system which is the next generation of programs stemming from Macsyma, Reduce, etc. Maple has libraries for calculus, statistics, linear algebra, and finite groups, just to mention a few. Due to the size of Maple and its libraries, a hard disk is a requisite feature. We will also make use of several smaller programs when appropriate; e.g.: Waits–Demana Grapher, Venn (a logic program), the True Basic Calculus Module, etc.

3.2. Logistics

The mechanics of converting a classroom to a laboratory are difficult. The necessary furniture, rewiring, and refitting the facility are small parts of the problem. In the large, it is much more difficult to find space that can be released from the classroom pool – at all universities classroom space is at a premium. Coordinating orders, storing partial shipments, and actually setting up the equipment takes much time and energy, and shows the benefit of our having an excellent support staff.

4. Laboratory Manual

Dr. J. Fiedler (Ohio State) and the author are preparing a laboratory manual in the style of chemistry / physics laboratory manuals that guide the student through experiments designed to elucidate the concepts. The manual focuses on the Macintosh computer running Maple. We have contacted several publishers who have expressed interest in the project.

5. Future Directions

The next phase is to expand the number of courses that the lab services – discrete math., computer science, and statistics are all natural targets. We also plan to connect the laboratory network to the university's VAX system through Ethernet and utilize the VAX as a file server to hold software and distribute assignment files, etc.

A second project is being developed by Prof. A. McEntire of ASU in a completely new direction. We are considering using the laboratory for summer workshops for in–service teachers and school systems' computer coordinators, both at the elementary and secondary levels, to work towards certification in computer literacy and to introduce them to software and methods appropriate to their classrooms.

6. Acknowledgments

The author would like to thank Oscar Knight, Computer Systems Coordinator of the Dept. of Math. Sciences, for much valuable assistance.

EFFECT OF COMPUTER GRAPHIC USE
ON STUDENT UNDERSTANDING
OF CALCULUS CONCEPTS

Charlene E. Beckmann
Department of Mathematics and Computer Science
Grand Valley State University
Allendale, Michigan 49401

The state of calculus course content and instruction has been called into question due to perceived student difficulties in understanding concepts and also due to the ready availability of computers and advanced calculators (c.f. Douglas, 1986; Steen, 1988). The literature concerning student difficulties in calculus suggests that students enter the course having a discrete concept of number. This concept of number is incompatible with the study of continuous phenomena (Confrey, 1980). A graphic representation, especially as displayed on a computer, holds promise for the development of understanding of the continuous processes and properties studied in calculus. Further, a perusal of the global features of functions represented by Cartesian coordinate graphs (Janvier, 1978) elicits further support for the use of graphs in the development of calculus concepts.

Based on the theoretical and empirical literature, student understanding of selected calculus concepts as developed through the use of a Cartesian coordinate graphical representation system were investigated. The purpose of the study was to compare the relative effectiveness of calculus instruction varying in the level of use of a graphic representation in developing the concepts of limit, continuity, and derivative on: (a) student facility with the use of a graphic representation in calculus; (b) student ability to solve applied, symbolic routine, and symbolic nonroutine calculus problems; (c) student attitudes toward mathematics; and (d) student attitudes toward the use and usefulness of graphs.

Subjects ($N = 163$) enrolled in four first-semester calculus sections at Western Michigan University participated in one of four treatment conditions: (1) Graphics (G), exposure to a computer-graphically developed conceptual course; (2) Graphics Plus (G+), exposure to the same course as G subjects plus provision of computer graphics software and related supplemental assignments; (3) Standard 1 (S1), exposure to a graphically-developed conceptual course without the computer; and (4) Standard 2 (S2), exposure to a traditional skill-oriented course.

In the experimental course presented to the G and G+ subjects, concepts were presented through real world applications which were modeled graphically, often through the use of computer graphs. The graphic results were then translated symbolically. A brief description of the development of the concept of derivative follows.

The Concept of Derivative as Developed
Through Computer Graphics

The graphical development of the concept of derivative begins with a dynamic visual representation of the relationship between two quantities and their corresponding rate of change. Through the use of the program, SPIDER (Beckmann, 1988), students observe a "spider" climbing a wall as a graph of his position with respect to time is drawn on the screen. Students observe that a graph can represent the relationship between two quantities and that the shape of the graph informs them of how the quantities change with respect to each other.

The spider situation is followed by a discussion of average velocity from the point-of-view of a commuter versus that of a police officer observing the commuter. Average velocity is determined similarly by the commuter and the police officer, determining the change in distance divided by the change in time over which the travel took place. If position is written as a function of time, students suggest that an expression for average velocity is:

$$\text{Average velocity} = \frac{f(b) - f(a)}{b - a} \ .$$

The commuter situation is modeled graphically. Students observe that the above expression is that of the slope of a secant line to the graph through the points (a,f(a)) and (b, f(b)). Asked to consider the police officer's perspective of this situation, students suggest that his interest is in instantaneous velocity at some time c in the interval (a,b). They further suggest that this is the limit of the above expression as the length of a time interval around c is reduced to zero. Asked to interpret instantaneous velocity graphically, students suggest that a secant would be drawn between two points which are very close together. They further suggest that the slope of such a line would better approximate the instantaneous rate of change as intervals of x became smaller.

Using *Master Grapher* (Waits & Demana, 1987), a variety of functions are investigated by graphing a function then magnifying the graph several times around a chosen point. Students observe that many continuous functions are locally straight on sufficiently small intervals. Asked to interpret the behavior of the displayed function with respect to a position vs. time situation, students suggest that the slope of the secant line containing the endpoints of the interval shown on the screen estimates the average velocity of the object whose position/time graph is displayed. When the graph appears to approximate a straight line, the average velocity closely approximates the instantaneous velocity over the interval displayed. If the graph of the function cannot be made straight over any interval containing a point (c, f(c)), then the instantaneous velocity cannot be determined at x=c.

Continuing the discussion, the program, SECANT (Beckmann, 1988), is used. SECANT draws successive secant lines through the points (c, f(c)) and (c+h_i, f(c+h_i)) where $h_i = h_{i-1}/2$. Students suggest that, as the length of the interval between c and c+h_i decreases, the resulting secant line appears to intersect the graph in one point. Students recall that they have seen a line with such a property, a tangent, in previous work with circles. The definition of a tangent is generalized to include tangents to any curve.

The definition of the derivative is then given as follows:

If y = f(x), then the ***derivative of y with respect to x at x = c*** is

$$f'(c) = \lim_{h \to 0} \frac{f(c+h) - f(c)}{h} .$$

if this limit exists. When the limit does exist, it is called the ***instantaneous rate of change*** of y with respect to x at x = c, or graphically, the ***slope of the tangent*** to the graph of y = f(x) at x = c.

As described, the concept of derivative evolves from a discussion of a graph representing the rate of change of the quantities position vs. time. The rate at which these quantities change determines the shape of the graph of the function representing them. With respect to the commuter situation, the expression for average velocity is interpreted graphically as the slope of a secant to the curve. As the width of the interval over which a secant line is drawn decreases, the resulting secant line closely approximates the graph of the function. Globally, this secant appears to be the tangent to the curve at a given point. This development uses graphic and natural language representations in parallel to build meaning for the symbolic representation of the concept of derivative. It is representative of the approach taken throughout the course.

To analyze the data gathered in the study, two investigations were undertaken. In each investigation, prior calculus experience was used as a blocking variable for cognitive measures on two levels: (1) prior experience and (2) no prior experience. Covariates used in the analyses with cognitive variables were scores on a pretest of precalculus competencies and responses pertaining to attitudes toward mathematics.

In Investigation 1, comparisons were made between the G and G+ sections on : (a) performance on routine applied, routine symbolic, and nonroutine symbolic questions; (b) performance on the departmental final exam and its subscales; (c) changes in attitude toward mathematics; and (d) attitudes toward the course. Multivariate analyses of covariance were performed for the graphic cognitive variables. χ^2 tests were conducted for the affective variables. These analyses revealed no significant differences ($p < .05$) between the G and G+ sections on any of the cognitive or affective variables. Scores on cognitive measures for G+ subjects were higher than for G subjects, but the differences were not significant. Such findings indicate the need for further study. Attitudes pertaining to the use of graphs were overwhelmingly positive.

For Investigation 2, multivariate analyses of covariance were performed for cognitive variables. χ^2 and LSD tests were conducted to detect changes in the affective variables. Analyses of variance were conducted to compare student attitudes toward the course. Significant differences favored: (a) the S1 section over the G+ section on routine symbolic questions; (b) the G section over the S2 section on nonroutine symbolic questions; and (c) the S1 section over the G and G+ sections on the final exam questions pertaining to limit, continuity, and derivative. Questioning the validity of covariate measures for the standard sections in the analyses of (a) and (c), analyses of variance were performed on cognitive variables. A significant ($p < .05$) difference was detected on symbolic nonroutine questions favoring the G section over the S2 section. No other differences were detected. Student attitudes were generally positive, both toward the course and toward mathematics. Retention rates were found to be much higher in the conceptually-developed sections than for the technique-oriented section.

Results suggest that developing the calculus concepts through the use of a graphic representation system, especially through computer graphics, can positively affect student understanding and interest without necessarily negatively influencing skill acquisition.

References

Beckmann, C. E. (1988). *Effect of computer graphics use on student understanding of calculus concepts*. Unpublished doctoral dissertation, Western Michigan University, Kalamazoo.

Confrey, J. (1980). Conceptual change, number concepts and the introduction to calculus. *Dissertation Abstracts International, 44*, 972A. (University Microfilms No. 80-20, 924.)

Douglas, R. G. (1986). *Toward a lean and lively calculus: Workshop to develop alternative curriculum and teaching methods for calculus at the college level* (Conference report). Washington, DC: Mathematical Association of America.

Janvier, C. (1978). *The interpretation of complex Cartesian graphs representing situations - Studies and teaching experiments*. Unpublished doctoral dissertation, University of Nottingham.

Steen, L. A. (Ed.). (1987). *Calculus for a new century: A pump, not a filter*. Washington, DC: Mathematical Association of America.

Waits, B. & Demana, F. (1987). *Master grapher* (Computer software). Reading, MA: Addison-Wesley.

MATHPERT: An Expert System for Learning Mathematics

MICHAEL J. BEESON
DEPARTMENT OF MATHEMATICS AND COMPUTER SCIENCE
SAN JOSE STATE UNIVERSITY
SAN JOSE, CA 95192
beeson@mathcs.sjsu.edu

Abstract[1]

MATHPERT is an expert system in mathematics designed explicitly to support the learning of algebra, trigonometry, and first semester calculus. This paper will describe what MATHPERT can do and how it is to be used by students. The internal design of MATHPERT will be described elsewhere. MATHPERT is intended to be used either alone or as a basis for more explicitly tutorial programs (which are not described here).

MATHPERT is an expert system in that it is capable of solving (almost) all problems in the stated subject domain internally and autonomously.[2] In this respect it is similar to existing computer algebra programs such as MACSYMA, MAPLE, or *Mathematica*. However, it differs from them in several respects, which are not only important but vital for education. The most basic of these differences are that MATHPERT is *glass box* and *cognitively faithful*; these terms are explained in the paper. MATHPERT produces not just "answers", but full step-by-step "solutions", with each step intelligible to the user and justified on-screen.

MATHPERT also incorporates a fairly sophisticated *user model*, and uses it to produce step-by-step solutions at a level of detail custom-tailored to the knowledge of the individual user.

MATHPERT is based on the analysis of the stated subject areas into several hundred *operators* which can be applied to mathematical expressions. When MATHPERT is operating in "menu mode", the user chooses which operator to apply next. The computer carries out the actual application of the operator. Operators which are "well-known" to the student, according to the student model, will be applied automatically, allowing the student to focus on the less well-known parts of the problem. At any time MATHPERT can be switched into "automatic mode", in which it will not only apply the operator but choose it, thus generating one (or more) steps of the solution automatically. The user can return to "menu mode" at will.

MATHPERT accepts arbitrary symbolic problems from the user; for example, a student might type in her homework. MATHPERT is designed for use with existing courses; whether or not the class is officially using MATHPERT, an individual student should be able to use MATHPERT beneficially. It is designed to be useful to students across the spectrum from those needing remedial work to the very brightest students.

MATHPERT includes a (two-dimensional) graphing package capable of every kind of two-dimensional graphing that might be useful for students, including facilities for rapidly changing the value of a parameter or the portion of a graph visible on the screen.

MATHPERT is written in Prolog and C, and runs on an IBM PC/AT with an expanded memory board and EGA or CGA graphics.

Glass Box and Cognitively Faithful

An expert system is called *glass box* if you can see how it arrives at its answer. MATHPERT can print out the individual steps of its solution, with their justifications. (We use the word "solution" to mean such a sequence of intelligible steps, whose last line is the "answer".)

An expert system is *cognitively faithful* if its own internal solutions correspond to the solutions a human would produce. MATHPERT solves math problems in the way we teach students to do, rather than using high-powered algorithms. Cognitive fidelity must be designed for from the beginning, as the demand for cognitive fidelity complicates the construction of an expert system considerably.[3]

---

[1] This work partially supported by NSF.

[2] The range of MATHPERT's capabilities is extensive. For example, it can solve problems in simplification, including all kinds of exponents and radicals, factoring, equation solving including transcendental equations, trig identities, limits, differentiation, and integration.

[3] The term *glass box* is in the literature, e.g. in Anderson [1988], Burton and Brown [1982] (where it is credited to a 1977 memo of Goldstein and Papert). The term *cognitively faithful* has probably been used, too: certainly the *concept* appears in Anderson [1988] and in Wenger [1987] (there under the name "psychological plausibility").

MATHPERT also incorporates an elaborate internal *user model*, or *student model* (but there may well be non-student users). This model contains (among other things), the information concerning which of several hundred pieces of knowledge are, for this user, *well known*, *known*, *learning*, or *unknown*. MATHPERT uses the model to tailor its output to the user. A naive user will receive more detailed explanations than a sophisticated one, and in particular ways tailored to the exact knowledge of that user; a generally sophisticated user with some gaps in her knowledge will still receive detailed explanations when her weak points are involved. This use of the student model to modify output results in MATHPERT's being "cognitively faithful" not just to some idealized student, but to the particular, individual user with whom it is dealing at the moment (provided, of course, that the internal user model is accurate).

The Operator View of Mathematics

MATHPERT depends on an analysis of its subject matter (algebra, trigonometry, and elementary one-variable calculus) into several hundred *operators* which can be applied to mathematical expressions. For example, one operator is called **collect powers**. It applies to an expression $x^2 x^9$ and produces x^{11}. The key to the solution of a mathematical problem, according to this view, consists in a correctly-chosen sequence of operators which are to be applied to the input. The "solution" itself is the line-by-line record of the result of these operator applications, together with their "justifications". The justifications are simply the names (or formulas) describing the operators applied.

MATHPERT operates in two "modes": in *menu mode*, the user directs the course of the developing solution by choosing the next operator from a system of menus. Since there are several hundred operators, it would not be practical to require the student to remember and type the names of the operators.[4] The menu system has been designed to show only those operators which might be relevant to the problem at hand, so that usually the student does not have to leaf through all four hundred operators looking for the right one. Moreover, even in menu mode, "well-known" operators will be applied automatically. Thus while doing integration by parts, for example, you need normally not search for the operator $-(-a) = a$, which should be well-known long before you are tackling integration by parts.

In *automatic mode*, MATHPERT will generate its own "ideal solution", step by step. The user can switch at will between automatic and menu mode. Thus you can start in menu mode, get stuck, switch to automatic mode for one or two steps for a hint; then, back on the track, you can continue in menu mode. When you choose "finished", MATHPERT will switch into automatic mode and see if you really are finished, according to its own internal algorithm. If not it will supply the last steps of the problem. If you switch into automatic mode and stay there, even at the beginning of the problem, MATHPERT will generate the complete solution for you. Thus in principle you could just type in your homework and have MATHPERT print out complete, step-by-step solutions to each problem.

Although automatic mode generates a single "ideal solution", menu mode permits "alternate solution paths": any correct sequence of operators will be accepted. At any point you can switch into automatic mode and MATHPERT will successfully complete the problem from that point, if possible.

Implications of the Operator View

The view of mathematics as consisting in the application of operators to expressions leads to the classification of mathematical knowledge into four kinds:

- *Knowledge Of or Knowledge That.* This consists in knowing what the effect of a given operator will be, and in knowing that this operator exists and can be applied when needed. For example, knowledge that $\sin^2 x + \cos^2 x = 1$.

- *Clerical Knowledge.* This consists in knowledge of how to apply a given operator. The boundary line between knowledge of what an operator does and the ability to apply it is hard to observe directly when students work with pencil and paper; if they fail to collect powers in $x^2 x^9$ you can't be sure why. However, MATHPERT takes over the clerical function for the student, leaving him (at least in menu mode) the responsibility to know (or discover!) the effects of operators.

- *Knowledge How.* This consists in the procedural knowledge of how to choose operators in the order necessary to solve a given problem. Very little explicit attention is given to this "control knowledge" in traditional

---

[4]Besides, students are generally poor typists. In MATHPERT, they will have to type only to enter their problem, not to solve it.

instruction, yet it is very important. MATHPERT permits the student to focus more attention on "knowledge how". In theory the correct sequence of operators could be chosen by some general planning process, based on the knowledge of the effects of operators. In practice students succeed in mathematics only when they learn a procedure for solving a given type of problem.

• *Knowledge That.* This consists in knowing the reasons why the operators are correct. For example, why is $\sin 2\theta = 2\cos\theta\sin\theta$ a correct operator? Why is collecting powers a legitimate operation? MATHPERT makes no contribution to help the student learn this kind of knowledge; the operators are just "given". However, the architecture of MATHPERT makes it easy for tutorial software running on top of MATHPERT to tutor students in "knowledge that". Note that "knowledge that" is almost never taught to students in conventional classes at the levels MATHPERT covers: although it may be mentioned in passing, it is never on the tests. Only the more elementary three kinds of knowledge are tested.[5]

Skills and Concepts

The guiding philosophy of my work on AI-based educational software for the personal computer is to make software that will help students in traditional courses. These courses are primarily skill-based, and I believe for good reason: Concepts grow, in my opinion, out of the use and mastery of skills, and not the other way around. This dictates the design of skill-based software as the starting point for computer-based education.

Let me offer an operational definition of "skill" and "concept", which I believe is helpful in clarifying these issues: How does one test whether a skill has been learned? By asking problems of the same sort used during instruction. How does one test whether a concept has been learned? By asking problems of a different sort from those used during instruction, but to which the concepts are applicable. If the student can solve only problems which follow a very strict template, then a skill has been learned.

Informal discussions with colleagues have made it clear that, to the contrary, many people feel the computer should be used to teach concepts, "liberating" students from the demands of skill-based learning. This is an issue that deserves a long discussion, which space here does not permit. I have only space to emphasize that MATHPERT has been designed on purpose as a tool for the learning of skills, not concepts; although it has been done in such a way to support tutorial software in the future which might teach concepts.

How will MATHPERT improve learning?

MATHPERT will relieve people of the clerical burden of mathematics, just as word processing relieves people of the clerical burden of writing. Instead of having to concentrate on the clerical details, people will be free to think about what they want to do to solve the problem.

This is, however, not the only effect of using MATHPERT. An important feature of MATHPERT is that it does not allow an incorrect step. It *will* allow "red herring" steps, which do not really lead in a useful direction, but are at least mathematically correct. But the fact that you can't really arrive at an incorrect formula with MATHPERT should prevent a lot of headaches with mathematics. When you realize that your "red herring" steps have led you "on a wild goose chase", you can use the "undo" facility to back up to where you went wrong.

Another way in which solving problems with MATHPERT differs from solving them by pencil and paper is this: MATHPERT's menu system permits you to peruse your possible moves at any time when you feel puzzled as to what to do. Simply seeing the possible operators laid out systematically before you will help you choose a useful one. Moreover, repeatedly seeing the operators while working problems is bound to make it easier to remember the range of possible choices.

Much previous work on ICAI (intelligent computer-assisted instruction) has focussed on the "buggy model of learning",[6] in which the computer is supposed to diagnose the student's incorrect ("buggy") operators and direct specific tutorial action to correcting these misconceptions. Generally speaking, the "buggy model of learning" will not be very useful with MATHPERT, since MATHPERT does not permit the student to take an incorrect step. The student will become disabused of buggy notions without the necessity of diagnosing these bugs. To put the matter another way: the "buggy model" describes learning in a traditional environment.

---

[5]Note also that justifications of operators are of two kinds: some operators, such as $\sin 2\theta = 2\sin\theta\cos\theta$, can be justified in terms of simpler operators. Others, such as the sum formula for sin, require a justification outside the framework of "operators".

[6]See e.g. Burton [1982] for an explanation and detailed example of the buggy model; and Matz [1982] for a discussion of the buggy model applied to high school algebra.

We do not believe it will describe how learning takes place in an environment based on MATHPERT. In this environment, we will "nip bugs in the bud". (Perhaps we should not mix metaphors: we will nip bugs in the larval stage.)[7]

A better theoretical foundation for learning in the MATHPERT environment is Maria Montessori's principle that educational materials should have "control of error", i.e. the student working alone with the materials should not be able to go very far wrong: the square peg won't fit the round hole.[8] Although Montessori had very young children and physical materials in mind, her principle (which for lack of space we have not fully explained) has much wider applicability.

The User Model in MATHPERT

The operators used by MATHPERT have been carefully chosen so as to correspond to cognitive skills, that is, to identifiable "chunks" of knowledge which can be taught and learned. Thus the "skills lattice" which is used in cognitive science to model learning can be directly correlated to a user model based on the executable operators of MATHPERT . Each operator is at the same time a procedure and a skill.

MATHPERT's internal model of its user consists in recording, for each of some four hundred operators, one of the values *learning, known, well-known,* or *unknown.* This section will describe how these values are used in MATHPERT and how they will be used by future software.

The values *learning* are used by individual operators. For example, we stated above that the operator **collect powers** will produce output x^{11} on the expression x^2x^9. This is not strictly true: if the student is still on record as "learning" this operator, it will produce instead x^{2+9}, after which the operator **arithmetic** will be automatically applied (if *it* is well-known) to produce x^{11} on the next line. A substantial fraction of MATHPERT 's operators are designed in this way, to produce more explicit versions of their output when certain operators are still recorded as "learning".

The values *well-known,* on the other hand, are used not by individual operators, but by the main MATHPERT system. Even in menu mode, well-known operators are applied automatically. This lets the user concentrate on the task at hand, relieving her of the necessity to search through the menus for an operator she learned two years ago and knows very well how to use. However, these operators will still be visibly applied on the screen. This provides the best of both worlds: the student does not have to think about applying $-(-a) = a$, but also can see explicitly that it was applied. Otherwise, the application of two or three well-known operators (invisibly) can result in confusion.

The values *known* and *unknown* are not explicitly used in the present version of MATHPERT, but are maintained to allow for the possibility that tutorial software may want to "grey out" unknown operators so that the student can't see or use them. This will be used particularly in the (not uncommon) case that one operator combines the skills represented by several simpler operators. For example, the operator **common denominator** is broken into five or six simpler operators intended for use while learning common denominators.

Evidently the user model is only useful if it is accurate. By definition, the process of *diagnosis is initializing and updating the student model.* A program called DIAGNOSER will initialize the student model interactively at the student's first serious session. DIAGNOSER will generate problems dynamically, based on the student's previous responses; since there are about four hundred operators, dynamic generation rather than a pre-stored test is necessary. At present (9-15-88), DIAGNOSER is in the design stage. Closely related to DIAGNOSER will be a program called EVALUATOR which will analyze the student's performance with MATHPERT and decide on the correct updating of the user model. At present EVALUATOR is in the early design stage, and we shall not comment further on it. DIAGNOSER and EVALUATOR are both components of an expert system MATHTUTOR which we plan to construct as a companion to MATHPERT.

MATHPERT can be used now, without DIAGNOSER and EVALUATOR, if a human tutor (or the user herself) adjusts the student model appropriately to the level of the student in question. It need not be absolutely accurate to be useful. The student model can easily be adjusted through the menu system, without altering the program itself.

---

[7]Of course, that doesn't apply to bugs that are already full-grown. The architecture of MATHPERT does permit it to accept a "buggy operator" as a new operator and simulate the mistaken behavior of a student, should tutorial software based on MATHPERT wish to do so.

[8]Montessori [1917], pp. 74-75, explaining the concept of control of error in the design of materials: "it is not enough that the stimulus should call forth activity, it must also direct it. The child should not only persist for a long time in an exercise; he must persist without making mistakes."

Graphics in MATHPERT

MATHPERT provides two-dimensional graphics; facilities are provided for graphing ordinary functions, parametric functions, and functions in polar coordinates. You can compare two functions on the same graph or on different graphs; you can choose the size, position, color, and other parameters of your graphs. You can pass from the symbolic manipulation part of the program to the graphics at will, for example first differentiating a function and then graphing it and its derivative on two graphs one above the other.

You can graph a formula containing a symbolic parameter, and change the value of that parameter at a touch of the '+' or '−' keys. You can "zoom" by a factor of two in the x or y direction by touching the arrow keys, or by several factors of two at once. In this way you can, for example, compare x^n with 2^x, alternately increasing n until x^n dominates, then zooming until 2^x dominates again.

The graphics of MATHPERT, however, is not cognitively faithful. There is no attempt to get the computer to sketch graphs as we wish students to do. It simply draws the graph in the usual computerized way.

Comparison of MATHPERT with Mathematica

These days one finds considerable enthusiasm for the application of computerized mathematics systems such as the new Mathematica (Wolfram [1988]) and its predecessors Maple and MACSYMA to mathematics education, particularly calculus. These applications fall into three categories, which largely correspond to the three major categories of abilities of Mathematica: computer graphics, symbolic manipulation, and numerical computation. The purpose of this section is to show that MATHPERT and Mathematica have different things to offer; one should not think of them as similar packages.

Graphics and numerical computation in MATHPERT are not different from graphics and numerical computation elsewhere, although the interfaces to them have been designed for education.

It is in the realm of symbolic computation that MATHPERT differs from Mathematica and its predecessors. Mathematica is similarly based on the concept of applying operators to expressions, but the operators in Mathematica do not represent skills that are taught to students. Cognitive fidelity is (intentionally) missing from Mathematica for two distinct reasons: First, the internal algorithms are not those taught to students, but rather the finest, fastest, most powerful algorithms known. Mathematica can factor $x^n - y^n$ for any value of n; MATHPERT, like the student, can only handle a few special values, and in many cases (for large n) can factor it only partially. This limitation of ability is a necessary correlate of cognitive fidelity.[9] Second, the operators provided with Mathematica are too powerful for use in unmodified form in education. For example, Factor is a single operator: it does not distinguish the many methods of factoring, and does not indicate the method or steps it uses.

References

Anderson, John R. [1988], The Expert Module, in: Polson, M. C., and Richardson, J. J. (eds.), *Foundations of Intelligent Tutoring Systems*, pp. 21-54, Erlbaum, Hillsdale, N. J. (1988).

Burton, R. R. [1982], Diagnosing bugs in a simple procedural skill, in Sleeman and Brown [1982], pp. 157-185,

Burton, R. R., and Brown, J. S. [1982], An investigation of computer coaching for informal learning activities, in Sleeman and Brown [1982], pp. 79-98.

Matz, M., [1982], Towards a process model for high school algebra errors, in Sleeman and Brown [1982], pp. 25-50.

Montessori, M. [1917], *Spontaneous Activity in Education*, reprinted by Schocken, New York (1965). (Page numbers in the text refer to the Schocken edition.)

Sleeman, D., and Brown, J. S. [1982], (eds.), *Intelligent Tutoring Systems*, Academic Press, London/ Orlando, Fla. (1982).

Wenger, E. [1987], *Artificial Intelligence and Tutoring Systems*, Kaufmann, Los Altos, Calif. (1987).

Wolfram, S. [1988], *Mathematica: A System for Doing Mathematics by Computer*, Addison-Wesley, Redwood City, Calif. (1988).

---

[9]Of course, one could always include an "oracle operator"— "appeal to methods beyond the scope of this course", which would result in the use of high-powered algorithms.

Dr. Dona V. Boccio
220-55 46 Ave.
Bayside, NY 11361

Using Computer Graphing Software as an Aid in Teaching
Maxima-Minima Problems

At the United States Merchant Marine Academy, the syllabus
for all sections of calculus includes five one-hour computer-
lab sessions per quarter. This time is devoted to using pre-
packaged software for problem-solving and to increase
students' understanding of the theory of calculus.

The speaker will present several examples of how a computer
graphing program can be used to solve maxima-minima verbal
problems that otherwise would be too time-consuming to
present in class. Problems encountered in using the software
will be discussed.

Examples will include minimizing the distance from a point to
a curve and maximizing the volume of a box created from a
rectangular sheet of cardboard. It will be shown how using a
graphing routine obviates the need for the instructor to
hand-pick the values for these types of problems.

Difficulties in choosing the "viewing rectangle" (that is,
the range of values of the independent and dependent
variables) for the desired graphs are well-documented ([1],
[2, p. 240]). The speaker will discuss several ways in
which the student can efficiently determine suitable regions.

References

1. Franklin Demana and Bert K. Waits, "Pitfalls in Graphical
    Computation, or Why a Single Graph Isn't Enough," The
    College Mathematics Journal 19 (March, 1988) 177-183.

2. Bert K. Waits and Franklin Demana, "Problem Solving Using
    Microcomputers," The College Mathematics Journal 18 (May,
    1987) 236-241.

Christine Browning

The Computer/Calculator Precalculus (C^{2}PC) Project and

Levels of Graphical Understanding

At The Ohio State University, a project group called the Calculator and Computer Precalculus (C^{2}PC) Project (Demana &Waits, 1988; Osborne & Foley, 1988) is moving toward increasing the attention given to graphing in the precalculus curriculum. The goal of the C^{2}PC project is to improve student understanding of functions and graphs by enhancing the precalculus curriculum through the use of computer and calculator based graphing. They are modifying the typical precalculus curriculum and applying computers and calculators to emphasize the correspondence between the numerical and algebraic representations of functions with their graphical counterpart. The number of examples, or the base for generalizing about functions and graphs, is increased by the use of computers/calculators. Students participating in the program, along with one precalculus class not in the C^{2}PC program that served as a control, provided the sample for this present investigation.

In order to determine the hypothesized increased understanding of functions and graphs provided by this modified curriculum, a measure of assessment needed to be developed specific to these areas of concern. Thus, the C^{2}PC project provided a motivation for the author to develop an instrument to assess student understanding of functions and graphs. The study (Browning, 1988) focused on developing a "Graphing Levels" instrument to determine if levels of understanding exist and can be characterized. The Graphing Levels instrument was then used to determine any growth in graphical "ability" in an attempt to assess any growth in understanding.

The title of the instrument indicates the author's belief that the understanding of functions and their graphs will occur in levels, perhaps hierarchical. This belief is based on research pertaining to the van Hiele levels of geometric understanding (van Hiele, 1958; Hoffer 1981, 1983; Usiskin 1982; Mayberry 1983; Fuys and Geddes 1984; Shaughnessy and Burger 1985, 1986; Crowley 1987). Although this research is based on levels of understanding related to geometric learning, a later work of van Hiele (1986) extended his ideas about levels to all mathematical learning. Very little research to date has focused on extending van Hiele's ideas to other areas.

Another major research project related to levels of understanding is the Concepts in Secondary Mathematics and Science (CSMS) program in England (Hart, 1980). Results of the CSMS study demonstrate the presence of levels of understanding in many mathematical topics (ratio and proportion, fractions, positive and negative numbers, graphs, vectors, algebra, etc). The research performed by the author was a "partial replication" of the CSMS project.

The research took place during Autumn Quarter, 1987 and Winter and Spring Quarters, 1988 at The Ohio State University, Columbus, Ohio. A 25 item instrument was designed, titled "Graphing Levels Test," and given to precalculus students at two city and two suburban high schools in central Ohio. The majority of the students were enrolled in the C^{2}PC precalculus classes with the remainder enrolled in a standard precalculus class. The Graphing Levels Test was used as both a pre- and posttest with only minor item modification on the posttest. Student interviews took place after both pre- and posttest administration. Pretest interviews were used as an aid in determining item modifications along with suggestions from mathematicians and mathematics educators. Results from posttest interviews aided in determining level characteristics.

Cluster analyses similar to that performed in the CSMS project was performed on pretest results from a random sample of 125 students from the original 211 to determine levels of graphical understanding. The group of 211 students was comprised of students from both groups in the original sample. Four clusters or levels were suggested from the analyses. The number of levels was based on the "facility band splits." A criteria span of not more than 20 percent was chosen, based on the CSMS project suggestions. A Guttman scalogram analysis was performed using the pretest random data to determine the validity of a hierarchy. A coefficient of reproducibility higher than 0.9 is considered to indicate a valid scale. The coefficient for the pretest random sample was 0.9975 indicating a hierarchy.

An examination of clustered items and review of posttest interviews suggested item characteristics for each level. The following is a sample of characteristics for each level.

Level 1: recognition of the graph of a parabola placed in different positions, simple interpretation of information from a graph, and development of initial vocabulary.

Level 2: translation of verbal information into a simple sketch of a graph, use of initial vocabulary learned in Level 1 and simple interpolation from a graph.

Level 3: usage of properties of graphs of functions to determine functions from non-functions, recognition of connection between a graph and its algebraic representation, and usage of properties of functions to construct graphs.

Level 4: usage of given information to construct a graph, usage of information from a graph to deduce more information, and recognition of what is required to find from given information.

As one advances from Level 1 to Level 4, it was noted the items became more complex, required more interpretation, required relating more ideas, depended on more knowledge, and required more complex problem solving strategies.

The pretest clusters were compared with posttest results from a split sample, that of the C^2PC group (n=125) and the control group (n=32). The level structure was essentially preserved for both groups, thus furthering the validity of the level construct.

Since the levels were found to be hierarchical in nature, a crosstabulation of the pretest data was done. Level scores were computed by alloting one point for each correct response to the items within each level produced by the cluster analysis. A criterion of correctly answering at least two-thirds of the items within each level was chosen as suggested by the CSMS project. The results of the crosstabulation showed the majority of students (77.5 percent) operating at Levels 0, 1 or 2 at the beginning of their precalculus year. A posttest crosstabulation showed the majority of the control group, 68.7 percent, remained at Levels 0, 1, or 2 while 73.1 percent of the C^2PC students were operating at the higher levels of 3 and 4. These results suggest the use of graphing technology within the precalculus classroom will provide for increased student understanding of graphs.

Another indication of growth in graphical understanding was found in comparing the posttest means of the five C^2PC classes with the one control class. All five posttest means were significantly higher ($p < 0.05$) than the control posttest mean of 13.7. The higher posttest means again provide evidence that the graphing technology makes a difference in the students' graphical understanding.

Results from this study indicate changes for the precalculus curriculum. Technology needs to be incorporated into the curriculum, specifically that related to graphing. If the precalculus curriculum is to prepare one for calculus, groundwork for calculus concepts and the corresponding skills must be laid. Part of this important groundwork is related to an understanding of functions and their graphs. As noted by the NCTM Commission on Standards (1987), among the most important of mathematical connections to be made in the curriculum are those between algebra and geometry. These connections can and should be developed in the precalculus curriculum. Graphing technology, as demonstrated by this study, is a definite aid to the student in developing the mathematical connections (see Pinker 1981, 1983 for further discussion on interpreting and making "connections" with visual displays). Technology provided C^2PC students with an increased example base and a greater number of graphing experiences than the control, furthering the C^2PC students' understanding of the connection between a function and its graph.

The hierarchical nature of the graphing levels suggests an order in the presentation of the topic . The greatest task for the educator is to allow students opportunities for making their own connections and not to provide shortcuts early on. For example, instead of telling the students what the graph of a quadratic equation looks like, allow them to experiment, a feat now possible with the aid of a graphing calculator. Not only does this allow the student to make their own generalization of what the graph of a quadratic looks like, it provides them exercises in creating quadratic equations.

In summary, results of this study show learning of functions and their graphs occurs in levels. The hierarchical nature of the levels implies an order to the presentation of topics related to graphing.

Results also imply the use of technology in the classroom improves student understanding of functions and their graphs by providing an increased example base. The interaction between student and computer also provided for making mathematical connections, connections necessary to understand and use graphs.

References

Browning, C.A. (1988). Characterizing Levels of Understanding of Functions and Their Graphs. Unpublished doctoral dissertation, The Ohio State University, Columbus.

Crowley, M.L. (1987). The van Hiele model of the development of geometric thought. In M.M. Lindquist, & A.P. Shulte (Eds.), *Learning and Teaching Geometry, K-12: 1987* (pp. 1-16). Reston VA: National Council of Teachers of Mathematics.

Demana, F., & Waits, B.K. (1988). *Precalculus Mathematics, A Graphing Approach.* In press (Addison-Wesley).

Fuys, D. & Geddes, D. (1984). An investigation of van Hiele levels of thinking in geometry among 6th and 9th graders. Research findings and implications. (ERIC Document Reproduction Service No. ED 245 934).

Hart, K.M (1980). *Secondary school children's understanding of mathematics.* London: Chelsea College, University of London, Centre for Science Education.

Hoffer, A. (1981). Geometry is more than proof. *Mathematics Teacher,74,11-18.*

Hoffer, A. (1983). Van Hiele-based research. In R. Lesh & M. Landau (Eds.), Acquisition of mathematics concepts and processes (pp 205-227). New York: Academic Press.

Mayberry, J (1983). The van Hiele levels of thought in undergraduate preservice teachers.*Journal for Research in Mathematics Education, 14, 58-69.*

National Council of Teachers of Mathematics, Commission on Standards for School Mathematics. (1987). *Curriculum and evaluation standards for school mathematics (working draft).* Reston VA: National Council of Teachers of Mathematics.

Osborne, A. & Foley, G. (1988). *Precalculus Mathematics, A Graphing Approach. Instructors Manual.* In press (Addison-Wesley).

Pinker, S. (1981). *A theory of graph comprehension* (Occasional Paper #10). Cambridge, MA: Massachusetts Institute of Technology, Center for Cognitive Science.

Pinker, S. (1983). Pattern perception and the comprehension of graphs. Cambridge MA: Massachusetts Institute of Technology, Department of Psychology. (ERIC Document Reproduction Service No. ED 237 339).

Shaughnessy, J.M., & Burger, W.F. (1986). Characterizing the van Hiele levels of development in geometry. *Journal for Research in Mathematics Education, 17,* 31-48.

Shaughnessy, J.M., & Burger, W.F. (1985). Spadework prior to deduction in geometry. *Mathematics Teacher, 78,* 419-427.

Usiskin, Z. (1982). *Van Hiele levels and achievement in secondary school geometry.* Chicago: University of Chicago, 1982. (ERIC Document Reproduction Service No. ED 220 288).

van Hiele, P.M. (1986). *Structure and Insight: A Theory of Mathematics Education.* Orlando, Fl: Academic Press, Inc.

van Hiele, P.M., & van Hiele-Geldof, D.A. (1958). A method of initiation into geometry at secondary schools. In H. Freudenthal (Ed.), *Report on methods of initiation into geometry* . Groningen: J.B. Wolters.

EPIC

James Burgmeier
and
Larry Kost

Department of Mathematics
University of Vermont
Burlington, Vermont 05405

EPIC (Exploration Programs In Calculus) is a microcomputer software package that permits users to investigate mathematical concepts. It includes mathematical style function input, an extensive function plotting routine, a module on limits, a module on integration, and a versatile polar coordinate and parametric equation curve sketching module.

EPIC supports the use of parameters in functions and easily shows how the graph depends on these parameters. Users may estimate the location of zeros, critical points, and inflection points, draw tangent lines and secant lines, and draw first and second derivative curves. Limits and difference quotients may be examined; the area under a curve or between two curves may be approximated by Riemann sums, the trapezoidal rule, or Simpson's rule. Intersection points of polar or parametric curves can be approximated and these curves can be "traced."

EPIC allows functions to be entered in multi-line textbook style, including restricted domain specifications and summations. Virtually any function in calculus texts can be entered and examined. EPIC can be used to enhance many topics in single variable calculus. Due to the required brevity of this paper, we will only discuss a few topics where EPIC may be useful in elementary calculus.

Continuity and Differentiability

EPIC supports the definition of functions involving more than one segment or section. For example, the function

$$f(x) = \begin{cases} \dfrac{x^2 + 1}{2x - 3}, & x \le \pi \\ a\operatorname{Tan}^{-1}x + b\sin x, & x > \pi \end{cases}$$

is entered into EPIC as shown in Figure 1. The primary difference between textbook display of such functions and their input into EPIC is the inclusion of the double line to separate sections.

Many functions included in the discussion of continuity or differentiability in calculus texts involve multi-section functions and these functions may be copied into EPIC exactly as they are written in the text. This formulation versus a one line, computer-language definition enhances readability and student understanding, minimizes error in function input and enables the student to focus on the mathematics of the task rather than the translation of the formula to a computer-language expression.

Figure 1

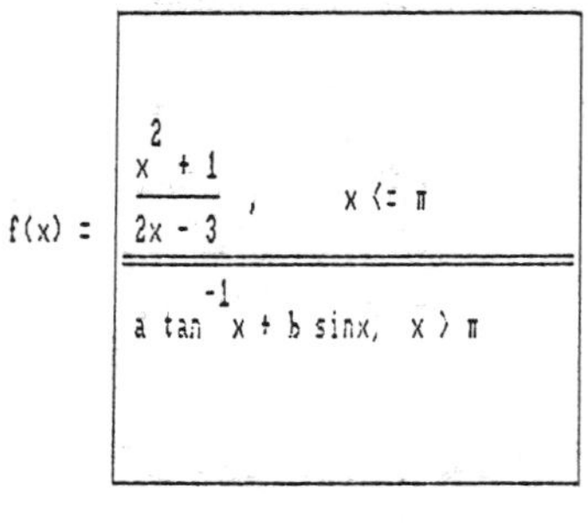

Students could be assigned several problems related to functions of this type. For example, what values should be chosen for the parameters a and b to ensure that $f(x)$ is continuous and differentiable at $x = \pi$? The proper value of a is $a = (\pi^2 + 1)/((2\pi - 3)\arctan \pi)$. EPIC supports algebraic expressions as input for numerical information, so the student could enter this value as $(\pi \char`^ 2 + 1)/((2\pi - 3)\arctan \pi)$.

EPIC will draw tangent lines and secant lines; the secant line feature may be used to illustrate that, when the choices for a and b differ from their proper values, the curve essentially has two distinct "near" tangent lines at $x = \pi$, and hence is not differentiable there.

Inverse Functions

The graph of $y = f(x)$ is the collection of points $\{(x,y) \mid y = f(x)\}$. The inverse *relation* corresponding to f is the set $\{(y,x) \mid y = f(x)\}$. EPIC will draw this "inverse" curve and frequently it will be a multi-valued function. A single-valued function may be obtained by restricting the domain of the original function. Students can be asked to suitably restrict the domain of several functions as an introduction to the restriction of the inverse trigonometric functions to their principle values. For example, the graph of $y = f(x) = x^3 - 3x + 1$ and its inverse relation are shown in Figure 2. Since the max/min occurs at $x = \pm 1$, one possible domain restriction is $-1 \le x \le 1$. This restriction to f and its inverse are shown in Figure 3.

Figure 2

Figure 3

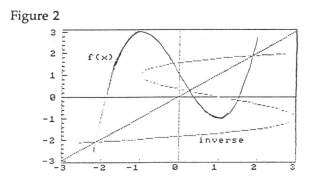

Relative Extrema and Inflection Points

Superimposed on the graph of f, EPIC will draw the graphs of f' and f''; the first and second derivatives may be computed by formula or approximated by difference quotients, at the user's discretion. EPIC will also find the zeros of f, f' and f'' (by Newton's method). The connection between a critical point and a relative extrema can be easily demonstrated as well as the distinction between *possible* inflection points and inflection points. In addition to drawing several graphs of a function by varying parameters in the function definition, EPIC allows two functions to entered and plotted at once, and then additional functions may be entered and plotted without destroying the screen. Thus many functions may be displayed at the same time.

The Computer May Mislead

EPIC plots functions by evaluating the function at equally-spaced points throughout an interval selected by the user. (EPIC starts with 128 sampled points, but the user may vary this from 2 to 600.) Because of this sampling, important behavior of a function may be missed. An extreme example is the graph of $f(x) = \cos 256\pi x$ on the interval $[0,1]$. Sampling at $x_i = i/128$ yields a value of 1 for each i. Consequently, the graph appears to be a straight line at $y = 1$.

Another example is the graph of

$$f(x) = \frac{e^x - 1 - x - \frac{1}{2}x^2}{x^3},$$

for $x \neq 0$ and $f(0) = 1/6$. The graph of f on a small interval about the origin produces inaccurate function values due to round-off error. The graph is "bumpy" when it should be monotone and nearly constant. See Figure 4.

Figure 4

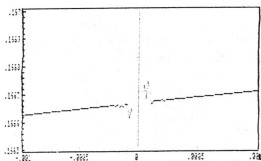

Implicit Differentiation

Consider the graph of the polynomial $f(x) = x^3 - ax + 1$. Figure 5 shows the curve obtained for $a = 2$ and $a = 4$. Increasing a seems to have moved the two leftmost roots left while the largest root has moved to the right. If we view $x^3 - ax + 1 = 0$ as an equation defining x as a function of a, then implicit differentiation yields

$$x'(a) = \frac{x(a)}{3x^2(a) - a}.$$

Figure 5

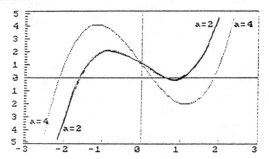

For the roots near -2 or 0.3, a quick calculation verifies that $x'(a) < 0$; for the root near 1.5, $x'(a) > 0$. This supports the observed movement of the roots. However, note that for the case $a = 2$, the function almost has a double root. For values of a smaller than that needed for a double root, there are no positive roots, leading to questions of when is a function $x(a)$ defined by an implicit relation. In this setting, EPIC may be used to motivate the need for more analysis of the situation.

Series

EPIC will plot functions defined by a series, including power series and Fourier series. Figure 6 shows the input screen for the series representation of Bessel functions, $J_n(x)$. Note the parameters in the upper limit of the summation and in the terms of the series. The first controls the number of terms computed to evaluate the function; the second parameter is the index of the Bessel function. Figure 7 shows the graphs obtained by choosing $m = 8$ and $n = 0, 1, 2$.

Figure 6

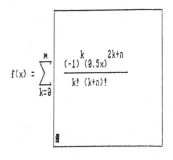

Figure 7

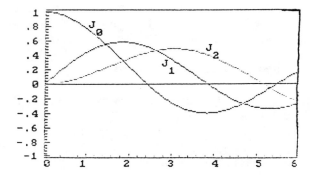

Additional Features

- EPIC enables the user to approximate area under a curve or between two curves by five methods: left, right and midpoint Riemann sums, the trapezoidal rule and Simpson's rule.

- EPIC contains a module that demonstrates the concept of limits. A small "window" (chosen by the user) of the graph of the function is magnified repeatedly, displaying details of the behavior of the graph.

- EPIC contains a module on polar and parametric equations. Some of the standard polar and parametric curves may be selected from a menu, or the user may enter the functions of his choice. A frequent question related to these curves is the precise order in which the curves are traced as the polar angle or the parameter varies. EPIC provides the answer to this question by slowly tracing these curves, with the tracing rate under user control.

- EPIC contains a calculator which "understands" most standard mathematical notation. The calculator may be invoked within every module. Parameters may be assigned to calculated values and then used in subsequent calculations. Up to ten formulas or calculations may be saved and recalled for later use.

A METHOD TO TEACH NUMBER THEORY USING A COMPUTER

Chris K. Caldwell

1. Introduction

Much of elementary number theory was discovered through computation, yet computation and discovery are almost nonexistent in the traditional number theory course. As a result this course often becomes stale and theoretical.

To allow discovery in our number theory course we have used muMATH by SoftwareHouse (augmented with our own number theoretical routines) as an unlimited precision calculator, making otherwise tedious computations trivial. Using carefully selected homework assignments, the students were led by to discover number theory for themselves. In class we developed *the student's* results and laid the groundwork for *their* future discoveries.

The homework problems not only set the pace of the course, but were chosen to teach the students how to experiment and conjecture. By using a text only as a reference we avoided the student's usual proof algorithm: (1) reread the techniques in the section containing the problem, (2) apply only those techniques.

In this paper we give examples of the homework problems and explain the philosophy behind their choice. We end with a brief discussion of the program and the text.

2. The Homework

To emphasize discovery, homework problems were chosen to be open ended. To make student discoveries central to the course, the lectures usually centered on the previous night's homework. Lectures also introduced definitions and concepts necessary for following problem sets. Consider the following homework problems:

1. Choose several primes p. For each evaluate (p-1)! modulo p. Make a conjecture. Prove your conjecture.

2. Choose a prime p and an integer a. Evaluate a^{p-1} modulo p. Repeat with several different a's and p's, be sure to try a = 0. Make a conjecture. Prove your conjecture.

While doing these problems the student quickly conjectures Wilson's theorem and Fermat's little theorem respectively, though they can not yet prove them. The proofs are given the next class day. Because we were proving results the student had discovered herself, the student was more interested, and remembered the results longer.

In real life we rarely know ahead of time if we can prove our conjectures, though we often have a gut feeling. To help develop this instinct in our students we did not indicate if they could succeed in their assigned proofs. The following problem, for example, leads to the unproven Goldbach's conjecture.

> 3. Choose any even integer larger than two. Can it be written as the sum of two primes? Make a conjecture. Prove your conjecture.

In the class meeting following this problem, we would discuss the progress that has been made towards a proof. More importantly, we would discuss why this result is harder to prove than the previous results.

Very little of undergraduate mathematics was done this century, so students rarely encounter unsolved problems. Number theory is an ideal place to correct this shortcoming. Here is one example:

> 4. Let $F(n) = 3n+1$ if n is odd, $n/2$ if n is even. Starting with any positive integer iterate this function until you reach one. For example starting with 13 we get the following sequence of numbers 13,40,20,10,5,16,8,4,2,1. Prove or disprove that the sequence always terminates (always reaches one).

(The function F is included as part of our computer software.) Shank's book [8] is an excellent source for this type of problem.

In the above examples the student could not have completed the proof herself. To avoid the danger that the student would not even try to prove her results, many of the problems had trivial proofs:

> 5. Take any three digit number and write it down twice. For example, 451 becomes 451451. Divide the resulting six digit number by 77. Make a conjecture. Prove your conjecture.

(Note abcabc = abc·1001 = abc·77·13.) It is surprising how many students can not tell whether proving a result will be easy or difficult. Hopefully students develop the necessary intuition as they see both easy and difficult problems side by side.

Two more pitfalls to avoid are the law of small numbers [5], and the student's desire to do only the minimum work required. We attack these problem with the following.

> 6. For $n = 1,2,...,20$ factor n^2+n+41. Make a conjecture. Prove your conjecture.

7. Use PRIMETEST to test the primality of the following numbers: 31, 331, 3331, 33331, 333331, 3333331. Make a conjecture. Prove your conjecture.

(PRIMETEST is part of our computer package.) Here the suggested numbers are all prime, so many students fell into the trap and conjectured that the results are always prime. Only about one half of the students looked look at unlisted terms in the second sequence. 33333331 is prime, but 17 divides 333333331.

Problems six and seven could be used to lead into a classroom discussion about Fermat and Mersenne primes, or about formulas for the n-th prime, or functions which take on only prime values... The student should be shown that there are vast unexplored frontiers.

We close this section with a follow up to problems six and seven.

8. We have seen several sequences whose first few terms are prime. Construct an interesting sequence of your own which begins with at least five primes.

This problem exemplifies our goals: it requires student discovery, it is open ended, and it can be used to lead into many classroom discussions.

3. The Program

All of the examples above are computationally intensive. Few students enjoy calculating 13^{72} modulo 73 by hand, or factoring $20^2+20+41$ by hand. So the computer was introduced as a calculator, making the calculation trivial - but leaving the student to interpret the results. We used muMATH by SoftwareHouse because it was cheap and ran on the IBM-PCs we had available. Any software package that had unlimited precision rational arithmetic could have been used. To this package we added the basic functions of number theory (phi, sigma, Jacobi symbol,...) along with routines to factor numbers (Pollard's rho and p-1 methods, ... [see references 2 and 7]). We also included the primality proving theorems of [1].

The computer was used as a discovery tool only, it was not the object of study and few students bothered to even list the subroutines. Less than fifteen minutes was necessary for most students to begin using the program. Many students enjoyed having the machine factor their phone numbers, social security numbers, even their names (viewed as integers base 36). The program greatly encouraged mathematical play.

As an example of the surprising power of this little system, consider the numbers in problem seven, $a_n = (10^n-7)/3$. It was known that these are prime for n =1,2,3,4,5,6,

7,17,39,49,59,77 and 100. Using the muMATH routines (and factorizations from [2]) we proved the primality of two of the next three probable-primes: a_{150} and a_{381}. Harvey Dubner showed a_{318} is prime.

4. The Text and Students

This course could have been taught without a text, or with a text like Shank's book [8], however, we chose to use a traditional text as a reference (Burton[3]). Next time we will probably use Eyden [4] or Rosen [6] because of their treatment of primality testing and RSA codes. Whatever text is chosen, do not follow it closely. Otherwise the students will look up the results rather than discover the results.   Not all students thrived under these conditions (a few were frightened by the problems sets and dropped the course immediately), but most students did very well, and a few even learned to love
number theory. The students' confidence increased. Many made mathematical discoveries for the first time in their lives. Several now regularly use the computer as a tool in mathematics.

Our small school has two number theorists and a biannual number theory course. So I have only been able to teach number theory once. I will definitely use this approach again.

REFERENCES

1. John Brillhart, D. H. Lehmer and J. L. Selfridge, "New Primality Criteria and Factorizations of $2^m \pm 1$", Math. Comp. 29 (1975), 620-647.
2. Brillhart, Lehmer, Selfridge, Tuckerman, Wagstaff, Factorization of $b^n \pm 1$, b = 2,3,5,6,7,10,11,12 Up To High Powers, Contemporary Mathematics 22, American Mathematical Society, 1988.
3. David Burton, Elementary Number Theory, Allyn and Bacon, 1975.
4. Charles Eyden, Elementary Number Theory, Random House, 1987.
5. Richard Guy, "The Strong Law of Small Numbers," Amer Math Monthly, 95 #8 (1988), 697-711.
6. Kenneth H. Rosen, Elementary Number Theory, 2ed, Addison-Wesley, 1988.
7. Hans Reisel, Prime Numbers and Computer Methods for Factorization, Birkhauser, 1985.
8. Daniel Shanks, Solved and Unsolved Problems in Number Theory, Chelsea, New York, 1978.

Department of Mathematics and Computer Science
University of Tennessee at Martin
Martin, TN 38238

THE MATHEMATICS LEARNING RESOURCE CENTER
AT OKLAHOMA STATE UNIVERSITY
Experiences with a Large Scale Mathematics Laboratory

by Pamala Çemen and Jerry Johnson
Department of Mathematics
Oklahoma State University
Stillwater, Oklahoma 74078

Oklahoma State University is a comprehensive land grant institution with a College of Engineering and an enrollment of approximately 21,000. The Mathematics Learning Resource Center (MLRC) was established here in 1984 as a comprehensive center featuring both electronic technology and conventional tutoring for supplementing and enriching the mathematics curriculum. The center occupies 4,354 square feet, has a full time director (Dr. Pamala Çemen) and a staff of 30 undergraduate and graduate student assistants. It is available to students 59 hours each week. Included in the MLRC are:

- 40 networked Texas Instruments Professional microcomputers,

- 12 networked IBM compatible Texas Instruments Business Pro microcomputers,

- 5 TI printers,

- 2 Control Data microcomputers,

- 19 combination VCR - television learning stations, and

- 5 caramate slide-tape players.

The two Control Data machines mentioned above are used for Plato algebra, precalculus, and calculus tutorial packages obtained as part of a beta test site agreement.

A wide range of commercial software is available for the Texas Instruments microcomputers including technical word processing, programming in the common computer languages, spread sheets, and a computer algebra system called muMATH™ *Soft Warehouse* used largely by students in linear algebra and differential equations. In addition, there are elaborate graphing packages and menu driven utilities that were written by faculty and graduate students here at OSU. One of them illustrates various concepts in differential equations such as direction fields and dependence of solutions on parameters.

Every attempt has been made to assure that a first time visitor to the MLRC with no knowledge of computers can quickly begin using much of the educational software. Everything from logging onto the Novell Network to calling up a program is done by simply following a sequence of menus. In addition, succinct documentation and on-line help has been provided.

Available for use on the VCRs are approximately 300 studio quality videos of mathematics lectures. Many were produced by Oklahoma State University faculty members through small grants and some were purchased from the Ohio Mathematics Association of Two Year Colleges. For the slide caramates there are similar sets of slides. The video tapes and slide sequences review the full range of topics covered in elementary mathematics through Calculus and Differential Equations.

In addition, the MLRC provides several support services such as conventional tutoring, a copy machine, electronic paper grading, placement testing, and a provision for instructors to place materials on reserve.

The problems of setting up and operating such a facility are enormous, but integrating it into the undergraduate mathematics curriculum is also difficult and we have made a vigorous attempt to do this.

Almost every student enrolled in mathematics from intermediate algebra through calculus, differential equations and linear algebra is assessed an $10 per semester fee for MLRC use. This semester (Fall 1988) that comes to about 5,800 students. Any other student at OSU may use the MLRC provided they pay the fee. On their first visit students fill out a card on which is kept a log of how they used the MLRC and what materials or software they checked out. The cards of students enrolled in a course requiring an MLRC fee are filed under the instructor's name so he or she may check a student's attendance if desired.

It is important to emphasize that the MLRC is strictly a supplement to the curriculum. No regularly scheduled classes meet there and no given amount of attendance is required. That is why the large number of clients we serve is manageable.

It is a fact of life that most students will do little academically that does not impinge on their grades. Therefore, ensuring that the MLRC is integrated into the curriculum requires the cooperation of individual instructors in giving graded assignments. While some are enthusiastic, others are less so. Part of the reluctance of the latter group comes from unfamiliarity with computers and part from philosophical disagreement with the whole idea, but most of it is simply due to the fact that OSU is a research institution where demands on the faculty's time make them unwilling to spend it in this way. It is therefore important that the instructor be required to exert very little effort in cooperating with the program. To guarantee this, the following steps have been taken:

At the beginning of each semester instructors are issued handouts for their students that describe the MLRC and provide a directory of the software and videotapes available. This directory is keyed not only to the subject matter but to the current textbook as well. Thus, if a student wants supplementary instruction on the material in section 5.3 of the text, appropriate videotapes and software are listed in the directory.

Throughout the rest of the semester instructors are issued self-contained MLRC assignments over various topics as they arise in the course syllabus. They may be used as the individual teacher wishes but every faculty member and teaching assistant is *strongly* encouraged to make them a part of the course.

Most of the exercises are designed to be fairly easy to grade and are self-explanatory with step-by-step computer instructions provided. Should the student require assistance there is always a staff member on duty at the MLRC to help. All this ensures that the instructor need know nothing about the computers and that little of his or her time is required to make the MLRC a part of the course.

While the program we have just described has been successful in most respects, there are improvements that still need to be made. One of the problems is creating *good* exercises. Most of the those mentioned above have been written by the first author and MLRC staff members in their "spare" time. It is very time consuming to design really meaningful assignments that teach concepts and reinforce ideas without the intervention of an instructor. A good way for a department to accomplish this is to provide interested faculty with release time to create such exercises, but this is not always a high priority.

Another aspect of this difficulty is that most of our assignments have only one version so that plagiarism is quite probable. It is likely that a student in a large section could copy all the MLRC assignments from a friend and never set foot in the lab.

The rapid change of technology is an inevitable companion in such an endeavor as well and the expense of staying up-to-date can be considerable.

Even with all its problems, we feel that the program is worthwhile. Without it there would be a lack of contact with the technology that is constantly changing our approach to teaching and learning.

THE USE OF SPREADSHEETS IN FINITE MATHEMATICS

Stephen D. Comer
Department of Mathematics and Computer Science
The Citadel, Charleston, SC 29409

This paper is a preliminary report on an experiment concerning the use of Lotus[TM] 1–2–3[TM] in a finite mathematics course for freshmen non–mathematics majors. The structure and purpose of the course and the rationale for using Lotus to do the modelling and calculations are described. In addition, the way in which Lotus is used in the course and the topics treated with the spreadsheet are discussed.

THE NATURE OF THE COURSE

A typical section of Finite Mathematics at The Citadel has approximately 30 students, mainly freshmen. They are mostly Business Administration or Political Science majors but the course is open to all non–mathematics majors. The course is used to partly satisfy the college's mathematics requirement. The purpose of the course is to increase a student's understanding of the value of mathematics through an exposure to different types of elementary mathematical models and to improve analytical skills by having a student solve problems using the various models. The course covers a standard set of finite mathematics topics: linear equations, matrices, geometric linear programming, quadratic functions, finite probability, and statistics. For the last several years a Finite Mathematics software package has been used with the course to assist students perform some of the tedious calculations, especially the computation of matrix inverses.

In at least one regard the past course has been a disappointment. Typically when ideas discussed in the Finite Mathematics course are encountered later in courses in other disciplines, a student exhibits total ignorance. We would like to overcome the attitude that mathematics is a sterile subject, only done in a mathematics class. One reason for introducing a spreadsheet as the vehicle for carrying out mathematical calculations is to improve student motivation for the course by increasing its impact on a student's career. The goal is to present the models used in the Finite Mathematics course in such a way that students will be able to build on these ideas when they are encountered in courses in their major.

WHY USE A SPREADSHEET?

Given the many software packages that are available to perform the calculations that are encountered in a Finite Mathematics course, why use a spreadsheet? There are two strong reasons — the wide availability of spreadsheets and their power. To some extent the "educational" software, geared to this particular course and useful for no other, contributes to a narrow view of mathematics' applicability. On the other hand, spreadsheets are widely available. Every computer in our PC lab is equipped with Lotus 1–2–3. Spreadsheet applications have become commonplace in businesses. It is felt that students, even those not majoring in business, will have many opportunities to build on the mathematical skills they acquires using a spreadsheet. Books by Deane Arganbright [1] and by Michael Kilpatrick [2] show how to implement a wide variety of common algorithms using a spreadsheet. A spreadsheet can be used to carry out every calculation needed in a standard Finite Mathematics course. Lotus 1–2–3 was the spreadsheet of choice because of the large number of built–in functions and operations available with version 2.0. Moreover, a student can begin to handle more complicated features after becoming proficient with the simple commands and functions. Those that master the use of macros can perform many sophisticated calculations. Lotus is a powerful calculating tool; however, it also should be emphasized that Lotus does not solve problems for students by magic. When using a spreadsheet to solve a problem or implement an algorithm, a student generally performs the same steps as if the problem were being solved by hand.

INSTRUCTIONAL USE OF LOTUS

To prevent the course from becoming preoccupied with computer skills, the spreedsheet topics to be introduced were carefully selected. For various reasons the topics were restricted to graphing, solving matrix equations with a non–singular coefficient matrix, and elementary statistics. These topics will be described in more detail below. Lotus 1–2–3 (version 2.0 or higher) provides an ample collection of @functions (not only statistical and financial, but also all of the standard trigonometric and logarithmic functions), the operations of matrix multiplication, matrix inversion, frequency distribution tabulation, and regression in addition to graphing. The one disadvantage of Lotus is the awkward procedure for printing out a graph. However, it was felt that the availability of so many built–in operations was worth the additional hassle with graphing. Lotus is used in the course in two ways.

1. It is used by the instructor to perform classroom calculations and to conduct classroom demonstrations.

2. It is used by students outside of class to complete assignments. To facilitate assignments that involve considerable data, Lotus worksheet files that contain the problem and data are stored on a hard disk for students to copy. This was done, for example, with a problem involving a Leontief model of the economy and problems involving the computation of probabilities using normal distributions.

Since it was assumed that a student in this course would have no previous experience with a computer, a certain amount of instruction in the use of a computer and the Lotus program was necessary. The discussion of basic computer operations included: how to format a diskette, how to create subdirectories, how to move between subdirectories, and how to copy and erase files. The first spreedsheet operations discussed were how to start Lotus, how to move around a worksheet, the types of data and how to enter data, how to save and retrieve files, and how to use the on–line help. The Command Menu was briefly explored. This introductory phase took approximately three hours of class, half of which was spent in a hands–on lab session.

During the course students are given hand–outs which, in a step–by–step manner, guide them through a "Lotus" solution to various mathematical problems and provide exercises. These exercises are coordinated with the classroom topics and discussions. Students are requested to turn in screen prints or the graph printouts of appropriate exercises. At mid–term each student was given a hands–on computer quiz.

As mentioned above three topics were chosen for spreedsheet implementation: graphing, solving matrix equations, and elementary statistics. Each of these topics will be discussed briefly.

GRAPHS

Ordinary XY graphs (as a set of ordered pairs) can be easily constructed using the Lotus 1–2–3 operations of /Data Fill, /Copy, and a formula for the function. Since functions such as @SIN, @LN, @EXP, and @ATAN are standard Lotus functions, students can quickly generate a large variety of graphs. The /Graph command is used to plot the ordered pairs on an XY axis for viewing. Up to 6 functions can be plotted at one time. After mastering the basic construction of a graph, a student was asked to recalculate a graph over various intervals. Once a graph has been created, if its domain is modified, the revised plot can be viewed with a single keystroke using the F10 key. Students were asked to examine various portions of a graph in order to identify the domain of a function, approximate zeros, and approximate relative maximum and minimums. For simple functions (eg., quadratic functions) students were asked to compare the approximate values they obtained with values obtained by hand calculations.

MATRICES

The algorithm for matrix addition is easily implemented using the /Copy command. Lotus contains commands for multiplying two matrices and for computing the inverse of a matrix. Students are still required to learn how to perform these operations by hand, but with problems that do not involve an excess of arithmetic. Using Lotus, students were asked to solve various matrix equations of the form $AXB + C = D$ where A and B are nonsingular. The use of Lotus makes it reasonable to experiment with various open Leontief models for an n–industry economy.

ELEMENTARY STATISTICS

For a given collection of data the Lotus /Data Distribution command allows one to construct a frequency distribution. Students can display the data as a bar graph (to approximate a histogram) or a frequency polygon by means of the /Graph command. To measure the central tendency and dispersion of data a student first learns to use the @AVG, @VAR and @STD functions from the library of Lotus statistical functions. For frequency distributions of grouped and ungrouped data the student is required to implement the standard formulas for the mean and standard deviation. When considering problems involving a normal distribution Lotus can be used as a substitute for looking up areas under the standard normal curve. A template worksheet that contains the cumulative normal probabilities is available for students to copy. With this worksheet problems involving normal probabilities can be worked using the @VLOOKUP function.

Topics, other than those mentioned above, could be incorporated in the course. For example, Lotus can be used to obtain binomial probabilities. Also, finite probabilities could have been treated, but this would have involved having students write macros. At present only one example, a macro for factorial, is included as a exercise.

REFERENCES

1. Arganbright, Deane E. Mathematical Applications of Electronic Spreadsheets. McGraw–Hill, New York, 1985.
2. Kilpatrick, Michael. Business Statistics Using Lotus^{TM}1–2–3^{TM}. John Wiley & Sons, New York, 1987.

STUDENT CENTERED DESIGN FOR EDUCATIONAL SOFTWARE
by Jere Confrey and Erick Smith
Cornell University

Introduction:

In this paper we wish to suggest that our current ventures into the design of computer software for the exploration of mathematical ideas ought to be viewed as tentative, but necessary first approximations. It is a plea for modesty that we might become aware of how impoverished our own knowledge is about student conceptions, learning, and the act of representing knowledge in a dynamic interactive medium.

In this spirit, we make four claims:

1) Software designed to in some sense give a literal "representation" of mathematical ideas will not be as effective as software designed to assist students in solving problems by providing a medium for them to express and record their mathematical ideas, allowing them the opportunity to reflect on these conceptions.

2) We should expect our first implementation to provide us with opportunities for revision. We need to examine how the students use the software and search out how the deviations from our expectation are evidence of alternative conceptions.

3) If we are open to it, examining these alternative conceptions may lead us to experience transformations in our own understanding of mathematics. These transformations are the result of our awareness of both the student as making viable mathematical statements within the constraints of his/her conceptual world, and of the computer as providing both a different means for mathematical action and for the communication of mathematical ideas.

4) Translating the idea for the program into a specific software program through interaction with our technical designers and programmers as they code it exerts another influence on the product. We can improve design by using tools such as Hypercard™ to storyboard the interfaces and thus reduce the distance between conceptual design and the actual program. This issue will not, however, be further addressed in this paper.

A Conventional View of Software Design:

To contrast our proposal for the design of software from that which seems often to be used, we present a hypothetical model of a typical design process:

1) Select a piece of mathematics to implement.

2) Choose which forms of representation (graphs, tables, equations, matrices, etc.) to use and the appropriate operations for each (numeric calculations, plotting, algebraic manipulation, etc)

3) Consider how to make this mathematics accessible and palatable to the student (user-friendliness) in such variables as speed and frequency of feedback, orientation to the system, attractiveness in the interface, motivational encouragement, etc.

4) Develop design documents for programmers to use to imitate as close as possible the mathematical representations, hiding to whatever extent possible the idiosyncrasies of the computer medium.

5) Complete the product and prepare documentation.

Such a process fails to take into consideration the importance of students' conceptions and the interactions between the student and the computer as an emergent process. Awareness and anticipation of these sources of design implementation can help both in the initial and later design stages. An example is provided to illustrate how these can have a useful impact on design.

Design of the Program *Function Probe* ™:

Function Probe™ is a piece of software which allows students to use multiple representations (graphs, tables, equations, and calculator) of functional relationships as tools in their problem-solving endeavors. In implementing the first version of this software, we had two working requirements:
1) Although certain basic functionality was required for each representation, this functionality was designed, modified, expanded, and implemented through an interactive process between educator and programmer in which observed student strategies were used as the development guidelines.
2) Each representation was required to be complete within itself, allowing students to freely follow any desired path in their choice and use of tools in the process of solving problems. In addition, the ability to transfer information among representations was required.

Our claim for modesty in this design is that it does not and cannot have the capability to allow students all the desired freedom and flexibility that would be ideal in problem-solving situations. However, the ultimate judge of how effective we are in approaching those goals and the determination for future modifications lies in the way the students actually use the software, not in how accurately the mathematics is presented or how attractively we create the interface.

Evidence for a Student-Centered Design Approach, an Example:

A single example will be provided of the evidence that students will not use the software as expected, and that these deviations from our expectations form the basis from which good design is made.

The traditional implementation of a logarithmic capability in a software program would be centered around the following curricular goals (precalculus level):
   a) translations from exponential to log form
   b) rules of logs
   c) change of base formula
   d) ln x as the inverse of e^x.
and would suggest the implementation of two log functions, log x and ln x. These are the tools and curricular areas which have been developed for working with logs and the most likely ones we would select as the "expert knowledge", "correct representations" and "traditional content" from our domain of mathematics. With the following example, we wish to demonstrate:
   1) how a tool can be used to transform our agenda and conception of how such a topic could be taught, and
   2) how putting the idea in context, where the problematic lies with the experience of the student, can allow him/her to see the representation as a tool to do what she/he wishes and can thus allow us to see how the tool can be a legitimate form of expression of their conception and a record of their actions in that process.

<u>The Problem</u>:

The current annual tuition(1987-88) for the Endowed College, Cornell University, is $12316. This is a result of annual tuition increases averaging 11.5% for the last 7 years. On the assumption that this annual rate of increase continues for the foreseeable future, the overall goal is to investigate expected future levels of tuition.

1. Create a table which predicts the tuition levels for the next three years.

2. Find a method by which you could <u>easily</u> find the expected tuition 10, 50, or 100 years from now, and add these values to your table.

3. Create a graph showing this relationship.

HOW MANY YEARS TO $250,000,000?

4. A letter to the editor in the New York Times claims that with present rates of increase, Cornell's tuition will exceed 1/4 billion dollars within our lifetime. You would like to check the figures, and also be able to create a method by which you could predict the following milestones:

> *a) the year tuition first exceeds 100,000*
> *b) the year tuition first exceeds 1,000,000*
> *c) the year tuition first exceeds 10,000,000*
> *d) Any dollar value you choose*

Can you come up with a method that will 'undo' your original process, allowing you to use the level of tuition to calculate the number of years until that level is reached? If you have a problem, what is it? What kind of a function would you need to accomplish this task?

When faced with the problem of how to reverse or "undo the process", the student is likely to look at the tool used for doing the problem. We initially used this problem in a classroom situation where only calculators were available. Since we had asked students to keep records of their calculator keystrokes, an individual record might look something like this:

Notice the y^x key is unusual, it is a binary key. Whereas our students have been successful in deriving inverses in previous (non-exponential) problems by "undoing" their calculator keystrokes, in this case they are stuck, since they have no way of "undoing" y^x. Although they will eventually find a way to solve their problem, typically by taking the log of both sides, this becomes for them an exercise in manipulation of logs, rather than an undoing of an operational process. We propose that:

> 1) teaching inverses on a calculator where procedures are carried out and recorded is a reasonable approach, and
> 2) inverses can be seen as undoing, a reversing of both the sequence and the operation of the keystroke record.

To assist in this process, we built in two features, a procedure builder and a binary log key. The procedure builder allows them to save a series of keystrokes under one calculator "button".

Thus in the example above, the given series of keystrokes could be saved as a procedure, and then used by entering first the number of years followed by a single keystroke that would carry out the set of operations above.

The binary log key, $\log_a x$ will allow them to directly undo the exponential operation a^x. In each operation, the calculator (as implemented in the software) will prompt the user to enter the desired base (any positive real $\neq 1$).

Although it is possible that we might have considered these implementations, without observing students, we think the procedures by which we made the decision to actually implement them are important:

1) we observed the students preference for and facility with calculators
2) we suggested to the students that they start keeping records of their keystrokes as an alternative representation of a functional relationship.
3) This, in turn, suggested a way to think of and implement inverse operations.
4) we observed, again, a relatively easy and intuitive ability to carry out inverses by our students
5) when exponentials were used, we observed students reverting to rule-based log manipulations, due to their inability to undo the exponential key.
6) we decided to implement both the procedure builder and the binary log key.
7) we expect this interactive process to continue as students use the first version of the software, leading to ideas for redesign.

Conclusions:

We suggest that we need to distinguish three conceptual sources that jointly interact in software design:

1) the concepts of the designer in his/her intended version of a mathematical idea
2) the concepts as understood by the programmers who must translate a design document into an interface and a program
3) the developing concepts in the student's mind which are expressed and recorded through his/her interactions with the software.

Together these result in the implementation of a product, a set of iconic shapes and a set of operations that form the tool. If it is an effective product, it will allow for the genesis and diversity of student representations of their ideas to be expressed, acted upon, recorded, and reflected upon in their educational process.

We must remember, however, that it is still early in the game and that, so far, the effectiveness of computers in mathematical education is only beginning to be demonstrated. We suggest that their continued viability will be based on the recognition of: 1) the three influences mentioned above in product design, 2) student deviation as legitimate expression of mathematical ideas which we must interpret as input for further design, not inappropriate student effort, 3) transformations in our mathematical ideas resulting from back-talk from programmers and students. Design, as conceived in this fashion, will be more effective and our products more provocative.

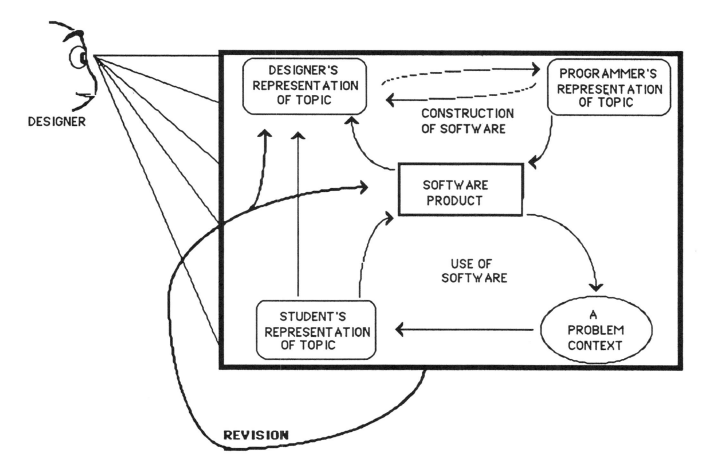

DESIGNER

DESIGNER'S REPRESENTATION OF TOPIC

PROGRAMMER'S REPRESENTATION OF TOPIC

CONSTRUCTION OF SOFTWARE

SOFTWARE PRODUCT

USE OF SOFTWARE

STUDENT'S REPRESENTATION OF TOPIC

A PROBLEM CONTEXT

REVISION

Model for Student Centered Design Process

Curriculum changes in calculus for scientists and engineers.
J. M. Anthony Danby,
Department of Mathematics, North Carolina State University,
Raleigh, NC 27695-8205.

This discussion concerns the teaching of calculus to
students majoring in science and engineering at a land grant
university such as North Carolina State University. It is not
concerned with the teaching of students majoring in mathematics.

In the excellent pamphlet "College Mathematics:
Suggestions on How to Teach it", prepared in 1979 by the
Mathematical Association of America, the Forward is taken from
an article written in 1971 by Professor Peter Hilton. He
suggests that the teaching of mathematics to future
mathematicians is effective. Then he writes: "However, the main
point to be made is that we are far less successful in teaching
effectively those who are not destined to become professional
mathematicians; and these, of course, constitute the vast
majority of our clientele as teachers of undergraduate
mathematics." Since that statement was made, computers and
their accompanying technology have revolutionized many aspects
of our lives; but the teaching of calculus has remained in a
backwater; virtually untouched, except by a few people on an
experimental basis. Apart from some cosmetic changes, our
textbooks could have been written many years ago. It is time
for radical change.

One of the consequences of today's technology is that
those in control of the curricula of engineering students are
demanding that the service courses in mathematics move more
rapidly, while containing more material. My own department has
aquiesced to such demands by simply concentrating further the
traditional syllabus, with less than admirable consequences.
Again, we need to change the syllabus.

The ingredients of the traditional syllabus, with the
order in which they are addressed, and their accompanying
"applications", congealed many years ago in an age when
calculation, if it took place at all, was limited to the slide
rule. The slide rule, physically, has gone. The mental
attitudes that accompanied its use are still with us. How can
we justify today such assignments as: "Solve an equation (non-
linear) correct to one place of decimals", or "Use differentials
to approximate the square root of 4.1"? Yet similar sentences
appear in all of our contemporary textbooks. (And I have seen
the approximation to the square root of 4.1 is described as
"good".)

"Applications" form a large part of our syllabus. One
response of authors to the need for change is to increase the
numbers and range of the applications: this is particularly
noticeable in introductory textbooks on ordinary differential
equations. Presumably we teach applications to demonstrate how

calculus can be actually applied. But some of my colleagues in engineering treat our "applications" with open contempt. I have heard these applications called "trivial", and unrelated to the material as it is used outside the calculus classroom; further, I have been told, they are presented by mathematics instructors who are unfamiliar with the subject matter. The engineers may have a valid point. For instance, many texts on differential equations, and nearly all texts on calculus include a discussion of orbital motion, including so-called "proofs" of Kepler's laws. I have yet to see a single presentation of this material that avoids fundamental error: an error that Newton himself corrected. It is clear to me that the authors are not competent to present this material, and that their textbooks are unusable for its presentation. At a time when the buzz word in curriculum review is "lean", it can be argued that "applications" represent fat.

Changes in the curriculum can be considered in various catagories:

1. Existing material can be dropped. Apart from many applications, some of the basic mathematics might go. Some examples: Most of the coverage of the convergence of infinite series could disappear without endangering the career of a single scientist or engineer. The same applies to the Wronskian. While "exactness" is important in calculus for future application elsewhere, "exact differential equations" are essentially unknown outside the classroom, and need not be taught. Many other similar assertions could be made.

2. Existing material can be modified. For instance, "curve sketching" is, in practice, mechanical drudgery for the student. But since curves can be graphed very easily, and almost instantaneously, on a computer monitor, visible to an entire class, or on a student's hand held calculator, we can now place our emphasis on the interpretation of curves rather than on their mechanical production. Newton's method for a single equation, which at present is applied with just two or three iterations, can be carried through to convergence to the full accuracy available: then qualities of the convergence can be observed and taught, and the method instantly compared with others, such as the secant method.

3. New material can be introduced. Newton's method for systems of equations, for example, or the method of steepest descent (easily implemented, and a beautiful application of the gradient vector). These topics cannot be tackled without at least a key-programmable calculator. Others will become accessible as programs for symbolic manipulation remove the drudgery from albegraic work. Power series are easy to manipulate on a computer, and can be applied to problems such as the numerical solution of non-linear differential equations. Other topics that are not part of calculus may be called for: those, for instance, traditionally found in the areas of difference

equations, linear algebra or finite mathematics.

4. The notion of "application" can be changed. The calculus and differential equations that we teach are, of necessity, so elementary, that only the most rudimentary physical systems can be discussed. Most non-trivial problems (especially if the field of differential equations) require computation. But if computational facilities, such as a PC are accessible, a freshman or sophomore can use calculus to discuss and solve problems that were, until recently, in the domain of graduate schools, and therefore never even seen by most students. Assuming that we teach applications to demonstrate how the mathematics can be usefully applied, we should use any available tool, and, especially, the computer.

5. The way in which we teach can change. Unless we ban calculators from the classroom, we must assume that students will have access to fundamental formulas through storage in calculators. So we must abandon putting a premium on the memorization of formulas. Instead, perhaps, we can concentrate more on technique and interpretation. Computers can be used in class to produce accurate figures. Three-dimensional figures, viewed with correct perspective, can ease the teaching and learning of multivariate calculus. Solutions of differential equations can be shown in animated form, emphasising the dynamic nature of differential equations. In all of these demonstrations, there can be interaction with the class, with students suggesting different parameters, such as the direction to view a three-dimensional surface, the height of a countour to be plotted, or initial conditions to a differential equation.

Curriculum revision requires experiment and argument. It also involves, ultimately, textbook revision, and this is a major problem, because of the crippling expense to the publisher. But perhaps the traditional type of textbook (all 1000+ pages of it) should become as obselete as the traditional curriculum: we should certainly consider alternatives. The chapters of a book cmight be published as separate modules, with options as to coverage, especially with regard to applications. If we can bear to dispense with glossy pictures, then a department, on the adoption of a text, would purchase rights to its reproduction from the appropriate diskettes. The material in a module could include software for computer aided instruction, supplementary exercises to be performed by student at the keyboard, and computer demonstrations that can be used in or out of class, with source code that can be understood and modified by the user.

For a presentation of this kind I can only apologise that the brashness of the comments cannot be balanced by more specific ideas. I have written further material in the following references:

"Learning Multivariate Calculus with the Help of the Computer." Collegiate Microcomputer, 1, 263-272, 1983.

"Computing Applications to Differential Equations."
Prentice-Hall, 1985.

"Computer Applications to Differential Equations." From
"Computers and Mathematics, MAA Notes, Number 9, pp. 73-78,
1988.

TEACHING CALCULUS WITH MACSYMA

A. DARAI
DEPARTMENT OF MATHEMATICS
WESTERN ILLINOIS UNIVERSITY

INTRODUCTION

In light of the recent developments in the area of applications of computer technology, from sophisticated computer algebra systems (CAS) to calculators capable of symbolic, as well as numerical computations, those concerned with the teaching of mathematics are faced with a serious question. This question is no longer whether or not computers should be used in the classroom, rather,it is how can the existing technology be most effectively used as an instructional tool.

The following is a brief account of how the CAS, MACSYMA, has been employed in the instruction of a three semester sequence of calculus. MACSYMA is a program which is capable of performing complex symbolic and numerical calculations and is available from Symbolics Inc. Computer Aided Mathematics Group.

FORMAT

Currently, MACSYMA is utilized in the instruction of one section of each part of the calculus sequence, which begins with the introduction of one variable calculus and ends with applications of multivariable calculus. There are four hours of lectures in calculus per week following a standard text. An additional one hour of discussion centers on the main theme of the topic in connection with the employment of the relevant capabilities of MACSYMA.

The computer sessions are usually conducted in a classroom which is equipped with a terminal and an overhead projector which is capable of projecting images from the monitor onto a large screen. Within this environment, the students are encouraged to fully participate in the process of experimentation by generating ideas and constructing interesting problems. Often the students' comments and questions bring about disscussions which help further explore the particular area of calculus being taught.
It is not uncommon for the instructor to be faced with questions concerning some of the capabilities of the CAS. However, direct answers to such questions may not be immediately apparent. Experience has shown that a teacher can frequently elicit worthwhile efforts from a large number

of students by posing the problem, with some guidance, as a challenge. This approach often leads to a continuation of the computer session discussion into the following lecture period, where students are eager to share their findings with the rest of the class and gain recognition for their efforts.

In addition to the usual quizzes and tests, short weekly assignments are given which concentrate on the main ideas of the topic under discussion. These assignments can be broadly divided into the following categories: 1) familiarity exercises--at the introductory stage of the course it is necessary to assign simple problems most of which can easily be solved without the use of the computer.

This will enable the students to achieve a basic understanding of the command structures of the CAS. For instance, one could require the use of appropriate commands in evaluating certain limits, derivatives or integrals. It has also proved interesting to give the students the task of constructing, within the limitations of the CAS, some elementary proofs which are given in the text. A typical case may be the proof of the quotient rule for differentiation. What plays a central role here is the process of translating a sequence of mathematical statements and operations, which are deductively related, from the language of the written text to the symbolism of the CAS. Inevitably, this process requires a firm understanding of the method of proof as opposed to the memorization of the intermediate steps. 2) Secondary problems-- among these one can include problems which may be conceptually easy to understand but cumbersome to compute (e.g."messy" inequalities or numerical integration with error calculations). This is where the students have to get their "hands dirty" with details and develop a sense for what is a "good" approximation in contrast to an exact solution. 3) Graphing and related applications-- under this category one can examine a variety of problems in relation to functions and their graphs. For instance, the salient features of graphs of large degree polynomials or of functions which exhibit rapid changes within a small interval (e.g. $\sin(1/x)$) can be extracted with relative ease. 4) Simple programming problems-- sometimes a frequently required sequence of computations cannot be performed by a single command from the CAS. In such cases the students may be asked to write a short program which will perform all the desired calculations.

It may be noted here that by requiring the students to include appropriate comments in their assignment outputs, whether they have been successful or not in obtaining a correct or a meaningful solution, one can encourage and stimulate the process of critical thinking as opposed to what we usually do.

OBSERVATIONS

One of the most difficult problems in teaching an undergraduate mathematics course where the students have access to highly powerful symbolic manipulators is to clearly define and impose a set of constraints on such usage. As a result the task of designing problems which are suitable for an interactive session with the computer is a delicate one. The skill related problems found in most standard calculus text books can easily be viewed as trivial by the students. This could easily lead to a greater misunderstanding of mathematics as a trivial manipulation of symbols which a computer can perform much more efficiently than we can. On the other hand a problem which involves a large number of calculations as well as some relatively difficult concepts, for the average student, can be so overwhelming that the student may easily loose all confidence in his/her mathematical abilities. Aside from choosing suitable problems one can counter-balance such negative aspects by frequently discussing and sharing with the students the rationale for employing a CAS in various situations. Therein lies the major function of the teacher. The teacher must select problems and pose questions that enable the student to not only see the rationale for employing a CAS but also to build a conceptual framework of the concepts of calculus. The question of an appropriate skill level necessary will be an issue debated for several years.

CONCLUSION

In this project, the student interaction with the computer has been conducted separately from the standard part of the course. Consequently, the changes in the existing curriculum constitute relatively minor shifts of emphasis over certain topics. In this context the basic concepts and applications of calculus require considerably more time than the techniques of differentiation and integration. The reason for adopting a parallel approach is two fold. First, it is believed that until there is consensus within the mathematical community on the merits of utilizing a program such as MACSYMA the students enrolled in these courses should have the option of transferring to a standard calculus course without being at a disadvantage. Secondly, after completing one part of a calculus sequence with MACSYMA some students may have no alternatives, due to timetable constraints, but to study the next part without MACSYMA (or vice versa). This problem is clearly dependent on the lack of the required funds which limits the student access to the computer equipment. The present technology indicates that microsystems are more viable than mainframe in projects of this kind.

Successful   implementation   of an integrated approach
with the appropriate curriculum changes requires the full
support of the academic institution.   Perhaps only then can
the merits of such reforms in the instruction of calculus be
fully assessed.

INTEGRATING GRAPHING CALCULATORS, PC SOFTWARE,
AND SYMBOL MANIPULATORS INTO A CALCULUS LAB

Robert Decker

John Williams

University of Hartford, Mathematics Department

West Hartford, CT 06117

Two sections each of Calculus I and II are being taught this year at the University of Hartford using programmable graphing calculators (Casio FX7000G), calculus software (EPIC), and a symbolic manipulator (muMath). The Calculus I part of the experiment will be discussed here. One hour out of four each week is spent in a computer laboratory, where 19 AT&T networked personal computers are available; each student is required to purchase the Casio calculator. Each week a laboratory assignment is given, which the students start during the in-class lab hour and finish out of class if necessary. The labs emphasize realistic problem solving, discovery and exploration; some of these labs will be discussed below. The calculators are used for in-class work, for homework problems, and can be used to finish lab assignments at home. We cover the same syllabus as the other four standard calculus sections.

There are several things which we accomplish using the lab format. One is to get students to attack realistic problems. Since they have muMath for finding derivatives, we can freely use any of the standard functions at a very early point in the course. We generally cover topics in the labs before they are introduced in the lecture. Thus the students get a chance to think about a topic before dealing with it formally. Sometimes this leads to discovering a property which is to be introduced later, and helps to cut down on lecture time. We also want to encourage group work; in the lab, there are two students per computer. The lab partners must interact with each other to get the lab done; watching the students trade and discuss ideas is a joy. Another goal is to encourage geometrical thinking, in addition to algebraic thinking. Finally, we require lab reports which demonstrate good communication skills. These must be written in correct English with each result explained in words.

The starting point for Lab I is a simple physics experiment. A mass is attached to a thick cord and suspended from the ceiling; the mass hangs at the level of the blackboard. The mass is displaced 20 inches from the rest position and released; every fifth return of the mass a chalk mark is made on the board representing the extreme position of the mass, and the corresponding time is recorded with a stopwatch. Due to the thickness of the cord, the motion of the mass is damped sufficiently so that the chalk marks end up being a few inches apart. The students are told that for small oscillations, a good model for the displacement y from the rest position as a function of time t is

$$y = A \ e^{-kt} \cos(wt).$$

The assignment is to adjust the parameters A, k, w to fit the data. This is accomplished in the lab using EPIC, although it can be done with the calculator almost as easily.

This lab gets the students to understand the difference between a variable and a parameter, and to discover the effect of each parameter on the shape of the graph. It introduces them to the experimental approach to mathematics; we want them to realize that there is more to mathematics than manipulating symbols. Each group comes up with slightly different values for the parameters, which leads to interesting follow-up discussions. For instance, is it better to fit the first two points, the first and the last or some other combination? What compromises should be made in trying to get a good fit?

Many of the other labs use the function developed in the first one. In Lab 2 the students are required to find the time values and velocities for the first time that the mass passes through the rest position ($y = 0$), and the first time the mass passes through the $y = 15$ position. To do this, they must solve an equation by successively enlarging the graph until the root can be determined to 4 decimal places. Then, the instantaneous velocity is estimated at these points using nearby time values and a difference quotient. This is done before the derivative is defined, but after an in-class small group session on estimating slopes of curved lines on a calculator. By graphing successive secant lines (performed by EPIC), one sees these lines approaching the tangent line. In a later lab, the velocity is found using muMath; this is still long before logs, exponentials and trig functions are formally introduced.

Lab 5 allows the students to discover the relationship between the graph of a function and the graph of its derivative. A polynomial is investigated first; both the function and its derivative are graphed together. The max's and min's can be approximated using the point locator of EPIC, and the roots of the derivative can be found using the root finder, whereby the above relationship is discovered. The student is then asked to apply this knowledge to the pendulum function and find its maximums and minumums on a small interval. This can be done with EPIC without using the formula for the derivative. It can also be done by taking the derivative with muMath and using the calculator to find the roots (we have given them a program for using the secant method to find roots on their calculators; see below).

Another lab involves solving equations with a combination of graphing and the secant method. In this lab, which is done on their calculators, we first give the students a program for solving equations using the secant method (we see this as somewhat preferable to Newton's method, since the derivative does not have to be programmed). Then they must find all roots of a given equation in a given interval. First a graph gives them approximate solutions, then the accuracy is improved using the secant method. Finally they are asked to find the last time value for which the pendulum from Lab 1 was 15 inches from the rest position. The graph must be magnified several times in order to find a good starting point for the secant method, since the last two roots of the equation they must solve are very close together.

Other labs involve discovering the product rule using muMath, finding areas through a sequence of better and better approximations, related rates problems using muMath, numerical-graphical exploration of differential equations (a supplementary program written by the instructors is needed here), and a discovery approach to the Fundamental Theorem of Calculus [1].

In addition to being available for use on the labs, the calculators are used on homework problems and in-class work. Besides the secant method for solving equations mentioned above, the students are given a program for Simpson's method for definite integration. These tools combined with the graphical capability of the calculators leaves the students with a powerful tool for dealing with problems not only in calculus, but in many other courses. One result is that the amount of time needed to cover the sections on max's and min's, curve sketching, and area finding is considerably cut down.

One effect of teaching calculus this way is the increase in active learning on the part of the students. They are continually challenged to come up with their own ways of solving problems. Sometimes what they come up with is not what was intended by the instructors; this is something we must learn to accept, and even encourage. For instance, after learning how to take derivatives, a student used finite difference quotients on one lab to find the equation of the tangent. This can be used as an opportunity to discuss exact versus approximate answers.

An advantage to using muMath is that equations with parameters can be used, such as the one used by the students in Lab 1. If particular values of the parameters are chosen, numerical experiments can be done. But if one wants to know what happens for any parameter values, muMath can be used to get certain information. Students come away with a better understanding of parameters, and the limitations of numerical approaches. Functions with parameters are rarely discussed in the standard Calculus I course.

The lab reports provide an alternative means of evaluation for the instructor. Although the students work cooperatively during lab time, each student must write their own lab report, where they have to assimilate and communicate the lessons of the lab. The instructor must emphasize this aspect in order to get good quality reports; math students are not used to using words to describe math problems.

We believe that all three technology components are important and each has a unique role to play. At this time there is no piece of software that has the graphics, numerical and symbolic capabilities needed, that is easy to use and that is available on PC's. Without the calculators, the students have nothing to use at home on homework problems, or to finish labs with. A scaled-down version of the lab approach could be used with the calculators alone.

The students gain an appreciation for the purpose of both calculus and the new technology by working on problems that would be hard to solve otherwise. Working with a function as complicated as the one used in Lab 1 without the technology would be difficult. Finding something like the velocity of the swinging pendulum would be hard to do experimentally without sophisticated equipment, hence the importance of calculus.

Reference

[1] Decker, Robert. "Discovering Calculus," to appear in The Mathematics Teacher.

TESTING WITH TECHNOLOGY

Scientific calculators, programmable calculators, and graphing calculators are in the hands of our students. Sophisticated software is available for students' own computers and in campus computer labs. Society in general has been much more receptive of modern technology than we in collegiate mathematics education have been. It is unlikely that technology will go away (even if we ignore it); therefore, we must take a leadership role in integrating calculators and computers into the classroom.

My philosophy on the role of technology in teaching mathematics (primarily precalculus and calculus) has evolved over the past several years as I have observed students struggling to make sense of the power they find in their hands. I believe that we must first teach students how to use calculators and mathematics software, then we must instruct them in the use of this technology for mathematics exploration and problem solving. Moreover, I believe that to achieve this objective, we must test to this objective.

Contrary to the belief of many educators and laymen, the use of calculators does not diminish the need for students to think or to acquire basic skills. Successful use of calculators and "toolkit software" requires a higher level of understanding than that required for rote computation or template problem solving. As an added benefit, the use of electronic tools in math classes allows the instructor to make a very strong case for the need to "look back" (Polya's last step in problem solving). "Does my answer make sense?" becomes ever more important, as the slip of a finger has the power to change the magnitude of an answer.

How can calculators and computers be used in student assessment? Simply allowing students to use calculators on traditional tests is not the answer. We need to make technology a viable component of the curriculum. We need to instruct our students in the use of calculators and computers and test them on their basic calculator and computer skills and on their ability to use calculators and computers to do mathematics.

The necessary skills for using technology effectively range from a basic knowledge of the operation of the calculator or program being used to a thorough understanding of the underlying mathematics. Students frequently have problems performing "chain calculations" on calculators, especially if they require the use of calculator memory. Issues involving round-off error and overflow gain importance as students use calculators for routine computations. Interpreting answers becomes vital. Finally, students must learn to recognize when an answer is unreasonable, either as a result of incorrect keyboarding or faulty logic.

A selection of calculator exercises on exams and quizzes stresses the importance of these skills to students. Problems can be approached from either of two perspectives: students can be requested to do specific calculations or they can be asked to interpret calculator results presented by the instructor. In the first case, apparently simple expressions such as $\sin(85\pi/4)$, $e^\bullet$, and $5,872,113^{37}$ can pose seemingly insurmountable problems on most standard scientific calculators; as such, they are a challenge to students' understanding of the mathematical concepts involved. In the second situation, questions such as "Why does the graph of this function appear to be just three vertical lines?" or "Why does my calculator show $\sin(\pi/2) = .0274\cdots$?" will lead students to discover the need to carefully analyze their procedures and results.

Accepting calculators and computers as standard equipment in the mathematics classroom necessitates a change in emphasis and approach for several topics. Techniques of integration, for example, become less important as calculators can do numerical integration with only a few keystrokes and computer algebra systems have the capability to do complicated symbolic integration. A topic that has previously received little emphasis in beginning calculus courses, but that now merits consideration, is error analysis. Important for the successful use of numerical techniques, error analysis has become feasible for beginning calculus students with the availability of sophisticated calculators and computer algebra systems. For example, the error bound for Simpson's Rule can be easily determined using a combination of computer algebra and graphics software.

Graphing calculators have the greatest potential of all of the technological innovations to impact on the way we teach precalculus and calculus. The ability to experiment with the graphs of functions enables students to gain a fuller understanding of the relationship between functions and their graphs, which results in a more meaningful concept of the functions themselves. This easy access to graphs aids in the understanding of such topics as absolute value, solving inequalities, tangent lines (hence derivatives), and limits. In addition, graphics calculators complement the study of curve sketching – generating a reasonable graph on a calculator frequently requires the use of graphing techniques from calculus along with the use of several screens for a complete picture.

The availability of graphs at the touch of a few buttons arguably has the greatest impact on the way we test in precalculus and calculus. "Sketch the graph of $y = (x-2)^2$" is not a challenging problem for students equipped with graphing calculators. However, the concept of transformations of graphs can be tested using a generic $f(x)$ and asking for the graph of $f(x-2)$, etc. Graphing techniques can be tested by requesting students to "Sketch the graph of a function having these properties . . ." or by supplying information about the function and its derivatives and asking for the graph.

Another effective use of technology for student assessment is found in the introduction of professional software (much of which is available in student editions), such as MathCAD and TK! Solver. In addition, this gives students a realistic look at how mathematics is used in the "real world" early in their studies. Projects using computer algebra systems and scratchpad software allow students to combine a variety of mathematical tools and techniques to solve non-trivial problems.

We teach best by example. If we expect our students to be proficient in the use of technology, we must be proficient ourselves. An excellent opportunity to use calculators and computers as our students use them is in the construction of exams. The most obvious example is the use of computers to generate graphs for exam questions. Furthermore, graphics capabilities make it possible to find the right function for a test question without having to search through dozens of textbooks.

Computers can also be used to randomly generate exams. There are commercially available test data banks and construction kits, as well as systems for computer generated and administered tests. For small classes, however, individualized questions can add variety to exams. A simple instructor-written program can randomly generate a different problem for each student in the class. Personal data can also be used to individualize questions. Personalized questions are especially suitable when a graphing or toolkit program is going to be used for the solution. Grading such questions often requires less time than is needed to give partial credit on a more traditional exam, as was the case when students were assigned randomly generated systems of equations to be solved using TK! Solver.

A technological revolution is happening now. As mathematicians, we have the opportunity to be at the forefront of this revolution. As educators, we find traditional values and methods being questioned. By combining the roles of mathematician and educator, we can utilize the available technology to enhance the teaching and testing of mathematics.

Gloria S. Dion
Department of Mathematics
Penn State Ogontz
Abington, PA 19001

October, 1988

Acknowledgement - I would like to thank Lawrence Harvey, a sophomore at Penn State Ogontz, for his assistance with this presentation.

RECOMMENDED READING

Douglas, Ronald G. (ed.). *Toward a Lean and Lively Calculus*, MAA Notes, Number 6, Mathematical Association of America, Washington, D.C., 1986.

Pedersen, Jean and Peter Ross. "Testing Understanding and Understanding Testing," *College Mathematics Journal*, 16(3), June 1985.

Smith, David A., et al (eds.). *Computers and Mathematics: The Use of Computers in Undergraduate Instruction*, MAA Notes, Number 9, Mathematical Association of America, Washington, D.C., 1988.

Steen, Lynn Arthur (ed.). *Calculus for a New Century: A Pump, Not a Filter*, MAA Notes, Number 8, Mathematical Association of America, Washington, D.C., 1988.

A Special Calculus Course at UNL
by J.A. Eidswick

1. <u>Background</u>. Last spring it was decided that UN-L should better
serve a special group of students: those entering the University after
successful completion of one full year of high school calculus. In the
past, such students had a difficult choice: enter at a lower level and risk
being bored stiff or enter at a higher level and risk getting blown away.

It was also recognized that such students, given proper enlightenment,
might develop into mathematics majors. At present, Nebraska's
undergraduate mathematics program is generally perceived by the mathematics
faculty as being close to nonexistent, so there's a definite need for
attracting such students into mathematics.

It was thus decided to offer a two-semester "enriched" calculus course
which would be tailored to the needs of this special class of students.
Information and forms were mailed out late last spring to all Nebraska high
schools involved in the teaching of calculus. In addition, records of
incoming freshmen who had taken a year of calculus were checked for ACT/SAT
scores, course grades, and other information. Ultimately, 123 students
were invited and 21 signed up.

The course was billed as one in which enrollment would be limited to
25 (noting that our usual Calc II classes have as many as 120 students) and
in which special emphasis would be placed on the concepts of calculus and
the meaningfulness of those concepts. At the same time, good use would be
made of scientific calculators, computers and other modern technology. A
background in computers would be helpful but not necessary. One special
feature of the course would be to introduce students to a "remarkable new
calculator, the HP28S". Students could purchase a calculator for about
$150, or if they did not wish to purchase one, they could use ones which
would be on reserve in the math library.

From the beginning, we didn't know whether to call this an "honors
course" because we didn't know if it would attract that kind of student.
As it turned out, it did and it now has that distinction.

Sometime before all the work began, I was asked by the chairman of the
department if I would like to teach the course. My response was that I
would be delighted to do it, but that it would be time-consuming and I had
a couple of requests. One was that I be given an assistant who could take
the class occasionally and help with grading and other chores. That
request was granted. The assistant that was assigned is a very able
undergraduate mathematics major with a good deal of experience helping
calculus students. A better choice could not have been made! [I had also
requested appointment of an advisory committee which would have included a
couple of key research people from the department. That request was denied
because at least one of those key research people was known to regard
activity such as this as counter to doing research.]

2. <u>The starting point</u>. The first problem was to determine the
starting point for the course. This was solved (as well as it could be) by
holding interviews. The entire second day of the semester was devoted to

that purpose and proved to be time well spent. Here's a summary of what was learned from the interviews and the resulting actions that were taken:

(i) Two of the original 21 students were found to be inadequately prepared. Their "calculus" courses had really been special topics courses in which some polynomial calculus had been taught. They were quickly transferred to a more appropriate class. All others, except one, had studied at least the equivalent of the first nine chapters of Howard Anton's "Calculus with Analytic Geometry". However, most felt a little rusty about the material and it was evident that a review would be welcomed. One student, who had grown up in Germany, had formally studied only the differential calculus but expressed confidence that he could pick up integral calculus on his own, and he was allowed to remain in the course.

It was decided that the course would officially begin with Chapter 10 of Anton. Chapters 1-9 would be regarded as review material and Chapter 12 (conic sections) would not be covered at all.

(ii) Computer backgrounds varied greatly. Two or three students were clearly in the "hot shot" category; others had not yet learned to appreciate wordprocessing. With the idea of keeping communication lines as open as possible, I had initially planned to have students keep "daily journals" of their reactions to the course. The idea was that they should keep a running record on a disk and submit a print-out every week. The plan was dropped when it was found that: (a) it would be a hardship for some to make daily trips to access a computer, (b) students seemed conducive to taking an unstructured course, and (c) all students indicated a willingness to be open about their feelings. More about this later.

(iv) A few of the students expressed a desire for longer, more challenging problems than the usual textbook problems. Because of this, I decided to offer a "smorgasbord" of special, nontraditional, problems and to require each student to solve and turn in at least two of them by the end of the semester. More about this later.

Of the 19 students remaining, 16 chose to purchase the HP28S calculator. (Two tried at first to get by on HP28C's but eventually found that the memory restriction of the HP28C was a serious handicap.) Two calculators and a printer are being kept on reserve in the math library exclusively for the use of the class.

3. <u>Calculus and calculators</u>. My plan was to carefully review the main concepts of Chapters 1-9 and quickly review the routine material (like differentiation and integration techniques). In addition, we would take whatever time was needed to explore what the class found interesting. We would take time for calculator activities whenever appropriate. From the beginning, I resisted the temptation to lecture explicitly about the calculator or about programming; when relevant activities were encountered, students were instructed to read designated portions of the "Owner's Manual".

The calculator got into the act right away -- and has been there ever since -- mainly because of its excellent graphing capabilities and because of the close connection between calculus and properties of graphs. These machines are truly remarkable, and wonderous discoveries are being made on almost a daily basis.

After only 8 weeks, the following four very effective ways of utilizing these machines have emerged:

(1) <u>To obtain approximate solutions</u>. The two most apparent areas of application: (i) analysis of graphs, including finding zeros, extreme points, intervals of monotonicity, intervals of concavity, and inflection points and (ii) evaluation of definite integrals (the ease at which these machines solve arc length problems is especially impressive).

(2) <u>As a pedagogical tool</u>. A moving picture is worth a thousand words -- especially if you can hold it in your hand!

EXAMPLES: Geometric interpretations of sequences, series, and convergence of Riemann sums.

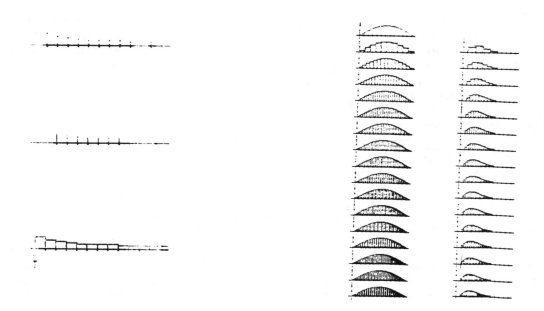

(3) <u>To make conjectures</u>.

EXAMPLES:

1. $1 + 1/2 + 1/4 + 1/8 + \ldots = ?$

2. $\lim x_n = ?$, where $x_1 = \sin a$ and $x_{n+1} = \sin x_n$ ($n = 1, 2, \ldots$)

3. $\lim y_n = ?$, where $y_n = n \cdot x_n^2$

4. $\displaystyle\int_0^{\pi/2} [(7x^2 - 3\pi x) \ln(2\sin(x)] \, dx = ?$

(4) <u>From limitations to concepts</u>. No matter how sophisticated calculators and computers get, there will always be limitations. The appropriate answer to the inevitable question, "what went wrong?" is that the machine simply didn't understand the concepts well enough. Conceptual understanding plus machine capability is a hard combination to beat!

EXAMPLES.

1. $\int_0^\infty e^{-t} t^{-1/2}\, dt = ?$ 2. $\int_0^{2/\pi} \sin(1/x)\, dx = ?$

3. $\int_0^2 [\text{IFTE } (x < .5 , 0 , .5) + \text{IFTE } (x > .5 , 0 , .5)]\, dx$

4. F(x) = (1-cosx^6)/x^12 ??   5. G(x) = x^x

What are the questions?

6. H(x) = .5 + (1-(1-.5*x^100))/x^100 ??

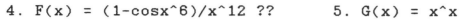

4. <u>Other technology</u>. Limited use in the course has been made of a computer/Data Display/graphing software combination. One interesting story: In an experiment with a graphing program, it was noted that the fifth derivative of x^x resembled a straight line. Obviously, the sixth derivative would look like a horizontal line. Wrong! Due to an inflection point that was hard to detect, the sixth derivative actually looked much more like a parabola than a horizontal line.

Limited use also has been made of MACSYMA to solve initial value problems which arise from pursuit problems.

5. <u>Adjustments</u>. About four weeks into the semester, it became clear that communication lines between me and the students were <u>not</u> as open as they should have been. Some students were apparently having doubts about their abilities, and tensions and frustrations seemed to be building up. So the decision was made to go back to the "daily journal" idea, this time allowing pencil and paper journals. The decision was the right one.

One source of anxiety turned out to be the "special problems". Even though these problems were within striking distance for most students, lack of success on one or two problems apparently had a rippling effect and resulted in general discouragement for some.

6. <u>Future plans</u>. Next semester UN-L will offer a calculator enhanced linear algebra course for students who have excelled in calculus. Next fall, we are planning a repeat of the present course plus, possibly, a special differential equations class.

Using the Group Experience Stack
in a Modern Algebra Course

Arnold Feldman and Marjolein de Wit
Franklin and Marshall College
Lancaster, PA 17604-3003

A major obstacle to the introduction of extensive computer use into mathematics courses has been the skepticism of the instructor that the computer has a genuine role to play in helping students understand mathematical ideas. Software is available for doing calculations in many areas of mathematics, but in most mathematics courses, either calculation is not the focus of the course, or the instructor is unhappy that it is the focus. Thus courseware, while taking advantage of the computer's ability to calculate and display information, should also advance the more theoretical purposes of a course.

We have designed a Hypercard™ courseware stack, called Group Experience, for this purpose. Group Experience can perform a large variety of calculations, but it is a teaching tool, not an all-purpose group calculator. What makes this software into courseware is a set of problems that helps the student abstract principles from calculations performed on built-in and student-created examples.

The plan to emphasize the purpose of the course through the use of the computer extends from the design of the program itself to the design of documentation. Thus in addition to explaining how the package works, we describe to the student how the use of Group Experience fits into the learning process:

> Usually a problem relates to an example group that already appears with the problem, but some problems guide you in using the software to create the relevant group. In either case the problem will direct you to do computations and make observations about the example group that will lead you to more general conclusions about finite group theory. Completing the problem will often entail making a guess or conjecture about all finite groups; sometimes you will prove your conjecture as part of the problem, but in other cases that conjecture may be a theorem your teacher will prove in class. A general principle you learn in one problem will often help you analyze examples in later problems.

Thus nearly every problem consists of two parts, the first involving calculation and the second a conjecture and/or proof for the student to do without the computer. The Hypercard™ environment has helped in several ways. First, it is possible to incorporate the problem into the example, in the sense that the student looks at the multiplication table of a group and a problem relating to that group at the same time. (Fig. 1) Second, the convenience of creating a visual display in Hypercard™ permits the student to make his or her own examples relatively easily. (Fig. 2) Also, we are building in a system allowing a teacher to tailor the courseware to suit the course by adding problems to those included already.

The particular choice of Hypercard™ has had a substantial impact on the design of Group Experience, both in what became possible and what became impossible. Finite group theory was, as we expected, an excellent topic to deal with in the Hypercard™ environment. The existence of interesting example groups with small numbers of elements suggested presentation methods appropriate for the topic--and convenient in Hypercard™--that would not apply to most subjects. On the other hand, computations are relatively slow in Hypercard™, making elaborate calculation impractical.

An important component of any Hypercard™ courseware design is the format for examples, i.e. the form of screen image (called a *card* in Hypercard™) on which the example appears. For Group Experience, we chose a single format, the group multiplication table, for illustrating every example. We made this choice after some experimentation, including consideration of graphical representation of the elements of certain groups. Eventually, we opted for the simplicity of a uniform format, but that choice could easily have been different. A crucial task in designing Hypercard™ courseware for other subjects would be the choice of format.

The general design of Group Experience would be appropriate for many mathematical subjects. The ingredients are:
• a set of problems allowing the student to use the computer to do calculations, then requiring an abstraction of principles from those calculations;
• integration of each problem into an example illustrated on a card of a Hypercard™ stack;
• the capacity for the student to create and manipulate examples just like those already built in to the stack;
• the capacity for the teacher to tailor the courseware to a particular course by adding problems as desired.

We are currently using Group Experience in the junior-level modern algebra course at Franklin and Marshall, taught by Feldman with de Wit as Help Session leader. Getting the students started using Hypercard™ and Group Experience has been relatively simple. One class period devoted to demonstrating the software (and simultaneously introducing some new material) sufficed to teach the students how to use it. There have been a few minor glitches, but not enough to discourage the students or embarrass the authors.

It is difficult to assess the impact that Group Experience is having on the course, for there is no control group; the students in this course have never studied group theory before, so it is not clear how their ability to relate examples and theory might have been different if they hadn't been using the software. Group Experience seems to have generated no great excitement among them, nor has it seemed to engender any resentment. They have matter-of-factly accepted it as an integral part of the course.

From the instructors' point of view, Group Experience has worked well so far in that it has successfully directed the students in the abstraction of group theoretic principles from exploration of small groups; i.e., the software is working the way it should and the students are using it properly. The computer really does seem to be allowing the students to look at examples in more detail with less tedium, without depriving them of experience in calculation by hand. It is too early to tell the extent to which this experience is enhancing their understanding of group theory.

Figure 1

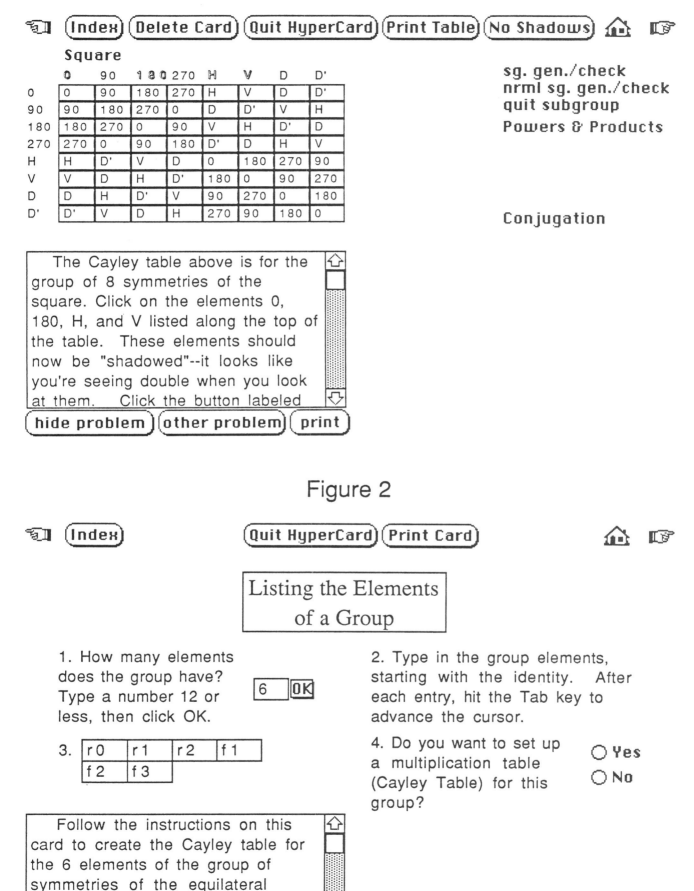

Square

	0	90	180	270	H	V	D	D'
0	0	90	180	270	H	V	D	D'
90	90	180	270	0	D	D'	V	H
180	180	270	0	90	V	H	D'	D
270	270	0	90	180	D'	D	H	V
H	H	D'	V	D	0	180	270	90
V	V	D	H	D'	180	0	90	270
D	D	H	D'	V	90	270	0	180
D'	D'	V	D	H	270	90	180	0

sg. gen./check
nrml sg. gen./check
quit subgroup
Powers & Products

Conjugation

> The Cayley table above is for the group of 8 symmetries of the square. Click on the elements 0, 180, H, and V listed along the top of the table. These elements should now be "shadowed"--it looks like you're seeing double when you look at them. Click the button labeled

(hide problem) (other problem) (print)

Figure 2

Listing the Elements of a Group

1. How many elements does the group have? Type a number 12 or less, then click OK.   [6] [OK]

2. Type in the group elements, starting with the identity. After each entry, hit the Tab key to advance the cursor.

3.

r0	r1	r2	f1
f2	f3		

4. Do you want to set up a multiplication table (Cayley Table) for this group?   ○ Yes   ○ No

> Follow the instructions on this card to create the Cayley table for the 6 elements of the group of symmetries of the equilateral triangle. Give this group the name

(hide problem) (other problem) (print)

Computers-in-Calculus: an Experiment at the University of Michigan-Dearborn.

by

John Frederick Fink, Margret Höft, David James

At the University of Michigan-Dearborn, three out of twelve sections of first semester Calculus offered in the fall semester 1988 use computers as a tool for teaching calculus. Preparations for the three sections started in March 1988 with a survey of mathematics departments where computers are used in calculus instruction. Information gathered through the survey and the current literature on the topic of computers in mathematics instruction led us to the conclusion that the use of the computer can strengthen the calculus course, and plans were made to use local resources to establish the pilot project Computers-in-Calculus for three sections of Calculus I. The instructors of the three experimental sections agreed that:

(1) Teachers should use the computer for classroom demonstrations of the concepts and applications of calculus.

(2) Students should use the computer in two ways: they should have access to computers in an unstructured setting to work on homework assignments that require the use of a computer, and to explore the available software. They should also be required to attend a supervised structured laboratory session, where specific computer exercises are completed. These computer exercises and problems should be designed to reinforce the concepts of calculus, and the computer should play an _essential_ role in their solution.

(3) One of the software packages used by the students should be in the form of a tutorial, encouraging exploration and experimentation, and one software package should have some symbolic manipulation capabilities. All software used by students should require no previous experience with computers, in particular no knowledge of a programming language, and it should require only minimal instruction for use.

Hardware The University of Michigan-Dearborn currently has only one classroom equipped with 30 IBM PC work stations. Blocks of time were reserved in this classroom for the supervised laboratory sessions. The University also has one microcomputer laboratory with a variety of IBM compatible machines that have recently been networked. This laboratory is accessible to any student who pays a fee of $7.50, and it is the principal site for unsupervised work with the calculus software. Engineering students can also use the laboratories in the School of Engineering. The School of Engineering maintains a Macintosh laboratory and a Zenith laboratory, and the calculus software is available in both.

Since no hardwired lines to the University's main frame computer are available in classrooms, the use of a microcomputer was the only reasonable choice for demonstrations in the classroom. All three experimental sections use the same computer, a Zenith ZW/158 with a hard disk, color graphics adaptor, and an 8087 arithmetic coprocessor chip. This computer belongs to and resides in the School of Engineering, is mounted on a cart, and is wheeled to each class session by the instructors. For classroom display, a Kodak Datashow LCD panel is used in conjunction with an overhead projector and a screen. These items belong to the University's Audio Visual Department and are delivered to the classroom by a student assistant from the Audio Visual Department. The equipment is then assembled by the instructor before class and disconnected after class, and the computer has to be returned to the engineering laboratory by the instructor. This is not an ideal classroom situation, where the equipment is securely installed in the classroom and no moving and assembling needs to be done.

Software   Software was chosen to meet the requirements in (3) above. An additional requirement was of course that it had to be compatible with the computers to be used in the project, and it had to be commercially available immediately.   To accommodate various configurations of hardware, three different versions of *MicroCalc* by Harley Flanders [1] were purchased and installed on the networks (site licences were obtained). This package has some symbolic capabilities and is very easy to use. As a tutorial, the students use *Exploring Calculus on the IBM PC* by John B. Fraleigh and Lewis I. Pakula [2] . For demonstrations in the classroom, the instructors use the same two packages, but occasionally other software (e.g. [3] and [4]) is brought in.

Laboratories   The experimental calculus sections are in design similar to a laboratory course in the sciences. Students in the experimental sections like those in the traditional sections meet with their instructor in a regular classroom setting four times weekly for lecture and discussion. In addition, the students in the experimental sections attend a weekly computer laboratory session supervised by their instructors and three student assistants. During this one and a half hour session, the students work through a four to six page handout of carefully selected computer experiments designed to explore in detail one of the concepts covered during the week's lectures. These handouts were written by the instructors, and topics include the derivative of a function, linear and quadratic approximations, Newton's method, the definite integral, etc. Students work in teams of two per computer work station, but discussion, interaction, and comparing of notes between teams is encouraged. Each team submits a laboratory report based on specific instructions at the end of the laboratory handout. The report is due one week after the lab session, and to increase the students' written communication skills, it has to contain a short discussion of what kind of mathematical questions were covered during the laboratory session. The report also contains solutions to problems, references to theorems used to obtain the solutions, and screen dumps of graphs as specified in the instructions for the lab report.

Each week the students are given a "discovery problem". These problems are meant to challenge the better students. They usually go somewhat beyond the material covered in the lectures and require the use of the computer and some ingenuity. Students use the microcomputer laboratories to work on the discovery problems and to finish up their lab reports. The three student assistants that help with the supervised laboratory sessions, are also available for help during the rest of the week. Two are senior students of mathematics, one is a senior student of engineering, and they spend up to ten hours per week in the microcomputer labs to work with the calculus students in the pilot project.

Midterm evaluation We have 71 students involved in the pilot project, and their initial response has been positive. When the students registered for courses, they did not know they were signing up for experimental sections, and during the first week of classes they were given opportunities to change to traditional sections if they wanted to. Only three students wanted to change, and many expressed enthusiasm for the project. Most students have attended the laboratory sessions regularly , and their lab reports have generally been of high quality. They seem to enjoy and discuss calculus more than their counterparts in the traditional sections. We plan a more careful evaluation of student reactions at the end of the semester.

The laboratory part of the pilot project is working well. The software turned out to be as user friendly as we had hoped. Even the students without any previous computer experiences had no difficulty understanding its layout, and they found the programs easy to use.

Computer demonstrations in the classroom are a problem. It takes too much time to assemble the equipment. Ideally, it should not take more than the flip of <u>one</u> switch to get the whole system started. LCD panels are not ideal projection devices. They are temperamental, do not project color, and the fluorescent lights in the classroom interfere with the image. The lights ought to be off for better visibility, but that makes lecturing on the blackboard impossible.

We are trying to get better projection equipment for next semester. We have also purchased copies of Maple and Mathematica, and plan to integrate them into the project next semester.

REFERENCES
[1] H. Flanders, "MicroCalc", MathCalcEduc, Ann Arbor, 1988.
[2] J. B. Fraleigh and L. I. Pakula, "Exploring Calculus with the IBM PC", Addison-Wesley Publishing Company, 1986.
[3] K. E. Petersen, "A Computerized Calculus Tutorial for the First Semester", Harper & Row Publishers, New York, 1988.
[4] B. K. Waits and F. Demana, "Master Grapher", Addison-Wesley Publishing Company, 1988.

Department of Mathematics and Statistics
The University of Michigan - Dearborn
Dearborn, Michigan 48128

Teaching Simple Linear Regression Using the Casio fx-7000G Calculator

Michael B. Fiske
The Ohio State University
Department of Educational Theory and Practice

In traditional elementary statistics courses, the teaching of simple linear regression by the method of least squares is complicated by inadequate student facility with arithmetic and their lack of understanding of slope. Textbooks used in these courses either devote minimal comments to linear regression (Moore, 1979) or substantial development to the fundamental concepts of slope and intercept (Freedman, Pisani, & Purves, 1978; Mosteller, Fienberg, & Rourke, 1983). The ready availability of calculators with linear regression modes and graphic displays both simplifies the calculations and enhances the opportunity for student understanding through wide exploration of linear regression models. This paper describes the linear regression functions available on the Casio fx-7000G calculator and discusses three teaching modules for exploring data sets in an elementary statistics course.

The Casio fx-7000G calculator uses the technique of least squares linear regression with the model: $Y_i = \beta_0 + \beta_1 X_i + \varepsilon_i$. The estimators of the regression coefficients, b_0 (constant term) and b_1 (slope), are calculated by the formula:

$$b_0 = \frac{\sum y_i - b_1 \sum x_i}{n}$$

$$\text{and} \quad b_1 = \frac{n\sum x_i y_i - \sum x_i \sum y_i}{n\sum (x_i)^2 - (\sum x_i)^2}.$$

Once the calculator is in the linear regression graphics mode and a satisfactory range for the data is set, data are entered in the form: independent variable, dependent variable (x_i, y_i). As each data item is entered, it is displayed on the graphics screen. The user can check that the correct number of items (n) have been entered and correct any incorrectly entered data. A drawback is that corrected points remain in uncorrected form on the graphics display. After entering the data, the user has the option of immediately graphing the estimated regression line or of obtaining the estimators of the regression coefficients by keying the appropriate function keys. If desired, the following values are obtainable on single keys: the averages $\overline{x}$ and $\overline{y}$; the sums $\sum x_i$, $\sum y_i$; the sums of the squares $\sum (x_i)^2$, $\sum (y_i)^2$; the sum of the cross product terms $\sum x_i y_i$; the biased and unbiased standard deviations $s_{x,n}$, $s_{x,n-1}$, $s_{y,n}$, $s_{y,n-1}$; and the coefficient of correlation, r.

The first teaching module uses data collected by Doll (Freedman et al., 1978) on the per capita consumption of cigarettes in 11 countries in 1930 and on the male death rate from lung cancer in the same countries in 1950. The module introduces students to the method of data entry and correction on the calculator and assists them in choosing and then entering an appropriate viewing range for the data set (0 to 1500 cigarettes per capita and 0 to 500 deaths per million men). Before proceeding to a regression model, students are asked to examine the scatterplot for any relationship they see between cigarette consumption and death rate from lung cancer. In addition, they are asked to discuss why different time frames were used for the data and why only men were studied. A simple keying sequence produces the graph of the estimated regression line (Figure 1). The horizontal axis is per capita cigarette consumption, and the vertical axis is death rate per million. Students are asked to examine the graph of the line to see how well it fits the data, identifying any points that lie far above or below the line. The estimated intercept (b_0) and slope (b_1) are obtained using function keys, resulting in the following estimated regression equation:

Deaths per million men from lung cancer = 67.56 + 0.2284 x (per capita cigarette consumption).

Predicted death rates based on the regression equation are introduced through the use of the $\hat{y}$ function. In order to see the effect of an outlying point on the regression line, the point for the United States (1300, 200) is removed and the new regression line plotted (Figure 2). Students can compare the suitability of each regression equation and be led to discuss why the United States had a substantially lower death rate.

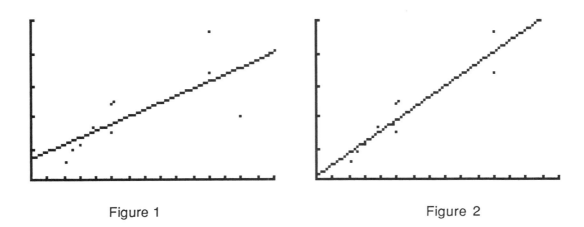

Figure 1   Figure 2

The second module uses data from the Olympic 100 m and 800 m running events to demonstrate how a logarithmic transformation of the dependent variable (running time) allows a comparison of the percent of decrease in running time between the two events (Cleveland, 1985). After entering the data in the form (Olympiad - 1, logarithm of running time), students use the estimated regression equations to predict future winning times. The graphs of the two regression lines (Figures 3 and 4) show that the percent of decrease in running times is almost the same for each event; that is, the slopes of the lines are nearly the same. The horizontal axes are the years (1896 to 1984, scale 12 years), and the vertical axes are the logarithms of the time (100 m run: 2.24 to 2.46; scale .1; 800 m run: 4.6 to 4.85, scale .05).

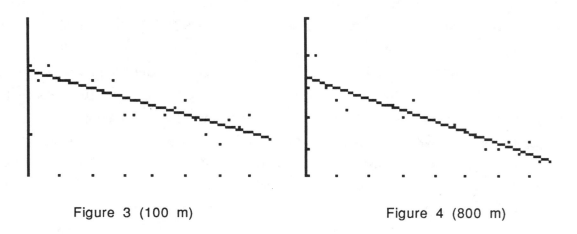

Figure 3 (100 m)   Figure 4 (800 m)

In the third module, the data of interest are infant mortality rates (deaths for children under 1 year of age per 1000 live births) in the state of Massachusetts during the years 1850 to 1970 (Travers, Stout, Swift, & Sextro, 1985). The initial regression line does not fit the data well. Thus, it is suggested that students break the data into two parts, the first 3 and the last 10 decades, fitting two regression lines (Figures 5 and 6). On the horizontal axes are the years, scaled by 10, and on the vertical axes is the infant mortality rate, scaled by 25 and ranging from 0 to 180 deaths per 1000 live births. Students are asked to explain why a downward trend began to occur in 1880. Furthermore, they are asked to explain the predictive value of the regression equations outside the range of the data. Additional explorations involve comparing infant mortality rates for social classes and with other industrialized countries.

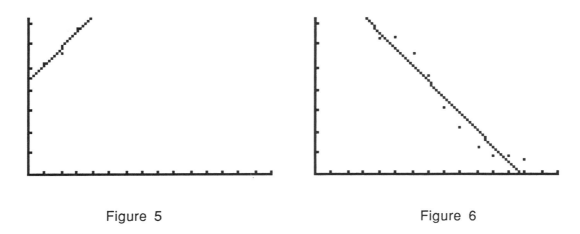

Figure 5   Figure 6

These three modules present the opportunity for expanded exploration within a traditional elementary statistics setting, including transformations of data and piecewise regression. Students are freed from extensive arithmetic calculation and graphing. They have the opportunity to ask and answer interpretive and evaluative questions regarding the data and the estimated regression line. The ease with which the calculator is used opens a new window on the world of statistics for the beginning student.

References

Biehler, R. (1985). Interrelations between computers, statistics, and teaching mathematics. *The influence of computers and informatics on mathematics and its teaching: Supporting papers* (pp. 209-214). Strasbourg: Institut de Recherches sur l'Enseignement des Mathématiques.

Cleveland, W. S. (1985). *The elements of graphing data*. Monterey, CA: Wadsworth.

Freedman, D., Pisani, R., & Purves, R. (1978). *Statistics*. New York: Norton .

Gordon, F. S. (1988). Computer use in teaching statistics. In D. A. Smith, G.J. Porter, L. C. Leinbach, & R. H. Wenger (Eds.), *Computers and mathematics education: The use of computers in undergraduate instruction*. (MAA Notes No. 9, pp. 79-83). Washington, D. C.: The Mathematical Association of America.

Moore, D. S. (1979). *Statistics: Concepts and controversies*. San Francisco: Freeman.

Mosteller, F., Fienberg, S. E., & Rourke, R. E. K. (1983). *Beginning satistics with data analysis..* Reading, MA: Addison-Wesley.

Travers, K. J., Stout, W. F., Swift, J. H., & Sextro, J. (1985). *Using satistics*. Menlo Park, CA: Addison Wesley.

RUBIK ALGEBRA

Charles G. Fleming and Judy D. Halchin

Rubik Algebra is primarily a tool for illustrating a variety of ideas from elementary group theory, using Rubik's cube. It is also very useful as a tool for someone trying to discover how to solve the cube. We will first describe the basic capabilities of the program and then a few short examples of how it can be used as an aid in a group theory course.

Section 1. Program features.

When the program begins, the screen is divided into two sections. On the left-hand two thirds of the screen is a picture of Rubik's cube, and the right-hand third contains a menu. This basic screen arrangement is maintained throughout the program, except when the user chooses to view the cycle decomposition of a move and when, on a few occasions, the picture of the cube is temporarily replaced by information the user has requested to see. Menu items are chosen either by using the arrow keys to move a highlighting bar to the correct choice and then pressing the enter key or by just typing the first letter of a menu choice.
The first choice the user may pick from the menu is to perform a sequence of face rotations. When this choice is made, the menu is replaced by a message asking the user to type in a rotation or sequence of rotations. The notation used for describing rotations is one commonly used in books about Rubik's cube. The six faces are called front, back, left, right, up, and down. The first letter of any of these six names denotes a 90 degree clockwise rotation of that face. A minus sign before the letter denotes a 90 degree counter-clockwise rotation of the indicated face. Rotations may be strung together to form a sequence, and numbers may be used to indicate that certain rotations should be repeated. For example, the sequence "r-f2b" indicates a 90 degree clockwise rotation of the right face followed by a 90 degree counter-clockwise rotation of the front face and a 180 degree rotation (two 90 degree clockwise rotations) of the back face. ("Clockwise" on the back face, for instance, assumes the viewer is standing behind the cube, looking at the back face.) A help screen is available at this point in the program to explain this notation for the new user. After entering a sequence of rotations in this form, the user is given the opportunity to indicate the number of times the sequence should be performed. The picture of the cube is then updated to show the cube after the sequence of moves has taken place the indicated number of times. The user can choose an option

of turning the entire cube right, left, up, or down to see
different sides of it.

The most useful feature of the program, from the point of
view of algebra, is its ability to decompose a sequence of
face rotations into cycles. By choosing "cycle decomposition"
on the menu, the user can see a list of the cycles that
comprise the most recent sequence performed. The notation for
displaying the cycles is similar to that used for describing
rotations. Corner positions on the cube are described by the
three faces that meet there. Thus, rfu indicates the corner
where the right, front, and up faces meet. Positions in the
center of an edge are described by a pair of letters, such as
br for the back, right edge. A cycle can then be given as a
list of corners or edges. For example, (rf bl) indicates a
two-cycle in which the the cubie (one of the twenty-six little
blocks that make up the cube) at the right, front edge is
moved to the back, left edge position, and the cubie in that
position moves to the right, front edge. It also tells us how
the cubies will be oriented. Matching up the first faces
listed in "rf" and "bl", we see that the right face of the
cubie in the right, front position will move to the back face;
likewise, the front face will move to the left side. If the
cycle had been given as (fr bl), we would have known that the
front face of the front, right edge cubie would move to the
back. The program also provides a help screen here to explain
these notational conventions.

Another useful feature that can be chosen from the menu
is the use of a library of moves. Upon making this choice,
the user is shown a secondary menu for the use and updating of
a collection of face rotation sequences stored in a disk file.
The user can choose to add the most recently performed
sequence to the file, display the sequences currently in the
file, delete sequences from the file, or perform one of the
sequences in the file. This feature is useful both for a user
wanting to unscramble the cube and needing to save useful
sequences, and for an instructor needing to save sequences to
be used in a classroom demonstration.

Other options on the main menu include a random
scrambling of the cube, unscrambling the cube to restore its
original condition, and an option to take back the most recent
operation performed. With this last option, a user trying to
unscramble the cube can easily undo a wrong move or any number
of wrong moves. After this choice has been made, a new menu
choice appears--that of restoring the move just taken back.
Thus, having made a series of moves, the user can move
backward and forward through the series of resulting positions
at will. When a new move is finally entered at any point, the
ability to restore "taken back" moves is then lost.

Section 2. Applications.

Some of the topics from group theory normally discussed
in a first course on abstract algebra are (a) the

decomposition of permutations into disjoint cycles, (b) even and odd permutations, (c) the order of an element in a group and, in particular, that the order of an element of a permutation group is the least common multiple of the lengths of its cycles, and (d) conjugation and the effect of conjugation on the cycle structure of a permutation. We will briefly mention how Rubik Algebra can be used to investigate and motivate these topics.

Of course one can calculate the cycle decompostition of a process by hand; however, for even a simple process such as rfu, this is tedius to do. For rfu, the cycle decomposition turns out to be (flu luf ufl), (fru ufr ruf), (dfr frd rdf), (dfl blu rbu rdb), (fu fr uf rf), (df lf ul ub ur br dr fd fl lu bu ru rb rd). One quickly tires of doing hand calculations when permuting the eight corners and twelve edges of Rubik's cube and also keeping track of orientations. Also, there are quite a few processes (elements of the cube group) that one wants to investigate. Rubik Algebra can take any process one wants to investigate and effortlessly generate its cycle decomposition.

Once cycle decompositions are available, a number of concepts and results can be illustrated. By looking at the cycle decomposition of f (or any other single face rotation), we see that f is an even permutation, and, since every element of the cube group is generated from face rotations, we find that all of the processes that can be applied to Rubik's cube are even. A consequence of this is the impossibility of swapping exactly two corners of Rubik's cube while leaving the edges fixed. Such a process would be a transposition on the two corners (an odd permutation) and the identity on the edges (an even permutation), and thus the process would be odd.

Another result that can be illustrated is the fact that the order of an element is the least common multiple of the lengths of its cycles. For example, by looking at the cycle decompostition of rfu, we find cycles of length three, four and fourteen, so the order of rfu is eighty-four. The student can apply rfu to the unscrambled cube eighty-three times (by choosing the repetition factor of eighty-three) and then apply rfu once more and see that the cube has returned to its initial state.

To illustrate the fact that the order of each element is finite, we ask the student to find a single process and the number of times that process must be applied to unscramble a scrambled cube. If P was the process used to scramble the cube, we use the program to obtain the cycle decomposition of P. From the cycle decomposition, we calculate the order of P. If the order is N, then applying the process P N-1 times will unscramble the cube.

The cycle decomposition of an element can also lead us to other interesting elements of the group. For example, the cycle decomposition of the commutator P = rf-r-f is (flu rfu ufl urf luf fur), (dfr drb frd rbd rdf bdr), (fu dr fr). P has order six, so 6P is uninteresting. However, we see that 3P is an interesting process since it has no edge cycles. In fact, the decomposition of 3P is (rfu luf), (drb rdf).

Finally, we mention how to use Rubik Algebra to illustrate the result that conjugation does not affect cycle structure. The process P = 12d-1-f2dfu-f2df12d-1-u has for its cycle decomposition (flu luf ufl), (fru ruf ufr), so the result of P is a clockwise twist of the flu corner and a counterclockwise twist of the fru corner. If we wish to apply this twisting action to the dfr and bdr corners, we can do this by conjugating the process P. The process -2f2rP-2r2f moves the dfr and bdr corners into the flu and fru corners, applies the twist, then move the two corners back to their original locations. This gives us a visual check that the cycle structure of -2f2rP-2r2f is the same as for P. We can of course also verify this fact by using Rubik Algebra to calculate the new cycle decomposition. The decomposition for -2f2rP-2r2f is (dfr frd rdf), (bdr rbd drb).

These examples show some of the ideas that can be easily illustrated to students in an elementary group theory course, using the program described here. Such demonstrations, much too lengthy to do by hand, are quick and easy when the proper tool is available and can make effective and memorable illustrations of fundamental ideas.

Computer Use in Calculus at WPI

W. J. Hardell and J. J. Malone*

Worcester Polytechnic Institute (WPI) is a college of science, engineering, and management located in Worcester, Massachusetts. It has an enrollment of about 2500 undergraduate and 650 graduate students. It operates on an academic year calendar of four seven week terms and most courses meet four hours a week. The basic mathematics sequence is four terms of calculus, one term of differential equations, and one term of linear algebra. Collectively, this material is referred to as "calculus."

Since many of the national concerns about calculus courses have relevance at WPI and since WPI was in the process of reexamining the freshman-year experience, in the fall of 1987 the department of mathematical sciences initiated a review which was to result in a revision of the basic math sequence. By the end of April a pilot program had been approved for implementation in September, 1988. At the beginning of the fall term we had 96 freshmen enrolled in three sections of the first course in the pilot sequence.

We identified four goals for action: change the orientation of calculus away from an emphasis on techniques to an emphasis on understanding the concepts of calculus; develop educational strategies so that calculus will conform to the students needs in the post-caclulus workplace; develop applications that will help motivate students to a better

*This work was supported in part by a grant from the Educational Development Council of Worcester Polytechnic Institute.

solving approach that will lead students to question and think;
integrate computers into the sequence for use by instructors and
students.

It was seen that topics in the basic sequence could be
profitably interwoven. For instance, there are topics in
differential equations that can be appropriately taught at
several places in the calculus; linear algebra can be employed to
present several differential equations topics.

The availability of calculators and computers with
symbolic manipulation capabilities raised questions about which
kinds of topics should be taught and how they shoud be taught.
We decided to utilize existing PC labs rather than require a
calculator with symbolic manipulation. We rejected the idea of
attempting to write our own software and decided to use
commercially available software.

An intensive review of text books and software began in
April. Over a period of three months we reviewed 16 calculus
books and 12 software packages.

Among the factors that led to a final text book decision
were treatment of basic concepts, exercises with an applied
flavor, exercises that could be adapted for computer use, and the
placement of numerical methods of integration in the integration
chapters. We wanted it close to the definition of definite
integral, not tucked away at the end of an applications chapter,
since numerical integration is taught before the Fundamental
Theorem is studied. By June we had narrowed our choice to
Calculus, 2nd ed., D. D. Berkey, Saunders College Publishing;

Calculus with Analytic Geometry, R. A. Hunt, Harper and Row; and Calculus, J. F. Hurley, Wadsworth Publishing. The final choice, heavily influenced by the applications included, was the Hurley text.

We are not programmers and did not want the course to involve programming. Desired features for a software package included: symbolic differentiation that could be used to find routine derivatives needed as parts of applied problems; graphics, both plane curves and surfaces in 3-space; modules that would encourage students to experiment with calculus concepts; adaptability for use in classroom demonstrations. No single package had all of these. There were three that had elements of what we were looking for. COMP-U-CALC, TrueBASIC Calculus, and Calculus-Pad. All three have been made available in the PC lab for student use, but COMP-U-CALC is the primary software tool for the first three courses in the sequence.

COMP-U-CALC has several attractive features that provide nice classroom demonstrations, and we are also using it for student assignments. Functions that are defined piece-wise can be entered, and limits and continuity can be explored using the plotting and tabulation module or the programs in the module on limits. There is a program that allows for epsilon-delta chasing. Also, we have used the bisection module in conjunction with the intermediate value theorem. Parts of the derivative module were helpful in the development of the derivative idea, but the symbolic differentiation program is not very useful for our purposes; it is slow and runs out of memory if the function is only modestly complicated. As we move ahead in the course sequence we will be using the the integration and applications of integrals modules. There are nice graphics, particularly for solids of revolution. In addition, it is easy for students to print the screen to provide hard copy for homework problems.

Later in the sequence, we expect to make greater use of Calculus-Pad. The surface graphing feature should be particularly useful in multivariable calculus.

TrueBASIC Calculus provides helpful classroom demonstrations and students should use it for some experimentation. The symbolic derivative program is fast, but does not simplify.

The text book and software selection process is not finished. We have to decide on a differential equations text and will be looking for a text that combines differential equations with some linear algebra. In software we will be looking for programs to handle the usual numerical methods for first order differential equation, second order linear differential equations, and systems of differential equations.

Reference has been made to classroom demonstrations. Although the AT&T 6300 is the PC available in our computer lab, in the classroom we use a portable Datavue 25 connected to a PC-Viewer. The PC-Viewer uses an ordinary overhead projector to project the content of the computer screen onto the classroom screen. The image is quite satisfactory and the classroom does not have to be darkened.

Worcester Polytechnic Institute
Worcester, Massachusetts

On the use of computer technology to teach
mathematics and computer science at a small black college

by
Andrew Hugine, Jr., Presenter
South Carolina State College
Orangeburg, SC 29117
and
Manuel Keepler

INTRODUCTION

Beating the odds is the story of the Mathematics and Computer Science Program at South Carolina State College. A profile of the institution in the 1987 edition of U.S. News and World Report's Guide to America's Best Colleges, describes a four year college with an enrollment of 4,037. Of this number, 95% are black, 4% white and 1% foreign. The combined SAT score of 670 is below the national norm and 85% of the undergraduate students receive some form of financial aid.

On the other hand, the South Carolina Higher Education Statistical Abstract, May 1987, indicates that 38% of the 140 Bachelor's degrees awarded during the 1985-86 academic year to blacks in South Carolina Colleges and Universities in the fields of Mathematics and Computer Science were awarded by South Carolina State College. The College, however, enrolled only 21% of all the blacks attending Colleges and Universities in the State. Students not only complete the programs but many go on to obtain advanced degrees at many of the prestigious institutions in this country such as the University of Illinois, Purdue, Michigan, Ohio State University and the University of South Carolina. Those not pursuing advanced studies are employed at top firms such as IBM, AT&T, Martin Marietta, State and Federal Government, and 10-15% of the graduates are commissioned into the Armed Forces.

What accounts for this phenomenal success?

The success is due to a large extent to a dedicated faculty, staff, and students. The other significant contributing factor is the curricula. The curricula is designed to overcome deficiencies which may exist in students' mathematical backgrounds while simultaneously responding to the recommendations of discipline organizations such as ACM and MAA.

THE CURRICULUM

The Mathematics and Computer Science curricula at South Carolina State College responds to the recommendations of ACM and MAA. The objectives of the first two years of the programs are two fold. First to provide the student with the prerequisite knowledge to undertake rigorous courses in calculus, modern algebra, geometry, and other higher mathematics. Secondly, to provide an introduction to the foundations of computer science, algorithms, data structures, and computer languages, software design and development, and properties of computer systems.

To achieve these objectives, students take seven courses in mathematics and four in computer science (29 credit hours). These courses are:

Mathematics 105-106	Pre-calculus I-II
Mathematics 203-204	Calculus I-II
Mathematics 112	Sets and Logic
Mathematics 212	Basic Proof Techniques
Mathematics 213	Discrete Mathematics
Computer Science 105	FORTRAN Programming
Computer Science 107	Computer Applications
Computer Science 201-202	Computer Programming I-II

Upon completing these courses, the student will have learned the elementary material of computer science, differential and integral calculus, and basic discrete mathematics; will be prepared for advanced courses in mathematics and computer science; and, most importantly, will be better able to decide on a chosen field. If the student wishes to switch majors after taking the courses of the first two years, he/she can do so with little or no loss of pace.

Relative to course requirements recommended by MAA, CSAB, or ACM, the South Carolina State College programs are not dissimilar from other mathematics or computer science programs at small four year colleges. The difference, we think, is in the novel way in which our curricula are designed to overcome deficiencies that may exist in the student's background. These dificiencies are identified through the college-wide testing of every freshman prior to placement in entry level courses. In mathematics, the test of the MAA are used. Reviews of tests results indicated specific weaknesses in the pre-calculus mathematics topics normally taught in a course equivalent to intermediate algebra. A formal study which investigated students' exit performance on the NTE further substantiated these initial findings.

The study correlated students performance in mathematics with their performance on the NTE. A multiple regression equation utilizing both the pre-college (SAT math score) and collegiate (course grades in calculus, statistics and linear algebra) variables for predicting NTE test performance, achieved an r squared of .6985 with the pre-college mathematics variable accounting for 55.76% of this total. The results of the study, as would be expected, indicated that the exit performance of students on the NTE may be improved by strengthening students' backgrounds in basic pre-calculus mathematics.

Based in part on the results of the study (not unlike many other colleges), students are offered a one year course in pre-calculus mathematics in an effort to overcome deficiencies which may exist. The structure and content of our pre-calculus sequence provide for an extensive review of intermediate algebra topics during the first semester. Emphasis is placed on the "algebra of calculus" with much drill and practice in exponents and radicals, solutions of quadratic equations, inequalities, etc. The first course concludes with a discussion of functions and their graphs. The second part of the course extends upon the notion of a function. In this course, attention is given to polynomial and rational functions, exponential and logarithmic functions, and the trigonometric functions.

USE OF COMPUTER TECHNOLOGY

The college is currently implementing two major projects designed to incorporate computer technology into the curricula. The first project, a federally funded Title III activity, is aimed at providing computer literacy for all students during their first year of enrollment. Partially in response to the Criteria of our regional accrediting body (the Southern Association of Colleges and Schools) the College has included computer literacy as a general education requirement. Students in all disciplines satisfy this general education requirement by successfully passing Computer Science 107, Introduction to Computer Science with Applications. In brief, CS107 is a one semester, three (3) credit hour course which explores the nature and history of computers, their impact on society, and their use in various fields and careers, including selected popular applications, such as spreadsheets, word processors, file and database managers, and graphics. The emphasis in this course is not on computer programming, but is on learning about and using computers. However, the course will cover the fundamentals of BASIC programming.

A computer lab with 24 IBM PS-2 computers has been established and the course was offered for the first time this year, Fall '88, with an enrollment of 75 students.

The second project, funded by the Sloan Foundation, has two principal phases. Phase I is designed to incorporate computer technology into the teaching of elementary statistics. As is true at most institutions, elementary statistics courses are offered in the mathematics department as well as other discipline areas. Examination of these courses will reveal that there is considerable duplication of effort. Our Mathematics and Computer Science Department has revised our elementary statistics course (M208) to cover essential elements of statistics and data analysis found in these courses. The revised course uses the computer as a tool to remove the drudgery from long statistical calculations. The SPSS-X (Statistical Package for the Social Sciences) computer package is used in this course.

The IBM computers in use also have the capabilities to generate "electronic" transparencies. This allows the instructor to produce professional visual aids that can compliment a somewhat technical presentation. Any text or graphical material can be produced on a microcomputer screen and shown using an overhead projector. The system allows you to project computer screens "online" or use the built-in editor to develop organized, multiple-screen presentations. Using this software, problem solutions, computer output, and other teaching aids may be created for each topic to be taught. For example, in the teaching of computer science, programming steps and procedures can be developed and projected on the screen in the classroom or an online demonstration can be conducted to illustrate programming exercises.

Finally, the second phase of the Sloan Project is in the exploratory state. As a follow-up to CS105 and CS107, programming assignments have been developed and will be introduced in freshman and sophomore level mathematics courses to improve programming skills and knowledge. When Phase II is completed, computer technology will be an integral part of all mathematics courses. Presently, review of available software to provide for computer assisted instruction in the following courses is being conducted:

M301	Introduction to Mathematical Logic
M401	Linear Algebra
M403	Differential Equations
M408	Introduction to Probality
M409	Mathematical Statistics

References

Association for Computing Machinery Curriculum Committee on Computer Science. **Curriculum '78: Recommendations for the Undergraduate Program in Computer Science. Communications of the ACM 22.3** (1979): 147-66.

Computers and Mathematics: The use of Computers in Undergraduate Instruction. MAA Notes, Number 9, 1988.

Criteria for Accrediting Programs in Computer Science in the United States. Computer Sciences Accreditation Board, January 23, 1987.

Educational Testing Service. National Teacher Examinations. Princeton: Author, 1947-1980.

Hugine, Andrew, Jr., and McDonald, S.N. "Predictors of NTE Mathematics Test Performance." **Bulletin of the South Carolina Academy of Science**, 1988.

Keepler, M. "Adapting National Curriculum Standards in Mathematics and Computer Science at a Small Black College". **Proceedings, First Southeastern Small College Computing Conference**, 1988: 74-79.

Mathematics Association of America Committee on the Undergraduate Program in Mathematics. **Recommendations for a General Mathematical Sciences Program.** MAA, 1981.

Report of the MAA Committee on Discrete Mathematics in the First Two Years. MAA, 1986.

SOME muMATH ACTIVITIES WITH LARGE CLASSES
OF DIFFERENTIAL EQUATIONS AND LINEAR ALGEBRA

by
Jerry A. Johnson
Department of Mathematics
Oklahoma State University
Stillwater, Oklahoma 74078

INTRODUCTION: There has been much attention given to the uses of computer algebra systems (henceforth abbreviated CASs) in undergraduate instruction. However, the bulk of it seems to have been focused at smaller schools where class sizes permit personal interaction with students. Some of these schools even have Honor Systems which free instructors from such mundane worries as plagiarism. Thus, activities designed for smaller schools are often not very helpful to those of us in large state universities where classes of 50 or more are common. In addition to large class sizes, the demands of research are usually heavy in big institutions (and small alike). Thus, left to themselves, many professors will not require graded computer assignments if it takes very much of their time.

This paper will discuss an attempt to respond to these problems by examining some computer assignments the author has incorporated into our linear algebra and differential equations courses.

We assume without apology that undergraduate mathematics students should be exposed to the power and utility of microcomputers in general and of CASs in particular as part of their education; but beyond just witnessing their power, students should see examples where CASs are integrated into the solution of a problem. We have tried to find ways to accomplish this in the face of the constraints mentioned above.

In summary we wanted each exercise to meet the following demands: It should (1) require the solution of a problem relevant to the course, (2) be impractical to solve without a CAS, (3) require very little of the instructor's time and (4) make plagiarism difficult.

FACILITIES: The setting for the exercises is the Mathematics Learning Resource Center (MLRC), a modern facility utilizing electronic technology and conventional tutoring to supplement the mathematics curriculum. Included in the MLRC are 40 networked Texas Instruments Professional microcomputers and 12 networked Texas Instruments Business Pro microcomputers. The MLRC is open 59 hours per week and has a staff on duty to check out software, take up assignments and assist students. This relieves the classroom instructor of some of the burdens of dealing with computer assignments.

A wide range of software is available at the MLRC including the CAS muMATH. It is relatively cheap and runs on any MS-DOS machine. Some of our graduate students and faculty (including the author) have created functions and programs in muMATH's lisp language muSIMP, along with documentation, that enhance the muMATH environment rendering it easier for our students to use. (I regret that I cannot provide examples of muSIMP code here due to space limitations.) In addition to programs that perform each of the three elementary row operations, are the following:

- ROWRED(M); puts a given matrix M into its row reduced echelon form.

- CHR(M); gives the characteristic polynomial of a matrix M.

- RATROOTS(P); searches a given polynomial P for rational roots.

- TSP(M); gives the transpose of a matrix M.

- Y'; gives $\dfrac{d\,Y}{dx}$.

Just as importantly, the author has written programs in muSIMP that facilitate the generating of numerous distinct exercises, along with solutions, that are provided to the instructor. Since each student's problem is different, plagiarism is deterred. Of course a student could ask a friend to help him in the computer lab, but he is at least being exposed to the technology and it has been our experience that slight differences in the exercises are enough to prevent deliberate copying. A brief help sheet is also provided that explains muMATH just to the extent necessary to solve the given problem. This frees the instructor from any extra effort other than simply matching up the problem with the solution.

This problem generating capability and the above enhancements make the following activities feasible for large classes.

Linear Algebra Exercises

Row reduction: One very simple exercise is to require each student to row reduce a matrix of size, say, 5 by 8 with non-trivial rational entries. This problem is one that the student will be familiar with but would be quite tedious to solve without a computer and liable to unacceptable round-off error without a CAS.

Constructing a list of examples and solutions is easy to accomplish: For example, begin with a 5 by 8 matrix A in reduced form and a 5 by 5 invertible matrix B. Let's say that column two of A is $[1, 0, 0, 0, 0]^t$ (V^t is the transpose of V). We employ a simple muSIMP program that prints the product B.A, increments the $(1, 2)$ entry of A and repeats. In this way we generate a list of complicated looking matrices which are alike except for the second column, and the solutions are easy to check using the original matrix A. For greater variety and complexity one may of course change A and B.

Diagonalization: Another good exercise is to ask a student to diagonalize a large matrix A, say 6 by 6, with non-trivial entries. Using muMATH it is quite easy to produce a variety of matrices A with integer eigenvalues and corresponding matrices P such that $P^{-1}AP$ is diagonal: One defines a diagonal matrix D and a matrix P each with one variable entry and uses muMATH to compute PDP^{-1}. Invoking the EVSUB command, one may substitute several values for the two variables in this expression, thus generating a list of desired examples with the accompanying eigenvalues, characteristic polynomials and diagonalizing matrices P.

muMATH may require a long time to compute PDP^{-1} with two variables. One way to speed things along is to define a function, say QUICK(R,S), that substitutes given values R and S for the variables in D and P respectively and then evaluates PDP^{-1}. Each invocation of QUICK takes very little time. If a class is taught in a large section of 100 or more, it is probably not necessary that each student have a unique problem. Twenty different matrices will probably suffice to discourage plagiarism if they are properly distributed.

To solve the problem the student invokes DET(xI - A) or CHR(A) to obtain the characteristic polynomial of A. Instructors have a choice of not divulging the existence of CHR or RATROOTS. Some are content to allow their use while others prefer the student to invoke DET and do a search for rational roots, using muMATH to help out.

Finding a matrix P such that $P^{-1}AP$ is diagonal is a tractable problem thanks to ROWRED. After invoking ROWRED(λI - A) for each eigenvalue λ the student reads the corresponding eigenvectors off the row reduced form of λI - A.

Finally we may insist that the student check his work by using muMATH to calculate $D = P^{-1}AP$. If printers are available hard copies of the results may be made, but hand written copies are not too burdensome.

ROOTS OF MATRICES: As an augmentation of the last exercise, the given matrix A has eigenvalues that are perfect cubes and it is required to find a cube root of A; i.e., a matrix B such that $B^3 = A$. Of course the heart of the problem is the above diagonalization procedure followed by $PD^{\frac{1}{3}}P^{-1}$. This is a very nice problem, for just as with the diagonalization exercise above, it uses a great deal of what the student has learned and would be virtually impossible to solve without *both* the theory he has learned and the CAS. I have also found that many students are fairly impressed by two aspects of it: First, when I show them the naive approach of letting B be a matrix with 36 variables and trying to solve the resulting 36 cubic equations in 36 unknowns they are appropriately amused. Secondly, when they are done and check the answer with muMATH by invoking B^3; they are impressed that it actually works and there is no excuse for a wrong answer!

POLYNOMIAL CURVE FITTING: In our current text *Elementary Linear Algebra* by Venit and Bishop the method of fitting a polynomial $Q(x)$ to data points $(x_0, y_0), \cdots, (x_m, y_m)$ is to minimize $\sum_{i=0}^{m}(Q(x_i) - y_i)^2$. The text observes that if $n \leq m$ then there is a unique such polynomial $Q(x) = \sum_{i=0}^{n} b_i x^i$; the vector $\mathbf{b} = (b_0, \cdots, b_n)$ satisfies $(U^t U)\mathbf{b} = U^t \mathbf{y}$ where $\mathbf{y} = (y_0, \cdots, y_n)$ and U is the matrix whose i, j entry is x_i^j, $0 \leq i \leq m, 0 \leq j \leq n$. For examples involving anything other than very small numbers of data points, solving $(U^t U)\mathbf{b} = U^t \mathbf{y}$ would be formidable. With a CAS, it's easy.

Generating lots of examples with solutions is done by a muSIMP program that gives the best fit to m data points by polynomials of every degree from 1 through $m - 1$. One can get several exercises from one such set of computations, by simply incrementing the y-coordinates of the data points. The polynomial solutions are the same except that the constant terms are also incremented.

One may also ask the student submit a graph of the solution if there is such a software package available. Because an elegant check of the answer here is unavailable except in the case of an exact fit by a polynomial of degree $m-1$, such a graph may provide an approximate visual verification. Students often have trouble with this exercise because they do not fully understand the matrix equation $(U^tU)\mathbf{b} = U^t\mathbf{y}$. If they attempt to find the exact fit first and verify it, they may be more likely to get the correct approximations.

Undetermined Coefficient Problems in Differential Equations

A differential operator $L[y]$ is contrived so that its auxiliary polynomial has a desired form; let's say a factor of $(r^2 + 1)^2$ and one real root, 2, of multiplicity 3. The corresponding exercises ask for solutions to $L[y] = g(x)$ for various functions $g(x)$. Of course muMATH quickly expands $(r - 2)^3(r^2 + 1)^2$ from which we get the operator. In this case, $L[y] = y^{(7)} - 6y^{(6)} + 14y^{(5)} - 20y^{(4)} + 25y^{(3)} - 22y'' + 12y' - 8y$. Letting $g(x) = x \sin x$, $\sin x$, $x \cos x$, $\cos x$, x^3e^{2x}, x^2e^{2x}, xe^{2x}, or e^{2x} gives us eight exercises. To make more exercises one may change the real root "2" to other values. There is an answer sheet generated by a muSIMP program that contains solutions for all the problems. Space does not permit elaboration.

To solve his problem the student is directed to use the method of undetermined coefficients. Thus the student must solve the homogeneous case and guess the appropriate form of a particular solution y_p with variable coefficients. The truly tedious part of the problem now appears when $L[y_p]$ must be computed and this is where the CAS takes over. On the 520K TI Business Pros it takes 3 minutes to evaluate $L[y_p]$ where $y_p = (Ax^6 + Bx^5 + Cx^4)e^{2x}$. After it is done, the equating of like coefficients produces a system of n equations in n unknowns with $n \leq 4$, depending on $g(x)$. This may be solved by using the CAS to invert the coefficient matrix or by hand, but using the CAS as a calculator. Finally, the student is required to check the final answer with the CAS.

Each of the problems presented here requires the student to integrate the CAS into the solution in an essential way, and it demands more than mindlessly pushing some buttons. A principle or method of solution must be understood and some of the work carried out by the student.

An Experiment to Evaluate a Calculus CAI Package

Mark J. Kiemele, Lt Col, USAF
Department of Mathematical Sciences
United States Air Force Academy

I. INTRODUCTION

Advanced technology has provided an onslaught of new instructional materials and modes of operation for use in the classroom. The mathematics classroom is no exception, with a myriad of mathematical packages available that will do almost anything that modern technology will allow. While mathematical software packages abound, relatively little has been done to actually determine whether or not the new technology enhances the educational process, and if so, how. Like many mathematics departments across the nation, the Department of Mathematical Sciences at the United States Air Force Academy has been heavily involved in both the development and implementation of software packages that could be described as computer-aided-instruction or computer-enhanced-learning tools. A major part of our mathematics technology implementation process is our quality assurance program. In addition to software validity, an important goal of the quality assurance program is educational validity, that is, to determine if the implemented technology contributes to the educational environment in the manner in which it was intended.

In 1984 the USAF Academy initiated a program called the Microcomputer-in-the-Dorm(MID)/Local Area Network(LAN) project. The goal was for each cadet to have his or her own microcomputer in the dorm and to connect all faculty and cadet computing facilities via a LAN. At the same time, the Department of Mathematical Sciences (DFMS) undertook a project to develop a software package that could make as big an impact as quickly as possible. Since every incoming cadet (approximately 1400 each year) took three courses in calculus, it was decided to develop a Calculus Problem Generator. This package was not designed to replace the instructor in the classroom but rather to supplement the classroom instruction with a tool that could help cadets throughout their entire four-year stay--i.e., something they could use on their own without any interaction with the faculty. Thus, the mathematical package under scrutiny in this experiment is a calculus problem generator, which is an interactive program that can randomly generate more than 4 billion problems in 11 different functional areas of differential and integral calculus. Each problem is randomly generated at run time based on a user-chosen objective and one of three levels of difficulty. Problems are presented in multiple choice format, and a detailed step-by-step solution is presented only if the cadet asks for it or if the cadet answers the question incorrectly. At the end of each session, the student is provided a performance summary which includes the number of problems attempted and the number answered correctly, for each objective and for each level of difficulty within each objective.

II. PURPOSE AND DESCRIPTION OF THE EXPERIMENT

The general objective of the experiment was to evaluate the DFMS-developed Calculus Problem Generator when used as a review tool. The regular track for cadets in core mathematics comprised three calculus courses during their freshman year. After a summer of military training, cadets took Differential Equations, also a core course, in the fall semester of their sophomore year. Typically, the first several lessons in Differential Equations constituted a review of the calculus which, for many cadets, seemed to have become a bit rusty after a summer's layoff. Specifically, then, the purpose of this experiment was to answer the following questions:

(1) Can the Calculus Problem Generator be used as a viable alternative to the traditional in-class review?

(2) Is there any particular aptitude group whose performance is enhanced by application of this software package?

(3) Do cadets want to or like to use the CAI package and, if so, which cadets?

Additionally, we wanted to obtain data and feedback so that the Calculus Problem Generator could be improved.

The procedure used to accomplish these objectives was a controlled, completely randomized experiment in which each cadet was randomly selected and then randomly assigned to exactly one of the following three groups:

(1) Control Group – these cadets were not allowed to use the package and had to attend a traditional in-class review.

(2) Experimental Group – these cadets had to use the software package and were not allowed to attend in-class reviews.

(3) Swing Group – these cadets were allowed to choose what they wanted to do: use the software package or in-class review, but not both.

Due to limited computer resources, the Experimental and Swing groups had to be limited in number. The primary purpose of the Swing Group was to determine the interest level among cadets in the computer package. The experiment was conducted in the fall of 1985 and involved two courses: Math 211 (regular track for differential equations) and Math 200 (differential equations for the less technically inclined). Hence, the Math 200 group represented a population with a lower aptitude for the material than the Math 211 group. Additionally, out-of-class study times were obtained for each cadet and a questionnaire was administered to each cadet in order to help evaluate the software package. The response or criterion variable was the score on a common calculus review exam which consisted of 16 multiple choice problems (each worth 6 points) and 4 workout problems (each worth 10 points).

III. EXPERIMENTAL RESULTS

Table 1 summarizes the results of each of the experimental groups with regard to the mean number of points scored on the common calculus exam. The total scores are broken down into the multiple choice and workout portions of the exam. Also shown are the results for each of the aptitude groups considered: Math 211, shown in brackets, and Math 200, shown in parentheses. A total of 808 cadets participated in this experiment, with the numeric breakdown as shown in the table. Using the sample mean data from the control and experimental groups, a simple F-Test for the equality of the two population means was conducted. The results indicate that the two means are not significantly different at the α = .05 level (P-value = .63). Another F-Test for homogeneity of variances was conducted, verifying the assumption that the variability within the two populations was the same (data for this test is not shown).

Table 1. SUMMARY OF MEANS*

	CONTROL	EXPERIMENTAL	SWING
n = sample size =	487 [373] (114)	158 [117] (41)	163 [127] (36)
Multiple Choice: 96	86.86 [90.23] (75.84)	87.19 [91.23] (75.66)	86.28 [89.48] (75.00)
Workout: 40	28.51 [31.42] (19.00)	27.29 [29.64] (20.59)	27.47 [29.98] (18.61)
Total: 136	115.37 [121.64] (94.84)	114.48 [120.87] (96.24)	113.75 [119.46] (93.61)

*Note: Math 211 data is shown in brackets, i.e., [Math 211 data]
Math 200 data is shown in parentheses, i.e., (Math 200 data)
The combined totals for both aptitude groups are shown without brackets or parentheses.

Table 2 summarizes the study time data that was collected from the cadets. The study times represent averages for the total amount of time spent studying outside of class. The data indicates that the Experimental Group had a 38% decrease in average study time, compared to their Control Group counterparts (59 vs 95). This difference is significant at the α = .001 level. Table 2 also shows that the lower aptitude group (Math 200) spent more time studying while still falling short of the performance measure of the higher aptitude group. This result was not surprising.

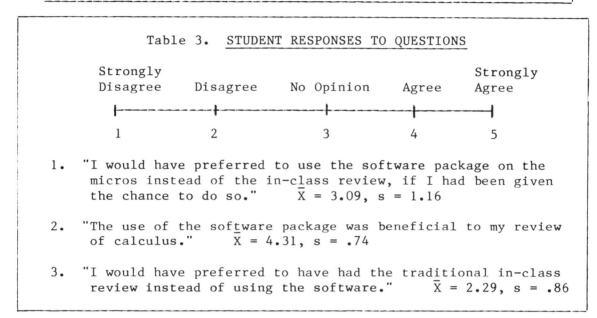

Table 2. STUDY TIME SUMMARY

By Group:	Control	Exp
Avg Study Time (min)	95	59

By Course:	Math 211	Math 200
Avg Study Time (min)	79	87
Avg Score (136)	121.05	94.91

Table 3. STUDENT RESPONSES TO QUESTIONS

Strongly Disagree	Disagree	No Opinion	Agree	Strongly Agree
1	2	3	4	5

1. "I would have preferred to use the software package on the micros instead of the in-class review, if I had been given the chance to do so." $\bar{X}$ = 3.09, s = 1.16

2. "The use of the software package was beneficial to my review of calculus." $\bar{X}$ = 4.31, s = .74

3. "I would have preferred to have had the traditional in-class review instead of using the software." $\bar{X}$ = 2.29, s = .86

Finally, a portion of the questionnaire responses is given in Table 3. The first question was asked of those cadets who were not allowed to use the computer package, while questions #2 and #3 were asked of all cadets who used the software package, be they in the Experimental or Swing Groups. The average response, as well as the standard deviation, is shown following each of the statements.

IV. CONCLUSIONS

Statistically, no significant difference was detected between the performance of the Control and Experimental Groups on any of the portions of the common calculus exam. However, the Experimental Group was able to achieve this same level of performance with about a 38% reduction in study time (95 vs 59). This is a significant finding, both statistically (as shown in Section III) and educationally, considering the magnitude of the time management problem imposed upon cadets at the Academy. Cadets liked to use the computer package. 72% of the Swing Group opted to use the software, and the questionnaire results indicate similar findings. While the experiment was too limited in scope to make wholesale decisions regarding the range of use of the software package, the Calculus Problem Generator does appear to be a viable review tool.

190

LARRY E. KNOP
Department of Mathematics and Computer Science
Hamilton College
Clinton, N.Y. 13323

A BETTER (Differential Equations) MOUSETRAP

As a teacher of mathematics I am primarily interested in educational technology as a source of tools that will enable me to better convey the concepts of my discipline. In one aspect of my differential equations course I have been able to develop a tool, a sequence of computer exercises, that has allowed me to go beyond anything I was able to accomplish using only words and chalk. Furthermore this success comes at no cost in class time, and at little cost in additional work for either myself or my students.

The course for which I developed the computer exercises is an introductory differential equations class; it has a year of calculus and a semester of linear algebra as prerequisites. The course serves about 35 to 40 students, in two sections, each spring. The students are sophomores, juniors, and seniors, with about two-thirds majoring in mathematics and the remainder drawn from physics, economics, etc. Most have not taken a college computer course. The text for the course is Boyce and DiPrima [1]; the course generally follows the text although with the linear algebra prerequisite I can do more with the linear operator point of view than the text does.

The specific aspect of differential equations that I addressed was the task of introducing students to the most basic ideas of the qualitative analysis of differential equations. In particular, I wanted a more effective way of teaching the material in Sections 2.6 and 7.5, 6, 7 of Boyce and DiPrima. This is a small part of my course, involving at most two or three lectures, and my expectations are correspondingly modest. I would like students to be able to sketch the direction field and the integral curves of a simple first order autonomous equation. Later in the course I want students to be able to draw the phase plane of a system of two equations with constant coefficients. Most of all, however, I want students to have an understanding of the meaning of the pictures. With only chalk and words I was not able to meet even these simple goals. Too many students became lost in the mechanics and details, and they never approached the conceptual understanding that I wanted.

Three years ago I turned to the computer for assistance, using a prototype program of uncertain antecedents that did what I required. The program (for a Macintosh) graphs the direction field of an entered equation or system, and draws in trajectories at the click of a mouse. Around this program I constructed two exercise sets that explore the mathematical behaviors of interest, and which bring in applications to give the results meaning. The sets are not perfect, even after several trials and many errors, but my main problem now is an embarassment of riches. The material has become quite easy for students. For example, I asked the following question on an exam last spring:

Let $X' = \begin{bmatrix} 3 & -2 \\ 2 & -2 \end{bmatrix} X$. The general solution is $X(t) = c_1 \begin{bmatrix} 1 \\ 2 \end{bmatrix} e^{-t} + c_2 \begin{bmatrix} 2 \\ 1 \end{bmatrix} e^{2t}$.

Sketch a representative selection of trajectories in the phase space (i.e. the xy-plane).

Label the trajectories with arrows in the direction of increasing t.

The exam was given in a standard setting, with only pencil and paper to use as tools. Every one of my 36 students answered this question correctly. Obviously what I need to do next is raise my expectations.

General Observations: The exercises I constructed are not models of ingenuity -- they aren't even particularly clever. They are, rather, very specific exercises based on a particular text and designed to work in concert with the text (and my lectures). The program I have is clever; it takes full advantage of the Macintosh capabilities to do exactly what I want it to do -- illustrate the graphs of solutions of a differential equation or a system of two differential equations. This is also all the program does, and the version I have has some serious flaws (for instance, attempting to print causes a system error). Thus the work I am reporting is not going to revolutionize education or even the teaching of differential equations; it is a limited experiment with limited relevance to other campuses.

The significance of my work, to me, is that of an existence proof -- it shows that "a better mousetrap" is possible. With technological assistance I have done a better teaching job than I was ever able to do before. I do not know how or to what extent this success can be duplicated in other areas, but at least now I have a rational basis for hope and some experience to go on.

An additional observation, learned through hard experience, I would like to share is that technology alone will rarely be a complete answer. When I first constructed my computer exercises I was not happy with the results. I felt the students had gained insight, but that insight did not show up on test scores. The problem, obvious in retrospect, is that there is limited transferability of knowledge and skills -- and I was giving the students computer work and then testing the results with pencil and paper questions. Since computer testing was not feasible, and since I was not interested in testing my students' facility with the graphing program anyway, the solution was to follow-up the computer work with pencil and paper homework and this combination was the successful one.

Finally, I would like to offer a few comments on educational technology, on what I would like to see, and on what I would use. The first comment is that, while revolutions may be exciting, revolutionary technology may not serve as intended. A classic example, one that no one knows what to do with, is the calculator. Simple numerical calculation is trivial these days, yet The Mathematics Report Card [2] found that only 26% of American high schools had calculators for use in 11th grade math classes. The public schools have an elaborate structure for teaching numerical calculations, and much energy is devoted to this task. Technology may have made (much of) this effort obsolete, yet it goes on with no end or even abatement in sight.

Personally, I am interested in revolutions but I am vitally involved in what I will be teaching next week -- and if technology is going to help me then there is where I would like the help. I want tools that will allow me to do my work better, or easier, or quicker. I want tools that require no training to use. My syllabii are already too full; I do not have time to teach the use of teaching tools, no matter how intrinsically interesting they may be. I also want specific suggestions and directions as to how to use the tools. I don't want a puzzle to figure out; my time is limited too. The appropriate model may be an exercise set. Give me a clear set of tools and let me select the combination to fit my work.

I don't know if such tools are economically feasible. This country has largely agreed on a single language, English, and textbooks can be marketed anywhere in the country. There are no such standards for hardware and software, so the educational technology market is fragmented. Obviously there are many problems, but there is also hope.

References

1.   W. E. Boyce and R. C. DiPrima, Elementary Differential Equations and Boundary Value Problems, 4th Ed., John Wiley & Sons, 1986.

2.   The Mathematics Report Card: Are We Measuring Up?, Educational Testing Service, June, 1988.

CALCULUS AND A COMPUTATIONAL LABORATORY

ERNEST J MANFRED AND GEORGE REZENDES

U. S. COAST GUARD ACADEMY

NEW LONDON , CONNECTICUT

The history of mathematics is replete with stimulations and challenges necessary to induce change. Today, one of the largest stimulations comes from the computer and computational science. The introduction of the computer in the calculus courses will give us a faster way of doing what we've been doing, with the additional opportunity that the student will be able to discover and explore concepts of the subject. The idea of an experimental approach to mathematics is not new. Inductive methods for generating a conjecture were used by Euler and Gauss (see Grenander 1982 Chap 1). The notion of a of a mathematical laboratory based on microcomputers to test ideas and conjectures based on numerical evidence is described in an article of the September issue of the SIAM NEWS 1985 by Bruer and Zwas. This paper represents an application of several ideas stated in that article

Calculus courses represent the core of the university mathematics programs in science and engineering. Content for these courses will not change , however the use of the computer will allow one to better understand concepts via computer graphics and computational analysis. Eight to ten computational modules have been introduced into each of the three calculus courses. These modules represent lab assignments to be completed. The assignments emphasize concepts of approximation, relative error and exercises to illustrate the more important concepts of the calculus. With the emergence of new software, the Macintosh is now a "supercalculator". For a list of available software used in the lab see Appendix 1. One side benefit of the introduction of the lab is its' usefullness in remedial work. This particular usage is enhanced since beginning with this years freshman class, all entering cadets are required to purchase a Macintosh . Software is available thru the department to help cadets improve skills necessary to succeed in the mathematics courses. For a more detailed list of computer usage for calculus courses see **The Influence of Computers and Informatics on Mathematics and its Teaching**, ICMI Study Series,Cambridge Press 1986.

A major concern of those teacching the calculus is the emergence of symbolic algebra computer systems. These systems are having an impact on the nature and scope of the traditional calculus courses. These programs differentiate , intergrate and do polynomial algebra together with many other operations. The day is not far of when this software will be available on hand-held calculators. More than ever, it will be important for the student to understand the concepts involved in the calculus.

College Algebra in a Liberal Arts Curriculum

James E. Mann, Jr.
Wheaton College
Wheaton, Illinois 60187

It is well remarked that the mathematical skills of college students have declined over the past three decades. Frequently, the fact is stressed that students need mathematics for later life, that mathematical training produces clear thinking and that our scientific future is being threatened by this lack of mathematical training. Nevertheless, many students come to a selective liberal arts college, such as Wheaton, with extremely weak mathematical preparation. Many of these students would like to avoid mathematics altogether, but because mathematics is required for those who have low test scores, they are forced to take mathematics of some type. At Wheaton, there are three options available: improve the test score, or take finite mathematics, or take college algebra. I wish to address the content and technological support of the college algebra course.

Here are the premises for the discussion:
1) College algebra courses offer little that is not in a second year high school course.
2) Most students have had two years of high school algebra.
3) Students think they will not use the material in their academic or vocational life.

If premises (2) and (3) are true, then students who take college algebra feel they are being penalized by having to repeat material that is essentially useless to them. How then, can we offer a course that will cover algebra without that course being too hard or repetitious? The answer depends on whether they have actually had two years of high school algebra. If they have, then a similar course in college is bound to be boring and repetitious. We can also assume that they did not like it the first time. For this group, it is possible to design a course that emphasizes the use of functions, the graphs of functions and some of the ideas of calculus. The functions used in the course will be polynomials or rational functions, but these must be made the prototype of all functions. The idea is to give the student plenty of opportunity for manipulating polynomials and at the same time give them something new by introducing the idea of the derivative. Though many of the technical ideas from algebra such as factoring may not be useful to liberal arts majors, the idea of a functional relation between variables and the rate of change of such a relation is likely to be helpful to many.

Since the algebra and calculus are to be based in graphical and intuitive ideas, it is important to have some examples of functions that the students construct from their own environment, e.g. number of words written vs. day of the week or number of pages read vs. day of the week. After making graphs of a few everyday functions, the students will make a transition to polynomials. Graphing a polynomial is a tedious job which is made much easier by the technology which is the subject of this conference. However, I think that it is better not to give the students at this level the most advanced graphing capabilities that are available. It is important to emphasize the relation between ordered pairs and the graph by having the students actually locate the points on the graph paper. It is important to have students think about the scale of graph so that the points fit on the paper. If they think about these things, we will reinforce the idea that many everyday events can be thought of in graphical or functional form. If the student is given a graphing engine that takes a whole polynomial and makes a graph, the student gets a graphical picture of the function that he learns to associate with the whole polynomial formula rather than with the individual point pairs. As I want to give the students the latter rather than the former picture, I argue against the use of the highest technologic solution for graphing. Nevertheless, calculating many ordered pairs for a fifth degree polynomial by hand would be considered inhumane in this decade; the calculation of the points is tedious even with a simple calculator. I opt for an inexpensive, programmable calculator. I insist that everyone have the same type of calculator or that the student himself be responsible for programming his model. Currently, I recommend (or insist upon) the Texas Instruments TI-60 because of its low price and ability to retain a general program for evaluation of fifth degree polynomials.

I do not attempt to teach the students to write their own programs; rather I take each calculator and enter the program for the student. The program is written so that all the student has to do is enter the coefficients into the appropriate storage registers, enter the value of x, push the R/S button and have the value of y calculated. Each student is provided with a written version of the program so that if the stored one is lost in the calculator it can be re-entered. Some instruction is given on how to use the program and how the polynomial is evaluated (Horner's method).

The calculator is used in several ways. First, the student is expected to make modest sized tables of xy-pairs and plot the graph of polynomial. Second, the student is taught to calculate the slope of a secant line, by graphical and analytic techniques. The TI-60 calculator has so few registers that it is necessary to write the values on paper to make the slope calculations (this is a serious limitation of this particular calculator). Once students can calculate the slope of the secant, they calculate sequences of slopes for x-values closer and closer together. A third use of the calculator is to locate approximate zeros of the polynomial by interval halving methods. Most of the calculations mentioned would be too time consuming without at least a programmable calculator. I claim that this piece of equipment is at once the minimum that will do the job, that it easy to teach the student to use, that it reinforces the proper ideas, and that it is cheap enough even for those who will not go further in mathematics. Therefore, I claim that the programmable calculator is the optimum equipment for the task (optimal claims do not extend to the particular calculator chosen).

Let us now turn to what calculus can be covered in a short course. The first task is to get the students to infer the derivative of x^n for several values of x and a particular value of n from the calculations that they can make with their calculator. The motivation given is that they are finding the rate of change of the function. Once they have made several inferences, something of a theoretical demonstration of the correctness of their result is given. The result is generalized to a polynomial and to rational values of n. By multiplying polynomials and differentiating the result, the product rule can be derived. The chain rule can be done the same way. As students do these things, it is important to remember that they are supposed to be practicing the algebra of polynomials; carrying out the above verifications gives ample opportunity for practice. Once several rules have been found, we can turn to the idea of extreme values of polynomials. With more graphing, the student can locate maxima and minima of several polynomials. At this stage, it is helpful to have some examples of polynomials that it would be interesting to maximize (profit, cargo load, area enclosed, etc). After looking for the relative maximum on a graph, someone will surely notice that this maximum occurs at the point that the derivative is zero. This observation, when coupled with looking for the derivative of a cubic, provides a reason to develop the quadratic formula. At this point students have a reason for wanting to know the zeroes of polynomials. Thus they can be taught to use their calculators to find zeros by the interval halving or other methods.

Dealing with integral calculus is also easy in that the calculator makes possible the evaluation of approximations for different subdivisions of the domain. The fundamental theorem of calculus can be illustrated numerically and supported by reasonable theoretical arguments. The harder part of treating integral calculus is to give practical examples of the use of integration that involve polynomials only. After all, it is seldom that one is interested in the area of some figure bounded by polynomial curves. A tentative solution is to introduce first order differential equations

(rate problems) for practical examples. Students are told
that it is frequently easier to characterize the rate of
change of a quantity than the quantity itself.   The difficulty with differential equations is that we must find problems whose differential equations are separable and whose
integrals are algebraic functions. Even though these constraints are limiting, there are enough problems to give
students insight into the use of integration.

Conclusions:

     College algebra can be taught from a numerical, graph-
ical point of view. A programmable calculator makes teaching from this viewpoint practical. It is also possible to
reinforce polynomial manipulation by doing the calculus of
polynomials. The value of this approach is that it goes
much farther intellectually than does algebra but it does
not require the bag of tricks associated with a full calculus course.

Using the Computer in Freshman Calculus

by John Masterson
The Michigan State University

1. Introduction

There are two separate and independent ideas addressed in this note. As a teacher of calculus I am interested in the manner in which computer/calculator technology can be used to help create a better learning model than that based on the traditional lecture. As chair of the Calculus Network in Michigan, a committee of the state section of the MAA, I am interested in the production of easily comprehensible textual materials that will assist the spread of this type of experimentation

The class I am teaching with the assistance of computers is a first semester calculus class of 38 students, mostly first year students. Class is held in a computer laboratory two of the five days a week. The students work in pairs, each pair having access to a Mac Plus. During the course of the term, we have used <u>Mastergrapher</u> by Waites and Demana and <u>True Basic Calculus</u> by Kemeny and Kurtz. This note will concentrate on the first few days experience with <u>Mastergrapher.</u>

In the spirit of the curriculum suggested in the report of the Tulane meeting, <u>Toward a Lean and Lively Calculus,</u> I used the computer to introduce the basic functions of calculus: polynomials, rational functions, trigonometric functions, ln(x) and exp(x). While all but the last two might be thought of as review, the freshman student's actual knowledge of any of these functions and their graphs (not to mention the function concept in general) does not support such a position.

Methodology

The computer (or smart calculator) significantly assists the utilization of the best kind of contemporary learning methodology: one in which the students take an active role in learning through experimentation with the large number of graphs that the computer can construct carefully and quickly. The cubic pictured below on the left was my starting point. The careful examination of graphs is aided significantly by the "zoom - in" feature of this program (a feature common to most graphing programs). On the right we see closer picture of the interesting part of the graph, one from which the students can make a better approximation of values.

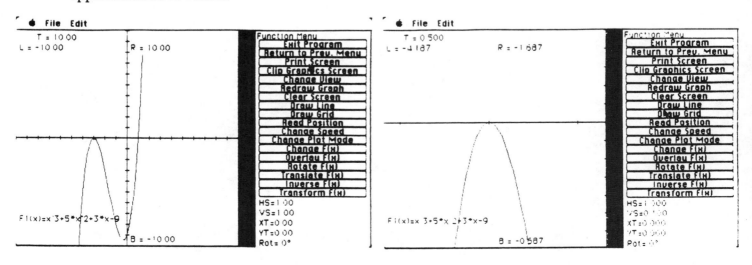

The following sample of the questions given in the written unit show the spirit of the experimentation being encouraged.

1. How many roots does F2 have and what are their x-coordinates?

2. On the basis of your answer to Question 1, can you write a factored form of the polynomial?

3. By changing only the constant coefficient of F2 can you create a cubic polynomial with 0, 1 or 3 roots? How or why not?

4. By changing any of the coefficients, can you produce a cubic polynomial with the basic shape of F2, keeping the double root at -1, but raising the height of the relative minimum?

5. What are the possible shapes of cubics?

Cognitive dissonance, in the form of the surprising graph, becomes more feasible in the presence of machinery. The students are instructed to graph the cubic pictured in the graph on the left : a most unique cubic polynomial hugging the x-axis as it does for several units length. The students are asked to zoom until they determine what is going on. Two astutely made zooms with long squat rectangles produce the significantly scaled picture at the right. This is one of many good exercises about one of the principal themes of modern scientific investigation: the distinctly different information one obtains from the microscopic versus the macroscopic view.

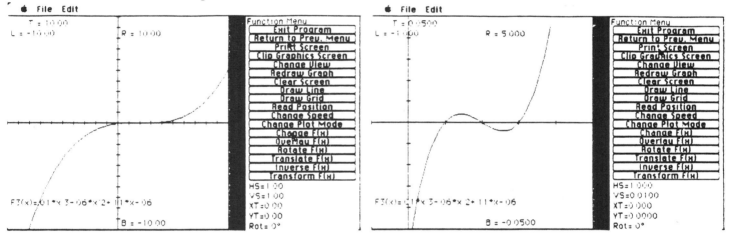

In a similar vein, examples suggested by Waites and Demana in studying rational functions display the macroscopic linearity of a rational function. Zooming out on the rational function $(2x^4 + 3x^3 + x^2 + 2)/(x^3 - x^2 + 12)$ pictured below at the left produces a dramatic vision of that "linearity" on the right. Our usual concern, the factored form of the denominator, can be studied better by zooming in near -2.

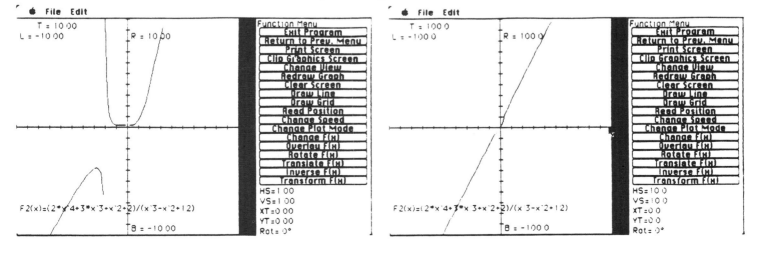

In a more organized fashion, the graphing units used with this program guide the student towards a constructive approach to graphing in general. Moreover, it gives the student a set of building blocks and processes from which graphs of many common functions can be produced easily and understood in a deeper qualitative sense. Translations, dilations, scalings, rotations and algebraic combinations of functions can be seen and practiced easily with computer. Space does not permit a more detailed presentation. Below are two examples that can be used to initiate the study.

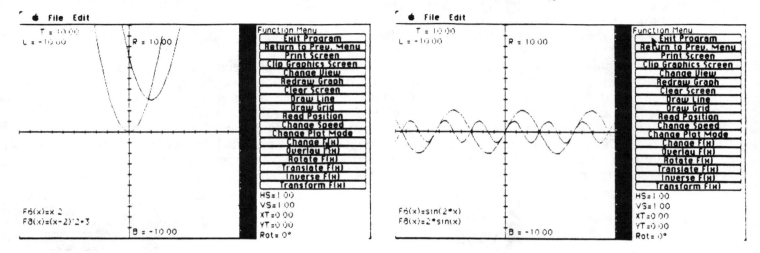

If this is thoroughly done, the students can be given the graph of a "black box function" f(x), for example, and they should be able to graph even such a general expression as $3 f(2x-4) + 1$. This brings us a lot further along the road to graphical understanding than we could ever have hoped for without the new technology.

Before leaving the topic of graphing I was able to present to these students who had as yet learned little calculus, some applied problems of a rather deep nature and discuss their significance in graphical terms. The function pictured below represents the time change of a first order chemical reaction. They cannot solve the differential equation but they can interpret the graph. After a discussion of the chemistry to elicit the asymptotic value of the function, a scaling change of the x-axis to allow us to see 100 units and a scale change of the y-axis to focus on the small vertical values, produces a dramatic picture of what the chemistry tells you should happen.

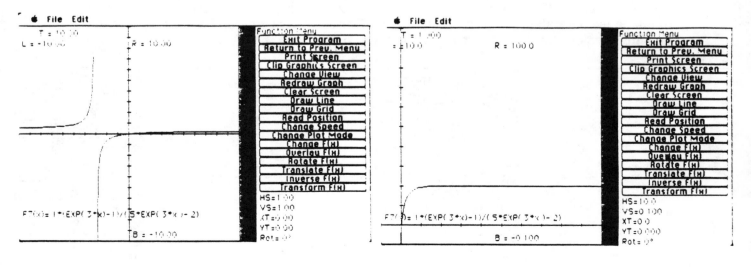

Writing Materials

If productive experimentation with a technology assisted curriculum is going to spread, it is important to write materials that allow a teacher with little experience with computers to see the simplicity of such an approach and be attracted to try it.

Materials which integrate the computer operations in a step-by-step manner with the mathematics provide, I think, the best possibility of attracting large numbers of teachers to make the attempt. By "step-by-step manner", I mean a set of instructions which tells the teacher exactly which buttons need to be pressed and when. Moreover, this should be done in the course of presenting the curriculum.

The initial page of my set of graphing units is printed here. A similar set of "simple minded" instructions for zoming in and out, setting the window, using lines and other computer techniques has been integrated into the curriculum in a similar fashion.

GRAPHING FUNCTIONS
USING MASTERGRAPHER

UNIT #1: Introduction to Graphing Functions

1. Developing Computer Procedure

There is an elementary sequence of steps which needs to be learned in order to make the Mastergrapher program easy to use. Once this sequence has been mastered it is easy to vary the procedure, introducing new techniques which make the program a powerful investigative tool. Perform the steps in the outline below.

1. Insert the Mastergrapher disk.
2. Double click on "Master Grapher 0.90".
3. Click anywhere inside the large box.
4. Select (Click) "Function Grapher" from the first menu.
5. Select (Click in circle) "Change or Remove Function" from next menu.
6. Select one or several functions from the function dictionary.

Select F1 as the first function we will investigate. If there is a black dot beside any other function click on it to erase it so only F1 is selected

7. Select "Previous Menu".
8. Select "Graph Function".

Notice that the cubic polynomial named as F1 is now graphed on a 20 by 20 unit grid centered at the origin. This grid can be changed in many ways. We will see this all later. There are values on this grid that are important for you to understand before going on. The values **T,B,R,** and **L** represent (respectively) "top", "bottom", "right" and "left". They give the dimensions of the grid which is represented on the screen. So this grid gives the set of points in [-10,10]x[-10,10]. In the lower right hand corner notice the values **HS** and **VS.** They represent (respectively) the horizontal and vertical distance between "ticks" on the horizontal and vertical axes. We will call this value the **tick value** . On this grid, then, each tick represents one unit value. The particular numerical values given for the variables in this picture are called the **default** values. When we change them later to do further investigations we will usually want to return to the default values at some point.

Do not perform the next step in the procedure since we want to get into the mathematics of graphing functions by asking some questions about the cubic polynomial which is in front of you. However, the last step completes the basic procedure so we list it here.

9. Click on "Return to Previous Menu" from the Function Menu to the right of the screen. This allows you to start the procedure again. Alternatively, click on "Change f(x)" to return immediately to the function dictionary.

THE INSTRUCTIONAL USE OF SIMULATION OF RANDOM EXPERIMENTS

DR. JAMES M. MEEHAN
DEPARTMENT OF MATHEMATICS AND COMPUTER SCIENCE
ILLINOIS BENEDICTINE COLLEGE
LISLE, ILLINOIS 60532

INTRODUCTION

In order to illustrate concretely the concepts encountered in a course in probability and statistics it is necessary to perform probabilistic experiments repeatedly. Simple examples are flipping coins and rolling dice. It is common that instructors have their students actually perform such experiments many times to illustrate a particular theoretical result. An obvious drawback of such procedures is that the collection of a sizable amount of data is tedious and time consuming. If fast, accurate methods for obtaining sample data were available, time could be saved, boredom eliminated and the learning process enhanced.

Here, of course, is where the computer is useful. All of the common probability distributions can be simulated by the computer via the random number generator. Furthermore, even today's inexpensive microcomputers are equipped with random number generators which are adequate for this purpose. The generator produces a sequence of independent observations which are uniformly distributed on the unit interval. (Henceforth, the notation $U(0,1)$ will be used for this distribution.) These in turn can be transformed, using results from probability theory, into independent observations from other distributions.

Given the capability of generating sample data from a desired distribution, the instructor can begin to design laboratory experiments which use the data to teach concepts. Elliot Tanis of Hope College in Holland, Michigan has written a laboratory manual [1] consisting of such experiments. The manual is used in a one credit laboratory which students take concurrently with a three credit lecture course. The purpose of this paper is to present examples of methods for simulating distributions, and to suggest techniques useful for designing experiments, like those of Tanis, which can be performed by students on their own computers or in the laboratory. It is intended that the experiments provide not only concrete illustrations but also the motivation to study and understand the underlying theory.

SIMULATING DISTRIBUTIONS

For the most part students are able to develop simple algorithms of their own for simulating the common discrete distributions, since most of these depend on repeated Bernoulli trials. The binomial, geometric, negative binomial and hypergeometric are examples of such distributions. Of the common distributions, the Poisson is an exception; it cannot be simulated (exactly) using Bernoulli trials. A technique for simulating the Poisson distribution is presented in the next section.

The theory behind simulating the continuous distributions is richer and more rewarding than that of the discrete distributions. One of the most useful theorems in simulating continuous distributions is the following:

<u>Theorem</u> <u>1.</u>   If the continuous distribution function F(x) is strictly increasing over 0<F(x)<1, and if random variable U has the U(0,1) distribution, then the random variable X=F^{-1}(U) has distribution function F(x).

The theorem is presented in several text books and tends to be difficult for students to understand. If the student has the opportunity to make use of the result, however, he/she may be better motivated toward an appreciation of the result. Two examples of how this theorem may be used to simulate distributions follow. In each case random variable X is expressed as a function of the random (number) variable V and X has the desired distribution.

<u>Example</u> <u>1.</u> The <u>Uniform</u> <u>Distribution</u>, U(a,b).

In this case F(x) = (x-a)/(b-a) for a$\leq$x$\leq$b. Solving U = F(X) yields
$$X = a + (b-a)U.$$

<u>Example</u> <u>2.</u> The <u>Exponential</u> <u>Distribution</u>, E(a).

Here F(x) = 1-e$^{-x/a}$ for x$\geq$0, a>0.

Again solving U = F(X) for X results in
$$X = - a \ln(1-U).$$

The simpler statement X = - a ln U can be substituted since 1-U also has the U(0,1) distribution.

In order to use this method the equation U=F(X) must be solved analytically for X in terms of U. Closed form solutions do not exist for some of the common distributions. The Gamma and Normal distributions are examples. A Gamma distribution with integral shape parameter, however, can be simulated as a sum of exponentials. The Normal distribution can be simulated using the Box-Muller transformation, which is discussed in the next section. Of course, from the Normal distribution can be obtained the t, F and Chi-Square distributions.

EXAMPLES OF EXPERIMENTS AND TECHNIQUES

One of the most illuminating techniques for studying the distribution of random data is the histogram. Consider an experiment in which a histogram of data from a Poisson distribution is constructed. The Poisson distribution is simulated by making use of the fact that, in the Poisson process, interarrival times have an exponential distribution. Simulation of the exponential distribution was discussed in the previous section. Exponential values are observed one at a time until their sum exceeds one. Subtracting one from the required number of observations yields an observation from a Poisson distribution with mean equal to the reciprocal of that of the exponential.

Appearing in Figure 1 is a histogram of 1000 Poisson values whose distribution has theoretical mean equal to three. A graph representing the expected number of observations has been superimposed in order to provide a comparison of observed versus expected. The sample mean and variance were computed and displayed for the purpose of comparison with the theoretical values. The same procedures can be applied to any distribution which can be simulated.

As another example consider the task of illustrating the Central Limit Theorem, one of the subjects most important results. As a limit theorem, it is difficult for undergraduate students to understand. Loosely speaking, the theorem guarantees that the sample mean from any reasonably well behaved distribution, when standardized, is approximately standard normal in distribution. The quality of the approximation (or the rapidity of convergence) is usually not discussed in undergraduate texts.

Let random variable W be the sum of n independent $U(0,1)$ random variables. If n=2 the density is triangular in shape. For n=k the density consists of k polynomial arcs, each of degree k-1, pieced together. Even for a value of n as small as n=4, the density of the sum is quite "normal" in appearance. Histograms representing 250,000 values obtained for the n=2 and n=4 cases are given in Figures 2 and 3. The data have been standardized and the standard normal density superimposed. The triangular nature of the density is clearly apparent in the histogram in Figure 2. The histogram in Figure 3 indicates the goodness of the approximation even for a very small sample size.

The importance of the normal probability distribution is evidenced by the Central Limit Theorem. Clearly it is desirable to be able to simulate this distribution. Data from the normal distribution can be used to design a variety of experiments for students of probability and statistics. Problems involving estimation, sampling, confidence intervals, regression and analysis of variance are some examples of topics whose concepts can be illustrated using normal sample data.

As mentioned in the previous section, normal sample data cannot be obtained by the method of inverting the distribution function. There is a method which transforms a pair of independent $U(0,1)$ random variables into a pair of standard normal random variables. The method is due to Box and Muller [2], and the transformation is given by

$$X = \sqrt{-2\ln U} \; \cos 2\pi V$$

$$Y = \sqrt{-2\ln U} \; \sin 2\pi V$$

Here U and V are the $U(0,1)$ variables and X and Y are the independent standard normals.

The problem of showing that the transformation actually works appears in several texts. It is a rather straightforward exercise using the Jacobian to obtain the bivariate density. Some texts may mention that the transformation is useful for simulation, but students often want to know how the authors came up with the equations.

Once it is pointed out that it is merely the polar coordinate transformation students seem satisfied. The random variable $\Theta = 2\pi V$ is uniform on the interval $(0, 2\pi)$. Since $X^2 + Y^2 = R^2$, the random variable R is the square root of a Chi-square random variable with 2 degrees of freedom. The latter distribution is exponential so that both $\bar{\Theta}$ and R are easily simulated.

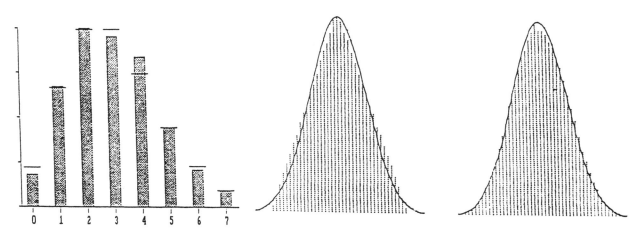

Figure 1. Poisson Histogram

Figure 2. $U_1 + U_2$

Figure 3. $U_1 + U_2 + U_3 + U_4$

CONCLUDING REMARKS

The previous example, although not actually a simulation experiment itself, reveals the primary purpose of the experiments. The need to find a transformation which yields data from a particular distribution is realized. A transformation which does the job is presented in class along with its proof. It is likely that the student will have a greater appreciation for a mathematical result and the underlying theory if he/she has direct experience involving its application. In general, carefully designed experiments can lead to desirable educational outcomes. Thoughtful instructors may find the time investment, both on their part and on the part of their students, yields valuable dividends in motivation and learning.

REFERENCES

[1] Tanis, E.A. (1978), Laboratory Manual for Probability and Statistical Inference, CONDUIT, Iowa City.
[2] Box, G.E.P., and N.E. Muller (1958), A Note or Generation of Normal Deviates, AMS 28, 610-611.

Using Computers to Teach
Series and Differential Equations

by Emily H. Moore
Grinnell College

At Grinnell College we teach "Series and Differential Equations" to our second semester sopho-mores. In this course we study notions of convergence and approximation — topics that are in-herently geometric. Traditionally these topics are taught with the aid of rough sketches on a blackboard. Some beautiful but static graphs appear in the texts. But students — especially those who are weak on the concepts — cannot produce correct sketches on their own. Computers have been used to teach isolated topics. But the computer programs have not generally been available for student use.

Computers, with their ability to perform numerical calculations and produce high-resolution graphics, can be used to present ideas of convergence and approximation. The problem is to choose software that is both powerful and easy to use. A general graphics package will graph any function you choose. But to graph, say, the partial sum of a Fourier series would require entering a very ugly function. On the other hand, special-purpose educational software often limits the functions that can be graphed. In some instances, the canned examples are little better than static drawings.

In the fall of 1987, I started work on a project to use the computer to teach concepts of convergence and approximation. My main goal was to provide an environment in which the student could experiment with a wide variety of functions. I wished to use the computer to enhance intuition through graphics, and to do some of the otherwise tedious or impossible numeric computations. The computer software must be so **easy to use** that attention is focused on the mathematical concept and not on the computer. It must provide students with as much **control** as possible over the output, yet make controls optional so reasonable graphs may be produced with little effort. And it must be accompanied by examples and exercises that encourage the student to explore examples, make conjectures, and back up intuition with rigorous mathematical argument.

I wrote and used early versions of some of the software and exercises in the spring of 1988. In the summer the Sloan Foundation supported work on improving the modules, implementing new modules, and writing supporting materials. These are being used this fall. Modules have been written to graph functions of one and two variables, to graph sequences of numbers (x_n vs. n), sequences of functions ($f_n(x)$), Taylor polynomial approximations to functions, partial sums of power series and Fourier series, and numerical solutions to equations, definite integrals, and first- and second-order differential equations. The computer modules are written in Matlab, a software package that provides matrix operations graphics calls. The software runs at Grinnell College on a network of Sun 3/50's. Computer monitors in classrooms allow use of the software for class demonstration. Students use the laboratory equipped with 18 workstations to run the software, producing graphs and numeric output on the screen and printing graphs and tables on the laser printer.

To illustrate how the software is used, I have chosen to focus on seven mathematical problems.

1. **Uniform convergence of functions:** Figure 1 shows a graph of the first 7 functions in the sequence $f_n(x) = nx \cdot e^{-nx}$. From the graph students can conjecture that each f_n has one critical point, and the maximum values for the functions appear equal. Calculus shows this critical point to be at $x = 1/n$, with $f_n(1/n) = 1/e$. The functions do not converge uniformly to 0 on $[0, \infty)$. But for $c > 0$, convergence is uniform on $[c, \infty)$. From the graph it is clear we must argue that past a point in the sequence f_n is decreasing on $[c, \infty)$. Graphing leads to conjectures which then must be proved.

2. **Taylor's Remainder Formula:** The sixth Taylor polynomial for $\sin(x)$ evaluated at x^2 is: $P_6(x^2) = x^2 - x^6/3! + x^{10}/5!$. Using Taylor's Remainder Formula, we find $|\sin(x^2) - P_6(x^2)| \leq$

$(x^2)^7/7! \cong .744$ at $x = 1.8$. The graph in Figure 2 illustrates the actual error at 1.8. Further, if we wish to approximate $\int_0^{1.8} \sin(x^2)\,dx$ by $\int_0^{1.8} P_6(x^2)\,dx$, the actual error is represented by the area between the two functions. Certainly an upper bound for that error is 1.8 times the maximum error. But a much tighter upper bound is the integral of the TRF bound, $\int_0^{1.8} x^{14}/7!\,dx$.

3. **Interval of Convergence:** In Figure 3 we graph the partial sum of the Maclaurin series for $f(x) = \ln(1+x)$ for $n = 3, 6, 9, 12$. The graph suggests the partial sums converge to the function for some values of x, but diverge for $x > 1$. Intuition motivates a more careful study.

4. **General solution to a first order differential equation:** In Figure 4 we have graphed the direction field for the differential equation $y' = 1 - 2xy$. Initial values for solutions were chosen by moving the mouse to a point (x_0, y_0) and clicking. From this simple graph the student gains an intuitive notion of the family of curves that make up the general solution to the differential equation. Also, clicking the mouse in the general vicinity of (2,0) shows the extreme sensitivity to initial values in some regions.

5. **Numerical solutions to first order differential equations:** In Figure 5 the solid line represents the solution to the initial value problem $y' = 3yx^2 \cdot \cos(x^3)$, $y(0) = 1$. The dashed line is the numerical approximation using Euler's method with $n = 100$ intervals. The graph vividly shows that the extreme curvature of the function makes it difficult for Euler's method to track. Euler's improved method and the Runge-Kutta method of order 4 do much better. The graphs in Figure 6 show the error in the Euler approximation with $n = 2000$ and the difference between the Euler approximations at $n = 2000$ and $n = 1000$. The similarity of the graphs helps to support the rule of thumb that the error at $2n$ intervals approximately equals the difference between approximations at $2n$ and n intervals, thus providing a stopping criterion.

6. **Series solutions to second order differential equations:** Computing series solutions to second order differential equations is a messy business. The power series solution to the differential equation $y'' = -xy' - y$ with initial values $y(0) = 1$, $y'(0) = 2$ is

$$\sum_{k=0}^{\infty} \frac{(-1)^k}{2^k \cdot k!} \cdot x^{2k} + \frac{2 \cdot (-1)^k}{1 \cdot 3 \cdots (2k+1)} \cdot x^{2k+1}$$

The behavior of this function is far from obvious. In Figure 7 we have graphed a numerical solution to the initial value problem together with partial sums of the series solution. The partial sum S_{49} is indistinguishable from the numerical solution. This computer tool not only gives the students a way to check their power series solution. It also provides an approximate graph of the function.

7. **Fourier series:** It is important to support the idea of adding terms in a Fourier series with graphical results. The module to graph partial sums of Fourier series allows the user to specify the period of the function $f(x)$ and to define the function over that period piecewise. In Figure 8 we have graphed the function that is periodic of period 2π, and is defined to be 0 on $[-\pi, 0)$ and x on $[0, \pi]$. Once the formulae for the Fourier coefficients have been entered, the student can graph the partial sum of the Fourier series for any reasonable n. The Gibbs phenomenon is observed at points of discontinuity. It is also enlightening to graph the difference between f and the partial sum.

Conclusion: Educational software can give students insight and power to solve a much wider range of problems than previously possible. The software must be both flexible and easy to use. Its success must be measured not only by its ability to demonstrate principles, but by its ability to put useful tools in the hands of the students.

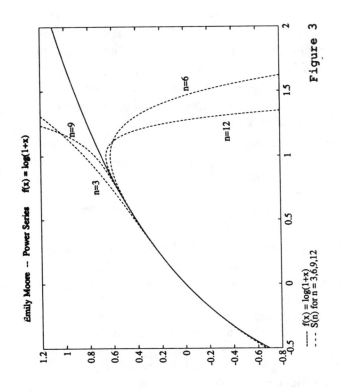

Emily Moore -- Power Series   f(x) = log(1+x)

Graphs for n = 3,6,9,12

—— f(x) = log(1+x)
--- S(n) for n = 3,6,9,12

Figure 3

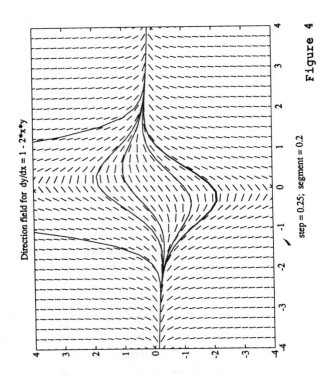

Direction field for dy/dx = 1 - 2*x*y

step = 0.25; segment = 0.2

Figure 4

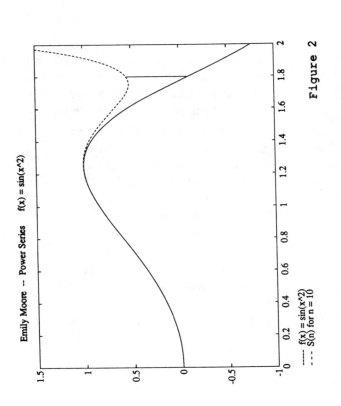

Sequence of functions: f(n,x) = n*x*exp(-n*x)

Graphs for n = 1,2,3,4,5,6,7

Figure 1

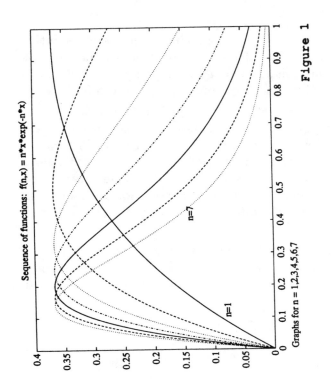

Emily Moore -- Power Series   f(x) = sin(x^2)

—— f(x) = sin(x^2)
--- S(n) for n = 10

Figure 2

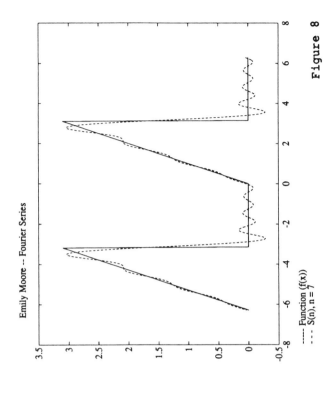

Emily Moore -- Second Order D.E. $y'' = -x*y' - y$

-- rungek, 50 intvls
-..S(9)

S(19)
-- S(49)

Figure 7

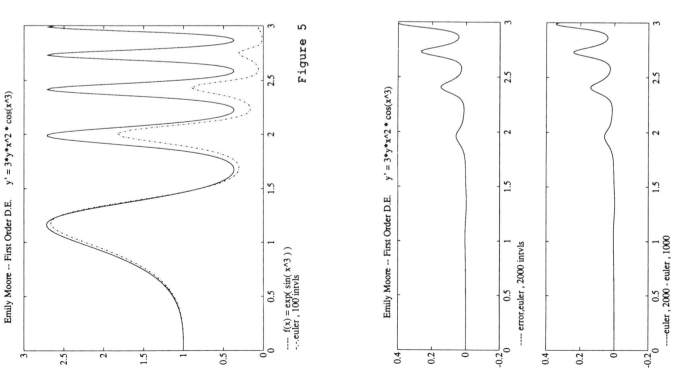

Emily Moore -- Fourier Series

—— Function (f(x))
--- S(n), n = 7

Figure 8

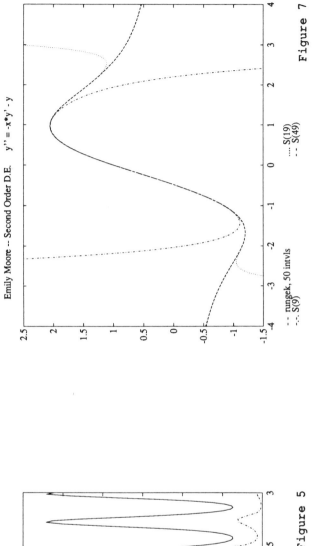

Emily Moore -- First Order D.E. $y' = 3*y*x^2 * cos(x^3)$

—— f(x) = exp(sin(x^3))
-..euler , 100 intvls

Figure 5

Emily Moore -- First Order D.E. $y' = 3*y*x^2 * cos(x^3)$

——error,euler , 2000 intvls

——euler, 2000 - euler , 1000

Figure 6

Nonlinear Differential Equations in *Mathematica* ™

by

T. D. Morley

School of Mathematics

Georgia Tech

Atlanta, GA 30332

(404) 894-2700

BITNET: MA201TM@GITVM1

Second and third year courses in <u>Advanced Engineering Mathematics</u> or <u>differential equations</u> traditionally cover models or equations that admit a simple closed form solution. Techniques such as perturbation and group analysis, while in principle no more difficult than some of the more traditional topics, cannot at present be effectively taught at this level because the intermediate computations are so complicated. The advent of *Mathematica* ™ and other such systems changes this. We give an example of how *Mathematica* ™ can be used in a classroom situation to study nonlinear ordinary differential equations. In particular, we show how *Mathematica* ™ can be used to study periodic solutions to the Duffing equation by means of perturbation .

In various courses, entitled <u>Advanced Engineering Mathematics</u> or <u>differential equations</u> various classical ordinary differential equations are studied. These include first order equations, second order linear equations with constant coefficients, and if time permits, some particular second order linear equations such as Bessel's equation. Increasingly, such courses include some numerical techniques for finding approximate solutions to those equations that one cannot solve explicitly. One such equation is Duffing's equation, which models a hard spring:

$$y'' + c^2 y + \mu h y^3 = 0,$$

subject to the initial conditions $y(0) = A$, $y'(0) = 0$. This equation models a hard spring, i.e., where the restoring force $c^2 x + \mu h y^3$ of the spring depends nonlinearly on the displacement y. When $\mu = 0$, this is just an ordinary linear spring. By making a change of variables, (and changing μ and A) we may rewrite this equation (with initial conditions) as

$$x'' + x + \mu x^3 = 0 \qquad (1)$$
$$x(0) = A \ , \ x'(0) = 0$$

Where x(t) is the (scalled) displacement. If $\mu = 0$, then it is well known that the solution is $A \cos(t)$. (Guess $C_1 \cos(t) + C_2 \sin(t)$. Given that $x(0) = A$ and $x'(0) = 0$, then the solution is $A \cos(t)$.) By numerically solving Duffing's equation (1) for values of μ near zero, we can get some idea of how the nonlinear term " μx^3 " contributes to the

---

Mathematica ™ is a trademark of Wolfram Research Inc.

qualitative behavior of the solution. But by doing this we really don't obtain any hard information, all that we get is our impression obtained by looking at several graphs. To obtain more information, we must use some of the analytical techniques for finding approximate solutions. Unfortunately, many of these techniques require extensive computation. However, it is this author's claim that some of these methods are no more difficult in principle than some of the methods currently in the curriculum. In what follows, we study, with the help of *Mathematica* ™, approximate analytic solutions of the the Duffing's equation. In particular, we obtain approximate solutions that are accurate within a given power of μ. This particular problem is chosen for at least two reasons:

1. The computations, while laborious, can be done by hand. When stating off with a new technique with a symbolic manipulator, it is important to be able to relate what is going on inside the computer system to what we already can do. I would not recommend actually carrying all of the details out be hand, but one can do some of the parts by hand.

2. The answer to this problem is (at least in some circles) well known. Thus the instructor will either know of this example, or can look it up in many standard texts, e.g., [1,2].

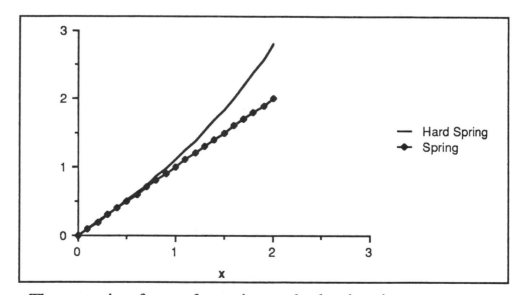

The restoring force of a spring and a hard spring --
$$\mu = 0.1$$

All right, here we go. Let N = 1 (or 2 latter on, or larger if you have the patience or computer time). We will look for periodic solutions. Of course the period of the solutions depends on μ. If $\mu = 0$ then solutions are periodic with period 2π. Let us guess that the period as a function of μ is $2\pi\omega$ where:

$$\omega = 1 + \mu\omega_1 + \mu^2\omega_2 + \mu^3\omega_3 + \ldots + \mu^N\omega_N + O(\mu^{N+1}) \qquad (2)$$

We will guess that x(t) can be written as a series on μ:

$$x(t) = u_0(\omega t) + \mu u_1(\omega t) + \ldots + \mu^N u_N(\omega t) + O(\mu^{N+1}) \qquad (3)$$

Where each $u_i(\cdot)$ is a periodic function of period 2π. Letting $\tau = \omega t$, then Duffing's equation (1) becomes:

$$\omega^2 \frac{d^2x}{d\tau^2} + x + \mu x^3 = 0 \qquad (4)$$

$$x(0) = 0 \quad , \quad x'(0) = 0$$

Using *Mathematica* ™ (or doing the computations by hand) we plug (2) and (3) into (4), keeping only the terms of order less than or equal to μ^N. We now collect terms based on powers of μ. The zeroth order terms give:

$$\mu^0: \quad \frac{d^2u_0}{d\tau^2} + u_0 = 0.$$

This is easily solved, using the initial conditions $u_0(0) = A$, $u_0'(0) = 0$, to give $u_0(\tau) = A\cos[\tau]$. This now tell us that:

$$x(t) = A\cos(\omega t) + O(\mu), \text{ where } \omega = 1 + O(\mu).$$

Plugging back this value for u_0 into the expanded equation, we get the first order terms:

$$\mu^1: \quad \frac{d^2u_1}{d\tau^2} + u_1 = \left(2\omega_1 A - \frac{3A^3}{4}\right)\cos(\tau) - \frac{A^3}{4}\cos(3\tau)$$

The solutions of this will not be periodic unless the coefficient of $\cos(\tau)$ on the right hand side is zero. So we set this coefficient equal to zero:

$$\left(2\omega_1 A - \frac{3A^3}{4}\right) = 0$$

and solve for ω_1. This gives $\omega_1 = \frac{3a^3}{8}$. Plugging this back into the μ^1 equation above, we get:

$$\frac{d^2u_1}{d\tau^2} + u_1 = -\frac{A^3}{4}\cos(3\tau)$$

If we find a solution of this with $u_1(0) = 0$ and $u_1'(0) = 0$, then $u_0(\tau) + \mu u_1(\tau)$ will satisfy the initial conditions $x(0) = A$, $x'(0) = 0$. We do this (in *Mathematica* ™) by the method of undetermined coefficients, and conclude that

$$u_1(\omega t) = \left(\frac{-A^3\cos[\omega t]}{32} + \frac{A^3\cos[3\omega t]}{32}\right),$$

where $\omega = 1 + \frac{3a^3}{8} + O(\mu^2)$. Remembering back to our guess for x, this gives

$$x(t) = A\cos[\omega t] + \mu\left(\frac{-A^3\cos[\omega t]}{32} + \frac{A^3\cos[3\omega t]}{32}\right) + O(\mu^2)$$

Note that this (approximate) solution of Duffing's equations tells us more than we could have obtained from studying plots of numerical solutions. It in fact tells us what the (approximate) frequency shift from the harmonic oscillator ($\mu=0$) case, and the presence of the "harmonic component" $\cos(3\omega t)$. Carrying out the second order calculation in *Mathematica* ™ we get:

$$x[t] = A\cos[\omega t] + \mu\left(\frac{-A^3\cos[\omega t]}{32} + \frac{A^3\cos[3\omega t]}{32}\right)$$
$$+ \mu^2\left(\frac{23\,A^5\,\cos[\omega t]}{1024} - \frac{3A^5\cos[3\omega t]}{128} + \frac{A^5\cos[5\omega t]}{1024}\right) + O[\mu^3]$$

where

$$\omega = 1 + \frac{3\,A^2\mu}{8} + \frac{21\,A^4\,\mu^2}{256} + O[\mu^3]$$

Note: The technical report version of this paper contains an appendix listing the *Mathematica* ™ notebook **Duffing**, in which the computations of this paper are done. This may be obtained from the author.

Bibliography:

1. W. F. Ames, **Nonlinear Ordinary Differential Equations in Transport Processes**, Academic Press, New York, 1968

1. W. J. Cunningham, **Introduction to Nonlinear Analysis**, McGraw-Hill, New York, 1958.

2. Stephan Wolfram, *Mathematica* ™: **A System for Doing Mathematics by Computer**, Addison-Wesley, New York, 1988.

Acknowledgements: The author would like to thank W. F. Ames for helpful discussions.

Two and Three Dimensional BASIC Graphics Programming and Its Uses

Umesh P. Nagarkatte and Shailaja U. Nagarkatte

<u>Introduction</u> Two dimensional BASIC graphics programming on IBM PC's or compatibles is extremely easy to use at any level starting from elementary algebra. It is flexible in the sense that it can be made as sophisticated as necessary depending on the student's mathematical background. The use of these programs requires no previous programming knowledge. It can be used to explain various difficult concepts in algebra, precalculus and calculus.

In elementary algebra, the concepts of consistency of systems of equations and in intermediate algebra, perpendicularity of lines, systems of linear inequalities in two variables can be explained using superimposing graphs. Students do not readily see the connection between graphs of functions and the real and imaginary roots of corresponding equations. These can be emphasized using graphics. Conic sections, and asymptotes to hyperbolas can also be easily handled. In precalculus, the concept of function - its domain, range and graph, the concepts of composite functions, inverse functions, graphs of rational functions, concepts of asymptotes, graphs of trigonometric functions and their inverses, effect of changing parameters on graphs of functions can be studied. In calculus, the concepts of limits, continuity and derivatives can be explained using graphing. The concept of Riemann Integral and areas between curves can be explained. Graphs of polar equations, which are time consuming to draw on the board, can be drawn very quickly. Taylor series approximation to functions can also be studied graphically. These activities can be handled effectively by using two dimensional graphics programming. In addition, derivatives can be applied as in shift operators to draw fast graphs of curves.

Students have a difficult time in visualizing three dimensional graphs. In this respect, a shell for three dimensional graphics based on easy BASIC instructions is helpful. Vectors, planes and surfaces, the tangent, gradient and normal to given surfaces can be explained using such a shell. But the ideas involved in 3D graphing are quite sophisticated.

<u>Implementation</u> We use 2D graphics programming in various classes as follows. Each student is provided a disk containing some DOS programs and a set of graphic programs. The instructor spends approximately 30 minutes in the micro computer lab explaining the procedure for using the graphics programs. Then lectures involving topics on graphing are conducted in the lab. As the instructor explains the concepts, the computers are used for reinforcement. Some exercises involving computer graphics are assigned in addition to regular exercises.

Quoted below are some of the actual assignments used in our classrooms. For more details see [8].
1. Given two lines with equations, y = 2x - 3 and y = -x + 2, complete a parallelogram with lines through (-5,-1). Students are provided a program containing two lines, which they modify to do this problem.

2. Given the program dealing with Riemann sums using the midpoint rule for $y = x^2 + 4$, write a program that draws proper rectangles and computes Riemann sums for y = (sin x)/x .

Two Dimensional Graphics Programming These programs consist of simple instructions. They can be easily modified. They use built-in functions in IBM or GW BASICA. These programs can be made interactive. They can incorporate other ideas such as shift algorithms, which are applications of derivatives for fast graphing. Only five BASIC graphics instructions are required to draw a graph. These instructions are as follows:

1. SCREEN - for medium or high resolution, activating color if desired.
2. WINDOW - for bringing the origin of the screen which is located at the upper left corner to the center. The "third quadrant" is brought to the lower left part of screen. This function also takes care of "line clipping" to disallow overflow or wrapping that might occur. By varying the scale, one can vary the size of the window to "zoom and pan".
3. LINE - for drawing the axes and tick marks, or for joining two given points.
4. PSET - for plotting points.
5. PAINT - for painting regions which are solutions of inequalities or areas between curves. Choose an interior point of the region for this instruction.

A typical graphics program is given below.

```
10 REM Draw a graph of the equation x^2 + y^2 = 25
20 CLS:SCREEN 1,0        'Clear screen, set med.res.,color on
40 x = 20               'Scale
50 WINDOW (-x,-x)-(x,x) 'Bring origin to center of screen
60 LINE (-x,0)-(x,0):LINE (0,-x)-(0,x)    'Draw x and y axes
70 LOCATE 1,20:PRINT "Y"                  'Name y-axis
80 LOCATE 14,20:PRINT "O"                 'Name the origin
90 LOCATE 14,35:PRINT "X"                 'Name x-axis
100 FOR x = -5 to 5 step .01
110      y = sqr(25 - x^2) 'Separate positive part
120      PSET (x,y)         'Plot the point
130      y = -sqr(25-x^2)  'Separate negative part
140      PSET (x,y)         'Plot the point
150 NEXT x
```

One can draw tick marks by inserting the instructions:
```
91  FOR t = -x+1 TO x: LINE (t,-.2)-LINE (t,.2):NEXT t
92  FOR t = -x+1 TO x: LINE (-.2,t)-LINE (.2,t):NEXT t
```

The above program draws an ellipse instead of a perfect circle. This distortion is due to the "aspect ratio" of the monitor screen. The aspect ratio is the ratio of the pixels along the horizontal axis to those along the vertical axis. Since this ratio is 4:3, the distortion could be corrected by using the instruction WINDOW (-4*x+1,-3*x+1)-(4*x,3*x).

Three Dimensional Graphing Three dimensional graphics programs use the same instructions as in two dimensions except that we replace WINDOW by WINDOW SCREEN. This instruction keeps the negative x and y coordinates in the upper left part of the screen. Three dimensional graphics consist of simulating the three dimensional objects on a two dimensional screen. Drawing a view of a surface depends on the location of the viewer. For this purpose it is convenient to take the origin at the viewing point of the viewer, the positive x-axis to the right, positive y-axis up and positive z-axis behind the viewer. The WINDOW SCREEN instruction facilitates this configuration.

The 3D graphics programming uses sophisticated ideas from projective geometry and linear algebra. The question which projective geometry directs itself is, "How do we see things?" or "What is the relation between a thing itself, and the thing seen?"[9] These problems are handled mathematically by using homogeneous coordinates. The regular 3D coordinates are transformed into view coordinates using rotations and translations and then into screen coordinates using projections. When we consider homogeneous coordinates, rotations and translations become linear transformations. Matrix multiplications play an important role in linear transformations. These transformations are discussed in [3,4,5,6,10].

Another difficulty is to remove the surfaces and lines which are hidden from view. Otherwise, the 3D effect is distorted. There are two types of methods for removing hidden surfaces and lines: Object-space methods based upon the object and image-space methods based upon the 2D image. Most books on graphics discuss the image-space methods. A failure proof method in hidden surface removal is the plane equation method, which is an object-space method. It uses calculation of the surface normal vector. The hidden surface routine expects the view coordinates to be provided in a consistent manner located counterclockwise around the outline of the surface. [2] When the surface normal points to the viewer, the surface is visible and when it points away from the viewer, the surface is hidden. In order to draw a visible surface, the area is cleared of all other graphics using "key matte approach". First, the required surface is drawn using an erasing color (cyan, for instance). Then the area is filled with the erasing color. Next, the outline is drawn in the correct drawing color (white, for instance). The area enclosed is filled with black. This is called the key matte apporach. Using this approach, the nearer surfaces will always cover surfaces which are further away, provided that the nearest surfaces are drawn last. [2] The authors are in the process of developing a shell of 3D graphing using the above ideas.

REFERENCES

1. Adams, J. Alan and Billow, Leon Descriptive Geometry and Geometric Modeling A Basis for Design: Holt, Reinhart and Winston, Inc., New York, NY (1988)

2. Adams, Lee High Performance Interactive Graphics: Tab Books, Blue Ridge Summit, PA 17214 (1987)

3. Artwick, Bruce Microcomputer Displays, Graphics, and Animation: Prentice-Hall, Inc., Englewood Cliffs, NJ 07632 (1985)

4. Mufti, Aftab A. Elementary Computer Graphics: Reston Publishing company, Reston, VA 22090 (1983)

5. Newman, William M. and Sproull, Robert F. Principles of Interactive Computer Graphics Second Edition: McGraw-Hill Book Company, New York, NY (1979)

6. Pavilidis, Theo Algorithms for Graphics and Image Processing: Computer Science Press, Rockville, MD 20850 (1982)

7. Rothenberg, Ronald I. Basic Computing for Calculus: McGraw-Hill Book Company, New York, NY (1985)

8. Sayrafiezadeh, Mahmoud Math Lab Manual for Graphics under preparation. Medgar Evers College, CUNY, Brooklyn, NY (1988)

9. Seidenberg, A. Lectures in Projective Geometry D. Van Nostrand Company, Inc., Princeton, NJ (Springer-Verlag, Inc., New York, NY) (1963)

10. Smith, David A. Three Dimensional Graphics: An Application of Linear Algebra An Apple Side Show Presentation Notes: Department of Mathematics, Duke University, Durham, NC (1984)

Medgar Evers College, CUNY
1650 Bedford Avenue, Brooklyn, NY 11225

Queensborough Community College, CUNY
56 Avenue and Springfield Blvd., Bayside, NY 11364

Henry C. Nixt, Associate Professor of Mathematics
Shawnee State University, Portsmouth, Ohio

Technology in Differential Equations

In this presentation I want to share some of my background experiences, humble beginnings, some mistakes made, and some student response to the numerical solutions portion of a differential equations course. Then, I shall conclude with some report on directions I hope to move and what seems to work for my teaching style. If there is extra time, perhaps some persons present will share their use of technology in college teaching.

Over the past eight years, I have taught one introductory course in differential equations six different times. Not having forseen the need for records of this course, I now wish that I had a more complete file of my experiences in differential equations classes.

The Early Years

I taught in a community college in central Illinois until 1987. Back in 1979-1980, the differential equations course on that campus had been taught as totally analytic. When I first taught D.E. the following year, I introduced some numerical solutions. The students and I used hand-held calculators, usually TI-55 or comparable models, to do Euler and the Improved Euler Techniques on first order equations.

Our college had one computer terminal on a TSO (time-share option) arrangement. I wrote some BASIC programs which were used for demonstrating numerical solutions. With only one terminal available, student assignments were necessarily short. The next time I taught the course I had a few students with experience writing BASIC and/or Fortran programs. I required all students to input my programs or write their own. From this experience I inherited improved programs for future use (Exhibit "A"). Shortly thereafter, our time share agreement was terminated.

Micro Computers

The community college established a lab of eight (8) TRS-80's (Radio Shack) computers. We adapted our BASIC programs to the TRS-80 set up. In a few years, we had a control center interfaced with the computers and three printers. At this stage the students no longer needed to input the programs since we could direct from the control station a program to various stations in the lab.

The interface and use of this lab for differential equations was often handled by one of my students who also worked in the computer lab. He was my first of a line of resident experts. [Each time that I have taught differential equations using technology--I have been blessed with a "resident expert" student. This quarter one of my

students is similarly gifted and industrious. I have his
permission to include his work (Exhibit "C").
    During one academic year, my students used TRS-80 s
interfaced with the control center, separate TRS-80 s, and
an Apple II computer housed in the Learning Resources Center
at the college.
    Several students in my class had Apple computers of
their own or had access to them at their local high schools.
The students had varied backgrounds. Some had experience
with course work in programming while some had no "hands on"
experience. Consequently, some were writing their own
programs while others used only those programs provided by
the instructor.
    Soon the availability of textbooks including flow
charts and/or sample programs showed that authors and
publishers were responding to the modern technology needs of
the student.

Conduit's Differential Equations Program

    Conduit's differential equations software package is
user friendly. After working with this package for two
classes over a period of two years, I have found this
program helpful to students. The package is available for
use on the Apple II line of computers. One of my more
recent "resident experts" managed to use his software and
programming ability to produce paper copy of some of the
Conduit results (Exhibit "B"). I have become aware that
Conduit has an "add on" hardware--Finger Print Plus--which
allows printing of the screen images. The hardware is
available for Apple II, IIe, and IIGS computers.

Hand-Held Calculators

    No matter what phase I have been working through this
eight-year period, I have always included some hand-held
calculator exercises for at least the Euler and Improved
Euler techniques. These exercises help the student toward a
better understanding of what goes into a numerical solution.
    Students must perform numerical techniques on many
differential equations--both those which are readily solved
analytically and those which do not lend themselves to
analytic techniques. From the former, students can compare
their analytic solutions to their numerical solutions to
begin to recognize the need for control of error. Those
which are not readily done by analytic techniques, of
course, demonstrate the power of technology in conjunction
with numerical technique.
    Some geometry helps the students appreciate their
results. Stepping across the interval using subintervals
and isoclines makes the solution more meaningful (Exhibit
"C"). The Conduit Program provides graphs and, if desired,
direction fields.

Trends

With the availability of software packages, I no longer teach programming. Time is valuable, and I prefer to use the time on traditional content of differential equations.

By and large, my students have been sophomores in pre-engineering curriculum. They have transferred into programs at the University of Illinois or Southern Illinois University. Most of the students were successful at these schools.

Former students' feedback indicates the need for more numerical work and work with orthogonal trajectories. My own feeling is that I need to pay more attention to existence and uniqueness--before "running to the computer." Also, error control must be watched more closely in the future.

Recommendations

Assign problems to be done numerically.

Suggest or require that students use a certain software package.

Spell out format to be used for hand-in work.

Require some problems be done with subinterval increments of n and n+1 evaluations.

By comparing results students can get an idea of what truncation error will do to results. Of course, students should compare the results when increments are $h = 0.1$, $h = 0.05$, and $h = 0.01$.

Students should compare various carefully selected problems using Euler, Improved Euler, Runge-Kutta methods, and some non self-starting predictor/corrector methods. (These comparisons are readily available in many recent differential equations textbooks.)

Textbook

In selecting a textbook, John Van Iwarrden, author of Conduit software package, has written a textbook specifically geared to the use of technology.

Other good texts with early introduction of numerical techniques could provide the base for an excellent course. [N.B. In selecting a book, a publisher's claim of independent chapters must be verified by reviewing the content and working some problems.]

In conclusion, even with modest equipment and modest investment, one can have a successful unit on numerical solutions in his/her differential equations course.

Textbooks

Finizio and Ladas. Introduction to Differential Equations.
    Wadsworth, 1976.

Guterman and Nitecki. Differential Equations: A First Course, Second
    Edition. W. B. Saunders, 1988.

Hagin, Frank. A First Course in Differential Equations.
    Prentice-Hall, 1975.

Kells, Lyman. Differential Equations: A Brief Course,
    McGraw-Hall, 1963.

Van Iwaarden, John. Ordinary Differential Equations with Numerical
    Techniques. Harcourt, Brace, Jovanovich, 1985.

Courseware

Van Iwaarden, John. Ordinary Differential Equations--Software
    Package. Conduit.

222

```
]LOAD IMEUL
]LIST
                "Improved Euler" Listing
 10   DEF  FN F(X) =  - (1 -  SIN (X) / 2) * Y
 20   PRINT "ENTER LEFT AND RIGHT ENDPTS,Y0,NMBIN,PSS"
 30   INPUT LEFTX,RIGHTX,Y0,NMBIN,PSS
 35   PR# 1
 37   PRINT "USING IMEUL": PRINT
 40  X = LEFTX
 50  Y = Y0
 55  Y1 = Y0
 60  H = (RIGHTX - LEFTX) / NMBIN
 70   PRINT "INTEGRATIONS:";NMBIN
 80   PRINT "H:";H
 90   PRINT " X"," Y"
 100   PRINT X,Y
 200   FOR I = 1 TO NMBIN
 210  F0 =  FN F(X)
 220  YP = Y + H * F0
 230  X = X + H
 235  Y = YP
 240  VLUF =  FN F(X)
 250  Y = Y1 + .5 * H * (F0 + VLUF)
 255  Y1 = Y
 260   IF  INT (I / PSS) = I / PSS THEN  GOTO 280
 270   IF NMBIN < > I THEN  GOTO 290
 280   PRINT X,Y
 290   NEXT I
 295   PR# 0
```

"A"

$$\frac{dy}{dt} = 3x - 2t \quad \begin{smallmatrix}-2<<<2>\\-2<x<2>\end{smallmatrix}$$

```
]LIST

 70  TS = 1 / 8:DT = (4) / 279:DX =
      (4) / 191:L = 10
 90   HGR : HCOLOR= 3
 100   FOR U1 = - 2 TO 4:M1 = 2 *
      U1: IF M1 = > 1 THEN L = 14

 101   IF U1 > 2 THEN M1 = 3: IF U
      1 > 3 THEN M1 = 5
 102   IF U7 < 7 THEN M1 = 0:U7 =
      7:U1 = - 2
 105   FOR U2 = 1 TO 2:M = M1: IF
      U2 > 1 THEN M = - M1
 106   IF M1 = 0 THEN U2 = 2
 107   FLASH : PRINT "M="M: NORMAL

 110   FOR TP = - 2 TO 2 STEP TS:
      T = 139 * TP / DT
 120  XP = (M * 2 + TP) / 3:F = 95
       + XP / DX
 130  CT =  COS ( ATN (M)) * L / 2
      :CF =  SIN ( ATN (M)) * L /
      2
 135   PRINT "T"T"...CT"CT"...F"F"
      ...CF"CF
 136   IF T - CT < 0 OR F - CF < 0
```

"C"

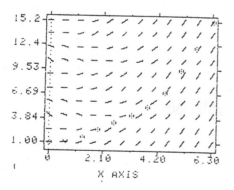

X	Y
0.0000	1.0000
0.6300	1.0803
1.2600	1.6341
1.8900	2.5172
2.5200	3.3660
3.1500	4.1800
3.7800	5.1588
4.4100	6.8430
5.0400	9.3881
5.6700	11.9629
6.3000	15.2135

$Y' = \sin Y + X - 1$
$Y(0) = 1.0$
Solution by IMEUL
on $[0, 6.30]$ with
$h = 0.63$

"B"

Using Numerical and Graphical Techniques to Teach Infinite Series

ARNOLD OSTEBEE, Department of Mathematics, St. Olaf College,
Northfield, Minnesota 55057

To understand the fundamental ideas taught in the first-year calculus course, it is helpful for students to see these ideas from several points of view. For example, a student's understanding of the function concept is deepened by working with functions represented as graphs and tables of numbers in addition to functions represented as algebraic expressions. Similarly, by examining other calculus concepts using graphical and numerical techniques, as well as the traditional algebraic methods, students can gain a deeper understanding of these ideas and a better appreciation for their applicability. Although graphical and numerical techniques are part of every calculus course, the computational effort involved in using these techniques is so great that only a very small number of examples can be examined in this way. These techniques can be given greater prominence in the calculus course if students have access to a computer algebra system.

The mathematics department at St. Olaf is currently in the midst of a three-year curriculum development project in which students and faculty use a computer algebra system (SMP) in first-year calculus courses. Support for this project has come from the National Science Foundation's College Science Instrumentation Program and the U.S. Department of Education's Fund for the Improvement of Post-Secondary Education. Students use SMP on Sun-3 workstations to complete locally written homework assignments that emphasize the development of conceptual understanding in addition to manipulative skills.

Most students in first-year calculus courses learn to apply convergence tests to infinite series, but many fail to develop a real understanding of what convergence or divergence means. For example, the idea that a convergent infinite series is equal to a *number* doesn't really sink in. Computer algebra systems can be used to improve this situation. We have used the capabilities of a computer algebra system to allow students to view the sequence of partial sums of an infinite series graphically. We have also augmented our discussion of convergence tests with exercises that require students to compute numerical estimates of convergent infinite series. The result has been a substantial improvement in students' understanding of the definition of the sum of an infinite series and the development of a better intuitive understanding of the idea of rate of convergence.

In this talk I will show some examples that illustrate how we use graphical and numerical techniques to help students learn about infinite series.

A GENERAL EDUCATION COURSE:
Mathematics,
BASIC Programming,
and Computer Literacy

Dr. Jerry W. Phillips
Professor of Mathematics and Education
Acting Dean of Academic Affairs
OAKLAND CITY COLLEGE
Oakland City, IN   47660

During a curriculum reorganization in a small midwestern liberal arts college, a faculty committee suggested that a general education course be developed which integrated mathematics and computers. This happened at the same time that an alumni group of former math teachers were asking about honoring a former professor. After meetings between the alumni, affected faculty members, and the Development Staff, a plan was formulated to memorialize the professor with a computer lab. The lab would be used for the general education course and other beginning courses in computing science. During this process, the college created both an associate degree in computing technology and a bachelors degree in applied math.

The course, Math 105, Mathematics and Computer Technology, was used as the vehicle to teach mathematics, BASIC computer programming, and computer literacy. It was also to be the hub of the new curriculum in computing science since limited resources made it impossible to offer a different course for applied Math/CS majors.

Two more factors influencing the development of the course were available hardware and student preparation. A network of TRS-80 model III and 4 computers has served well but has been limiting. Our student population has been such that several students have found that their lack of mathematical background and no computer experience has necessitated remedial courses before enrolling in Math 105. Remedial courses available are Developmental Mathematics, Beginning Algebra, and Introductory Programming. This approach has been very helpful for marginal and nontraditional students. We have found that for students with some computer literacy and average secondary mathematics backgrounds, the course is challenging but realistic.

In addition to the three primary objectives listed above there have been two supporting considerations during the course's development: 1) reducing math and/or computer anxiety, and 2) improving oral and written communication skills. The instructors have found that it is possible to develop the kind of caring atmosphere where the instructor personally interacts with the students and is able to reduce math and/or computer anxiety. Faculty offices are located next door to the lab which facilitates the answering of student questions. Also mathematics work-study students are available to assist students in the lab. Communication skills are encouraged by the careful evaluation of written assignments such as essays and computer programs. Students are expected to participate in class discussions and present homework solutions on the chalkboard. Students know they are expected to speak and write properly.

FIGURE 1

Relative Point Value of Course Requirements

Requirements	Possible Points	% of Total
3 Exams	300	46
3 Programming Projects	150	23
Miscellaneous Quizzes	100	15
5 Written Reports	50	8
Participation & Attendance	50	8
totals	650	100

Evaluation for the course is determined by the criteria listed in Figure 1 above. Ample opportunity is provided for students to garner points to achieve at a level appropriate to their ability. The three 100-point exams are approximately 40 points on mathematics concepts, approximately 30 points on tracing programs and supplying the output, and the remaining 30 points on flowcharting or coding mathematical algorithms. Usually ten 10-point quizzes are given during the semester to encourage attendance and completion of home-work, to let students demonstrate mastery of the content, and to find areas of misunderstanding before an examination. The programming projects are assigned during the last half of the semester to give the students an opportunity to apply the programming concepts in a meaningful manner. These projects may be mathematical in nature, the entire class may do the same programming exercise, or students may choose to develop their individual applications. These projects are evaluated as to originality, style, completeness, and error-handling of inappropriate data. Flowcharts for each project are also considered in the evaluation.

The five written reports are usually assigned relatively early in the semester before students are working on the programming projects. This requirement of the course encourages students to read current articles related to computers and provides for evaluation of written communication. Students must identify at least one source with a bibliographic entry. Students are encouraged to relate the computer to their college major or explore topics such as the history of computing, particular software or hardware, technological limitations, and effects of computers on society. Students are encouraged to use a word-processor and write at least two pages. The final 50 points are available for the instructor to reward hardworking students for their diligent efforts or to penalize students who have not attended or put forth noticeable effort. This system of evaluation has been found reasonable in that marginal students who are willing to put forth considerable effort have been able to achieve a passing grade.

The content of the course closely follows the text, Using Computers in Mathematics by Elgarten, Posamentier, and Maresh. Mathematics areas covered include algebra, geometry, number theory, probability, and statistics. A parallel development of BASIC programming concepts are introduced as the mathematics topics are being covered. See Figure 2 for specific topics of the two disciplines and their parallel presentation. (A deviation from standard outlining is used in this figure.)

The text was written as a first course in BASIC programming with the intent to teach BASIC programming and mathematics at the same time. The text is not hardware specific and is readily adaptable to common microcomputers. Although it does not deal with any storage procedures, after the first few weeks we include instruction on cassette tape storage due to its ease of use and availability. Students with diskette experience are permitted to use diskettes. Structured programming is emphasized throughout the course by encouraging a minimum number of GOTOs and illustrating modularity. Other good programming practices encouraged are the use of remarks, meaningful use of variables, user-friendly messages, proper error-handling, debugging practices, and efficiency considerations. These programming practices are also used in program evaluations. Students with previous programming experience often find this portion of the course the most frustrating because it requires them to break sloppy programming habits. Students have to be encouraged to use good programming practices and not to just get a program to run.

The course, Math 105, Mathematics and Computer Technology, appears to be meeting the needs of our liberal arts students. It seems to be teaching the fundamental mathematics, BASIC programming, and computer literacy foundations. The anxieties toward mathematics and/or computers seem to be helped by the informal, caring, "hands on" atmosphere of the class and laboratory environment. As we look to the future literacy needs of these liberal arts students, we especially want to increase the software packaging capability so that more instruction can be included on word-processing, data base, and spreadsheet packages.

The above description is merely a summary of the course developed for general education at our institution. If an instructor is interested in more background information, the overall computing science curriculum, student preparedness for the course, or a more in-depth analysis of the course, please see an article by Sermersheim from The Journal of Computing in Small Colleges. Mrs. Sermersheim has been one of the instructors for this general education course.

Bibliography

Elgarten, G.H., Posamentier, and Moresh, Using Computers in Mathematics, Addison-Wesley, 1983.

Sermersheim, Robin A., "Killing Three Birds with one General Education Requirement," The Journal of Computing in Small Colleges, Consortium for Computing in Small Colleges, P.O. Box 329, Evansville, Indiana, Volume 3, Number 2, November 1987.

FIGURE 2

MATH 105 COURSE OUTLINE

NOTE: Numbering Code is as follows:
Roman Numerals indicate title of unit.
Capital letters indicate mathematical topics.
Hindu-Arabic digits indicate programming concepts.

I. INTRODUCTORY PROGRAMMING CONCEPTS
A. Solving Simple Equations
1. Computer Programming and Operations
2. Preparing a Problem for Computer Solution
3. Variables
4. Order of Operations
5. Assignment Statements
6. INPUT, PRINT, and END instructions
7. Tracing Programs
8. Entering and RUNning a program
9. Formatting Output
10. Initialization of Variables
11. EDITing Programs
12. Flowcharting Algorithms
13. Unconditional Branching (GOTO)
14. Conditional Branching (IF-THEN)
B. Perimeter and Area Calculations

II. ELEMENTARY ALGEBRA CONCEPTS
A. Fahrenheit and Celsius Conversions
1. User Prompts with INPUT
2. Statement Separators
B. Divisibility
3. Greatest Integer Function
C. Absolute Value
4. Absolute Value Function
5. TAB Function
D. Newton-Raphson Method for Approximating Square Roots
6. Truncation
7. Rounding
8. Square Root Function
E. Solving Systems of Linear Equations by Reduction
F. Determinants
G. Cramer's Rule for Solving Systems of Linear Equations
9. Avoiding Endless Loops
10. Use of Menus
H. Galileo's Gravitational Formula
11. Use of a Counter
I. Invoicing Problems
12. READ-DATA Statements
13. Use of an Accumulator
J. Finding the Arithmetic Mean
14. FOR-NEXT Loop
K. Factorials

III. GEOMETRY CONCEPTS
A. Triangle Inequality Problem
1. Logical Operators AND and OR
B. Pythagorean Theorem
2. String Variables
3. Nested Loops
C. Classification of Triangles
4. Exchanging Contents of Two Variables
D. Heron's Formula for Area of a Triangle
E. Distance Formula in Two and Three Dimensions
F. Congruency
* G. Circle Intersection Problem

IV. OTHER ALGEBRAIC TOPICS
A. Solving Quadratic Equations
B. Classifying the Roots of a Quadratic Equation
* C. Euler's Method to Solve Diophantine Equations
* D. Arithmetic Sequences and Their Sums
* E. Harmonic Sequences and Means

V. NUMBER THEORY
A. Prime and Composite Numbers
B. Euclid's Division Algorithm
C. Perfect, Abundant, and Deficient Numbers
D. Amicable Numbers
1. Subroutines
E. Finding the Greatest Common Divisor
F. Finding the Least Common Multiple
G. Operations with Fractions
2. One-dimensional Arrays
3. Replacement Sort
4. Bubble Sort
* 5. String functions (MID$ and LEN)

VI. PROBABILITY AND STATISTICS
A. Fundamental Counting Principle
B. Permutations
C. Combinations
D. An Introduction to Probability
E. The Birthday Problem
F. Using Random Numbers
1. The Random Number Generator Function
G. Measures of Central Tendency: Mean, Median, Mode
H. Data Frequency Charts
I. Summation Notation
J. Measuring Dispersion: Variance and Standard Deviation

* Optional Topics

CONFERENCE ON TECHNOLOGY IN COLLEGIATE MATHEMATICS
Oct 27-29,1988 The Ohio State University

BORIS D. RAKOVER,Dept. of Math and CS
St.John Fisher College,Rochester New York
THE IMPACT OF COMPUTER PROGRAMMING ON TEACHING
COLLEGIATE MATHEMATICS.

The growth and development of computer technology during the last four-five decades has stimulated new approaches in teaching Collegiate Mathematics.

The purpose of this paper is to present a teaching model which deals with various classes of solvable uniform mathematical problems from Collegiate math courses.

Computer programming processes involve techniques and approaches being convenient tools in developing abilities for problem solving, logical reasoning.Investigating the process of solving mathematical problems by college students we found that more than half of all math students experience difficulties in solving math problems.To estimate the degree of difficulty we used evaluation functions,that is,how long one is wandering before the right direction is found.Our conjecture is:

THE INTELLECTUAL ACTIVITIES OF SOME MATHEMATICS STUDENTS CAN BE PARTLY GUIDED DURING THE PROCESS OF SOLVING MATHEMATICAL PROBLEMS.

Our teaching model reduces the difficulties in cases when a problem belongs to a class of uniform problems.

Creating and using a problem-solving model for a set of uniform mathematical problems from the undergraduate math curricula is perhaps the most fascinating aspect in mathematics teaching process.It should be realized that the difficulties which one meets in the process of mathematical thinking during a problem-solving situation can be critical and it is obvious that the problem on hand will never be solved without an outside impulse.It is our conviction that the most efficient impulse is an algorithmic prcedure created and formulated for the class of problems being considered at a given time.

We define an algorithmic procedure (AP) as a
SEQUENCE OF STEPS LEADING TO THE SOLUTION FOR A CERTAIN SET OF UNIFORM PROBLEMS WHERE DETAILS OF SOME STEPS CANNOT BE DETERMINED BEFOREHAND.

The knowledge of an AP is not a garantee to find the solution of a problem from a given class,it provides a proper path,a general direction subdivided into a number of steps.To perform the actions inside each step one should have the appropriate abilities and skills in order to support the mathematical activities until the problem is solved.That clarifies the concept of an AP: it is only a frame to be used in critical situations,it is free of a detailed description of steps which cannot be subdivided any more.

There are problems arising during the process of forming AP.We will specify just a few of them:
1.HOW SHOULD BE DESIGNED THE CRITERION OF DECOMPOSITION A PROCESS INTO SIMPLER SUB-PROCESSES?

2.WHAT SHOULD BE ACCEPTED AS THE FIRST-LEVEL ELEMENT (WHICH CANNOT BE DECOMPOSED ANY MORE) IN THE PROCESS OF SOLVING MATHEMATICAL PROBLEMS?

3.WHAT ARE THE MATHEMATICAL ABILITIES AND SKILLS OF MATHEMATICS STUDENTS WHICH WILL SATISFY OUR REQUIREMENTS?

4.IS THE LIMITED TIME PREDETERMINED FOR A CERTAIN TOPIC COMPARABLE WITH THE AMOUNT OF KNOWLEDGE TO BE ACCOMPLISHED BY A MATHEMATICS STUDENT?

Our findings show that it does not exist one and only one way to decompose an entire process into less complicated subprocesses. In many cases it is impossible to predict and to determine in advance the size and the content of a certain subprocess. Moreover,a certain subprocess can be considered as a very simple one from person's A point of view and very complicated from preson's B point of view.

There is an additional problem to be cosidered:WHAT ARE THE DIFFERENCES ON SPECIFYING A PROBLEM-SOLVING PROCEDURE BY A HUMAN AND BY A COMPUTER?

We are considering only one aspect in copmarisons:THE PROBLEM-SOLVING PROCEDURE.It is appropriate to observe only a few of them:

1.In general a human needs only the major steps of the procedure, having always enough room for creativity,while a computer program must be designed in such away that all first-element steps must be indicated.

2.A human has a common sense,is constantly in a thinking "mode", acts in a reasonable way,while the computer cannot exercise judgement unless it has been provided with explicit directions for making a decision.

3.Specifications can be reduced to a general description when they are designed for a human use,while the computer accepts procedures being described in a certain language related to an exacting technique requiring attention to details.

Our approach is based on the logic of mathematics itself and on the accessibility (in terms of common sense) of the structural, logical,reasonable construction of the AP.That approach has a tendency to raise the opportunities of the cogitative activity on a higher level,namely to establish favorable conditions for independent creativity.The training to create and use AP provides a tool to solve mathematical problems,to develop one's critical type of thinking and abilities for insight.

Consider now four classes of problems from College mathematics.

1.Find an inverse function $f^{-1}(x)$ for a given monotonic function f(x) determined in an interval I.The corresponding AP can be formulated as it follows:

Step 1. Express x in terms of y: $x = f^{-1}(y)$

Step 2. Exchange x and y: $y = f^{-1}(x)$i.

It is obvious that step 2 is precise and clearly determined. Meanwhile, step 1 describes only tha general action,that is,the requirement to express x in terms of y. In some cases the inversion is very difficult or even impossible (for example,when f(x) is a polynomial of degree five or more).

2.A rational function of trigonometric functions R(sinx,cosx) is to be integrated.The task is to examine the given function in order to establish an appropriate method of integration and to describe as much as possible the corresponding AP. The analysis of that problem results in four available types of substitution,each of which will convert the given trigonometric function into rational function with respect to t: a) cosx = t, b) sinx = t, c) tanx = t, d)$\tan\frac{x}{2} = t$.

To be able to make the right choise we will investigate different possibilities for the expression R(sinx,cosx)

2.1 Let the function R(sinx,cosx) be odd with respect to sinx, that is, R(-sinx,cosx) = -R(sinx,cosx). In that case the given function can be represented as $R(sinx,cosx) = R_0(\sin^2 x,cosx)sinx$ The substitution cosx = t transforms the function R(sinx,cosx) into a rational function of t.

Example. Integrate:

$$\int \frac{\sin^3 x}{cosx - 3} \, dx.$$

The integrand is a rational function with respect to sinx and cosx and   an odd function with respect to sinx,therefore the substitution cosx = t converts the given integrand into a rational function of t:

$\sin^3 x = \sin^2 x \ sinx$. If cosx = t,then -sinx dx = dt, $\sin^2 x = 1 - \cos^2 x$. Now the given integral is transformed to the form:

$$\int \frac{\sin^3 x}{cosx - 3} \, dx = \int \frac{t^2 - 1}{t - 3} \, dt.$$

It is obvious that the same substitution can be applied in case when $R(sinx,cosx) = R_1(sinx)cosx$.

2.2 Let the function R(sinx,cosx) be odd with respect to cosx, that is, R(sinx,-cosx) = -R(sinx,cosx). In that case the given function can be represented as $R(sinx,cosx) = R_0(sinx,\cos^2 x)cosx$. Referring to the case 2.1 it is easy to see that the substitution sinx = t can be applied in order to convert the given rational function with respect to sinx and cosx into a rational function with respect to t.

2.3 Let the function R(sinx,cosx) be even with respect to both sinx and cosx,that is, R(-sinx,-cosx) = R(sinx,cosx).We will prove now that the substitution tanx = t converts the given function into a rational function with respect to t.Indeed,if tanx = t,then

sinx = t cosx,and $\sec^2 x \ dx = dt$.Knowing that

$$\sin^{2k} x = (1 - \cos^2 x)^k, \text{ and that } \cos^2 x = \frac{1}{1 + \tan^2 x} = \frac{1}{1 + t^2}$$

we conclude that all expressions involved in the integrand and dx are rational functions of t.

2.4 If conditions in cases 2.1-2.3 are not holding then it is appropriate to apply the universal substitution $\tan\frac{x}{2} = t$.

3. Consider the general form of a linear second order homogeneous differential equation with constant coefficients: $a_0 y'' + a_1 y' + a_2 y = 0$

The AP to solve that differential equation consists of the following steps:

Step 1. Seek the solution in the form: $y = e^{kx}$.

Step 2. Find the first and the second derivatives for y and set up the characteristic equation: $a_0 k^2 + a_1 k + a_2 = 0$.

Step 3. Compute the discriminant of the characteristic equation:

$D = a_1^2 - 4 a_0 a_2$.

Step 4. If $D > 0$ then k_1 and k_2 are real and distinct numbers. The solutions are:

$$y_1 = e^{k_1 x}, \quad y_2 = e^{k_2 x}$$

The general solution in that case is:

$$Y = C_1 e^{k_1 x} + C_2 e^{k_2 x}.$$

Step 5. If $D = 0$ then $k_1 = k_2$ are repeated real roots.

The solutions are:

$$y_1 = e^{kx}, \quad y_2 = x e^{kx}.$$

The general solution in that case is:

$$Y = C_1 e^{kx} + C_2 x e^{kx}.$$

Step 6. If $D < 0$ then a + bi(a and b are real numbers) is a complex root of the characteristic equation. Since the original equation has real coefficients the conjugate complex root is: a - bi.

The solutions are:

$$y_1 = e^{ax} \cos(bx) \text{ and } y_2 = e^{ax} \sin(bx).$$

The general solution in that case is:

$$Y = e^{ax} [C_1 \cos(bx) + C_2 \sin(bx)].$$

4. The main part of the solution of a system of linear equations using Gauss-Jordan method is devoted to the conversion of the given matrix (augmented) into a special form. Following is the AP to perform this part:

Step 1. Find the column j that contains non-zero entries.

Step 2. Find the row i which contains a non-zero entry in the intersection with column j. Let that entry be a_{ij}

Step 3. Divide row i by a_{ij}.

Step 4. Add suitable multiples of row i to all rows 1,2,...,i-1,i+1,...,n,so that all entries in column j excluding the i row become zeros.

Step 5. Consider a new matrix not containing row i and column j, and go to Step 1. Continue in this way until it is impossible to follow the steps futher.

A considerable part of mathematical applications are related to verbal problems which arise in different areas of human activities. The history and nature of mathematics itself is a chain of findings and discoveries related to various scientific verbal problems.

The process of solving a word problem at any level of mathematical education is creative and not predictable. We cannot formulate in advance a prescribed set of steps, that is, an algorithm which will solve any given word problem. On the other hand it is possible to illustrate some patterns of solving word problems and to emphasize the common parts showing the general approach of this type of mathematical thinking. The AP reflecting that type of problems is:

Step 1.Read careful the given verbal problem (more than once).

Step 2.Write down the given numerical values and the corresponding scientific notations.

Step 3.Write down all unknown and required values,introduce your own notations.

Step 4.THE MAIN PART OF THE PROCEDURE: use your knowledge (refer to Guidebooks if necessary) to create relationships between the known and unknown values resulting into an equation or system of equations (linear,quadratic,trancsendental,differential). That step forms the MATHEMATICAL MODEL of the given problem.

Step 5.Solve the pure mathematical problem obtained in Step 4.

Step 6.Interpret the solution.Try to explain completely the obtained results (some of them could be senseless).

As a rule the process of working out a apecific AP is preceded by an ALGORITHMIC FLOW CHART MODEL (AFCM),which indicates a computational and logical procedure.The entire process as applied to a certain class of uniform problems exhibits each of the individual parts of the solution together with all the interactions among them.These parts and interactions are indicated by blocks and connecting arrows.Details depend upon the complexity of the class of problems.

Conclusion.

An AP cosists of generalizations,logical sequences,in some cases branching processes.There is a significant distinction between an algorithm and an AP. An algorithm which separates the solution of the problem into very small portions giving exhaustive information inside the steps,and exact sequence of execution so that both,a human and a computer are able to work out all steps in order to obtain the solution,is an absolute predescribed,predetermined, precise instrument,while an AP permits various paths of design, freedom in choosing a decomposition,it presents an application of knowledge,intuition,skills,and scientific background.

A MATHEMATICAL MODEL
SOLVING A LINEAR NONHOMOGENEOUS DIFFERENTIAL EQUATION
OF THE N-th ORDER WITH CONSTANT COEFFICIENTS

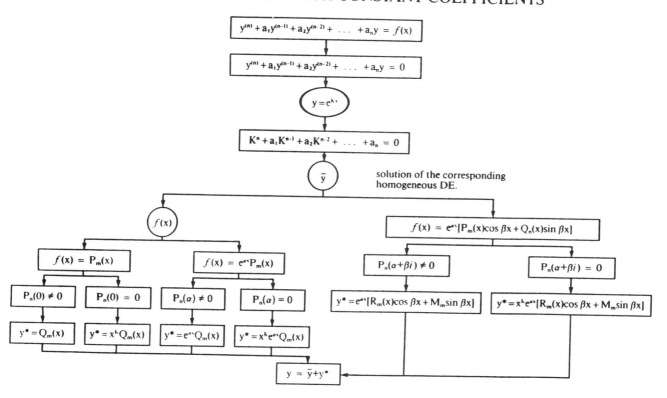

LINEAR ALGEBRA USING SPREADSHEETS

V.S.Ramamurthi
University of North Florida
Jacksonville, Fl 32216

This paper briefly describes a first course in linear algebra which was taught using a spreadsheet as basic software with no matrix calculation facilities built-in to start with. The following topics were handled using the spreadsheet:

1) Gaussian Elimination
2) Arithmetic operations on matrices
3) Calculation of inverse
4) Calculation of determinant
5) calculation of eigenvalues and eigenvectors
6) Gram-Schmidt Process

Each topic was passed through three stages. First, the students had to work typical problems by hand. This is to enable them to learn the principles. Secondly, they solved similar problems using the spreadsheet. Here they had to know the appropriate formulas but the spreadsheet will do the arithmetic and the repetitive portions of the task. This highlights to the student the advantage of using the computer to solve the problem. The third step was the creation of a template for solving any problem of a particular type. The use of this template was restricted in the following way: a template for a problem of type X should not be used to solve a stand-alone problem of the same type. But if a problem of type Y involves the solution of a problem of type X, the template can be applied to the problem of type X and the solution used in the Y type problem. For instance, the template for Gaussian Elimination is not allowed to be used to solve a problem on systems of equations. But it can be used to solve the systems of equations which arise when calculating eigenvectors.

The class met once a week in the computer lab where they used the computer to solve problems which they did by hand during the other two meetings in the regular lecture room. Each test was of two parts (i) a classroom part involving definitions, theorems and problems of small magnitude (ii) a computer lab part involving a larger number of problems of bigger magnitude.

The spreadsheet was tried as the basic software for this course due to the following reasons:

1) The software will only do the arithmetic and the repetitive tasks. The student will have to know the principles and the formulas. The software is not a "black box" facility.

2) The student can see all the steps of the solution on the screen including the formulas used. These formulas can be modified any time if necessary without having to start the problem from scratch.

3) There are very few rules to be learnt in order to use the software for the purpose of solving linear algebra problems. Further, the software has a lot of other built-in mathematical functions. Therefore it can be used in other courses like calculus and discrete mathematics too.

4) Spreadsheets like Lotus 123 have a Pascal-like language facility which can be used to write programs which will reside in the particular workfile and aid calculation or other required tasks. Writing such programs which are called "macros" is more easily learned than regular programming languages like Pascal or Basic.

5) In many cases, solving one problem of a particular type automatically creates a template for solving all problems of that type and this template can be used again and again.

Course Material

We give below two computer lab instructions as samples of course material prepared - one for Gaussian Elimination and the other for calculating a determinant. These are based on the spreadsheet Lotus 123, but can be modified easily to fit any other spreadsheet.

Gaussian Elimination

Problem: Solve the system $2x-2y+z=3$
$$3x+y-z=7$$
$$x-3y+2z=0$$

Procedure: Enter the augmented matrix in the cells b13 to e15.

$$
\begin{array}{rrrr}
2 & -2 & 1 & 3 \\
3 & 1 & -1 & 7 \\
1 & -3 & 2 & 0 \\
\end{array}
$$

To reduce this, we start by making the entry in b13 equal to 1 by dividing the whole row by 2. To do this, go to b20 and type the formula +b13/2. Then, press the keys

/, c, <enter>, -->, ., -->, -->, <enter>   ...(1)

This has the effect of dividing the whole row by 2. The idea is that the formula is applied to the first element in the row and just copied to the other elements. For all row operations, the same procedure can be applied. This saves time and the possibility of arithmetic mistakes. We will signify the sequence of keystrokes (1) by "replicate" henceforth. The following table describes the further procedure.

To do row operation	go to cell	enter the formula	do
r2	b21	+b14	replicate
r3	b22	+b15	replicate
r1	b24	+b20	replicate
r2-3*r1	b25	+b21-3*b20	replicate
r3-r1	b26	+b22-b20	replicate
r1	b28	+b24	replicate
r2/4	b29	+b25/4	replicate
r3	b30	+b26	replicate
r1+r2	b32	+b29+b29	replicate
r2	b33	+b29	replicate
r3+2*r2	b34	+b30+2*b29	replicate

r1	b36	+b32	replicate
r2	b37	+b33	replicate
r3/0.25	b38	+b34/0.25	replicate

---

r1+(0.125)r3	b40	+b36+(0.125)*b38	replicate
r2+(0.625)r3	b41	+b37+(0.625)*b38	replicate
r3	b42	+b38	replicate

This produces the matrix

$$\begin{array}{cccc} 1 & 0 & 0 & 2 \\ 0 & 1 & 0 & 0 \\ 0 & 0 & 1 & -1 \end{array}$$

From this you can read the solution $x = 2$, $y = 0$, $z = -1$.

===

Determinant of a 3 by 3 matrix

Enter the matrix

$$\begin{array}{ccc} 1 & 2 & -1 \\ 3 & -1 & 0 \\ 2 & 5 & 1 \end{array}$$

in the cells b5 to d7.

To calculate the determinant of this matrix, enter the following formula in cell b10 :

    +b5*(c6*d7-c7*d6)-c5*(b6*d7-b7*d6)+d5*(b6*c7-b7*c6)

The determinant is displayed immediately. Now whenever you want to find the determinant of a 3 by 3 matrix, just enter the matrix in the location b5 through d7. Then the determinant is automaticlly calculated and the result is displayed in cell b10.

Exercise: Make a similar template for calculating the determinant of a 4 by 4 matrix.

The Natural Language Mathematics Testing System

Peter Rice
Mathematics Department
University of Georgia

In the summer of 1985, the Mathematics Department of the University of Georgia launched a project to design a computer system that would create, administer, grade and record all of the regular exams given in College Algebra. Since then, the project has grown to include the testing of Precalculus and has been sold to other schools where it is in use in a variety of courses.

The original idea was to create a system similar to the one designed by Stephen Franklin at the University of California, Irvine. It utilized a database of questions stored on a mainframe computer and administered exams with terminals. There were two types of questions: questions that required a numeric answer and multiple choice questions.

The questions requiring numeric answers were administered in a straightforward manner. Sitting at a terminal, a student would work out the answer to a problem using pencil and paper, then enter the answer. Checking the answers to such problems amounted to comparing the response to the correct answer.

The major innovation of Franklin's system was the "rolling" multiple choice problem. The question was presented along with one of the possible answers. If that answer was rejected it was removed and the next possible answer was shown. The student continued replacing old possibilities with new ones until a satisfactory answer appeared. However, no answer could be viewed or chosen once it was rejected. This feature defeated the common practice of reviewing all of the answers to select the most likely answer and required the student to know absolutely which answers were correct and which were incorrect.

At Georgia, the designers were convinced that a more intelligent testing system was within the scope of modern technology. Specifically, it should be possible to have the student type in an expression and have the computer analyze it to determine whether it was an acceptable answer to the question posed. For example, if the problem was to factor a quadratic expression in x, the student could type in the answer using the normal keys on a computer keyboard. The process of determining whether it was a correct factorization would first check to see if the student's response was functionally identical to the correct answer, then to see that it was factored.

Determining whether two expressions are functionally identical can be a difficult or impossible problem, depending on the class of functions allowed, but the class of functions that is normally encountered in elementary mathematics courses is so small that a practical test of identity is not difficult to construct. To determine whether an expression is factored we compare the two factored expressions, the student response and the correct answer, to determine whether their factors are the same. Using procedures of this sort, we were able to produce grading procedures for common problems.

Using a database of problems causes certain difficulties. The database has to be large enough to ensure that the questions do not get passed around by the students, or the tests would be compromised. If it is large enough to avoid such problems, then it is difficult to construct and takes up a large amount of space. A better method of preparing a large number of problems is to use algorithms that generate specific problems using randomly chosen values for parameters. One algorithm can generate hundreds or thousands of individual problems with unique answers and it could be designed so that a single procedure would be used for grading. Since between one and two hundred algorithms would suffice for a course, the problem database could be reasonably small and easier to construct. The major difficulty with this approach is that each algorithm has to be individually coded, requiring programming expertise in addition to the ability to make up good mathematics problems.

Another feature of Franklin's system that did not attract us was the use of a mainframe computer. Not only was there a large cost associated with setting up such a system (or even renting space on an existing computer), there were the problems of access and reliability. When a large machine is used heavily, the response can become unacceptably slow, making it almost impossible to arrange timed tests. Also, when a mainframe computer goes down, all of the testing system goes down with it. For these two reasons, we decided to try to create the system on microcomputers, which offered the advantages of low cost and freedom from catastrophic breakdown.

The major problems associated with developing a large system on microcomputers turned out to be the speed and graphics capacity of the machine and the existence of an easy to use but powerful programming language. The display of algebraic expressions called for graphics displays, and the processing of graphics causes most small machines to slow down considerably. Luckily, we were able to develop a system that was acceptably fast using assembler language on a PC clone. But most smaller machines, such as the Apple II series, Commodore, etc. were too slow or offered unacceptably low graphics resolution to be useful.

The other problem was the choice of a programming language. While the developers were competent in several languages, including assembler, we were planning to code the algorithms for the problems, and hoped that this task could be distributed among several people, not all of whom could be assumed to be so competent. At that time, BASIC was not well suited to system development because it was not a procedural language and had a 64K code and data restriction. Turbo Pascal 3.0 was much better because, although it had the 64K size limitation, it allowed the use of overlays and, using locally developed hooks into the operating system, we were able to have a Turbo Pascal program load and execute another program. Also, Turbo Pascal was becoming common and had an easy to use programmer interface. It took about two years to reach the limits of the language, and development slowed down considerably before the introduction of Turbo Pascal 4.0.

At the present time, major portions of the system are being rewritten in C both for speed and portability. An additional benefit of this conversion is the existence of many highly competent professional C compilers that can push the PC to its limits. However, the problem algorithms are still being written in Pascal for ease of support and because C will never be a common language.

A number of other problems came up and had to be solved in the course of the development of the system. Some were connected with the sheer mass of students that had to be tested, some were associated with the general problems of computer support and repair, and some have to do with the mechanization of the processes that we as teachers perform one at a time for our students. Here are three examples of such problems and a brief description of the solution.

1. In testing 1000 students per week on 35 computers, it is necessary to have the computer lab open at virtually all hours. Instructors cannot be expected to be in the room at all of these hours, so it was necessary to hire student assistants to man the labs. As long as the system was designed to be simple to use and had good security measures, this system proved workable.

2. Computers do break down. Whereas a system designed to be run on micros can sustain the breakdown of a testing computer, the file server must be working at all times. By having a spare machine available and keeping good backups of data, it is possible to avoid long delays or disasters. However, it is still helpful to have a ready source of computer repair expertise. We discovered that routine repair tasks on PC clones can be performed by untrained people, so equipment repair became a minor issue.

3. When we test in class, we regularly accommodate students with disabilities. With a computer testing system, such accommodation has to be built in to the system. We designed the system to allow the operator to create a printed version of a test so that students with special problems can be served in a more traditional way.

Because we were introducing technology to replace a traditional teaching function, it was inevitable that the question of suitability arise. Our colleagues immediately asked whether the system can test students as well as traditional methods. The best answer to that question is the personal testimony of teachers who have used the system and can compare it with traditional testing methods. After two years of experience, their testimony is that the system is as effective at testing students as the traditional method and, in some ways, does a better job.

However, complete evaluation of the system is more complex because it does more than just replace a traditional function of the teacher, it fundamentally alters the teaching environment. For example, it is designed to be used by students at their convenience, so all the students in a class do not take tests at the same time. This makes it possible for students who take the test later to gather information on the tests that have been given to their friends. Does this availability of extra information compromise the test? Also, since there is quite a bit of information about tests floating around, do the students concentrate on learning how to pass the tests and not on learning mathematics?

We have explored one statistic that sheds light on these questions. We have some data on the grades of students who took a computer graded precalculus course (all tests, including the final exam, given on the computer) and those taking a traditionally graded precalculus course along with their grades in the subsequent first quarter of calculus. After normalizing the precalculus course for grade average (in non-computer graded courses, class averages differed considerably), both groups had the same expected performance in calculus. That is, the computer graded precalculus student can expect his calculus grade to be .79 lower than his precalculus grade, while the student having the traditionally graded course can expect a drop of .78. This statistic indicates that the computer testing system did not promote rote learning of exam questions to a greater degree than the traditional testing system.

The system is being extended, expanded and continually evaluated, and has now been installed in other schools. Short term development goals include enhancing the system so that it may be used in calculus courses and looking at applications of the system in schools.

Using Technology to Implement a Constructivist Approach to Calculus and Abstract Algebra

John Selden and Annie Selden
Department of Mathematics, Tennessee Technological University, Cookeville, TN 38505

This is a report on two yearlong experimental courses and an analysis of the technology needed. The research aim of the project is to gather information useful in the development of future courses that produce students who can use their knowledge in flexible and creative ways to solve nonroutine problems. In a study of first calculus students done in Spring 1988, we concluded that average students coming from small, well-taught traditional classes cannot work nonroutine problems [2].

In a traditional calculus course many students rely heavily on remembering sample solutions. This produces inflexible students, unable to understand problems and applications they have not previously seen. In addition, such students are often deficient in conceptual understanding, remember the material poorly, and tend to find calculus boring.

The courses described here are meant to be similar to traditional ones in content, credit, and manpower requirements, while allowing for a constructivist approach to their teaching. By a constructivist approach we mean that the teacher observes and uses a student's conceptual base in guiding his construction of further concepts and skills [5]. The opportunity to understand a student's thinking arises naturally in one-on-one tutoring but is uncommon in traditional lecture courses.

DESCRIPTION OF THE COURSES. Although there are many possible ways of implementing constructivist ideas, most would involve considerable interaction between the teacher and the students. In our approach, students present their own solutions to problems in class. It is best if the problems are *cognitively nontrivial*, i.e., if the students have not previously seen solutions to essentially similar problems. For a discussion of problem solving in this sense, see [1]. Since adding a significant amount of this sort of work to a traditional lecture course could more than double its length, we decided to remove many of the explanations, sample solutions to problems, etc., and instead, converted this material into problems for the students to solve. Students must explain and justify their solutions when presenting them to the class and we provide criticism.

In both first calculus and graduate algebra our goal is not only that students should learn the usual course content but also that they should be able to use it effectively in new contexts. In calculus this means students should be able to solve concrete, but cognitively nontrivial problems and provide arguments for the validity of their solutions. In graduate algebra, students should be able to solve problems and prove theorems not covered in the course. Often these goals are not met by the traditional lecture method. Lecturing can work well for more sophisticated students in graduate algebra, but more naive students often obtain only superficial knowledge which they are unable to apply. For an analysis of misconceptions and errors made by undergraduate abstract algebra students when taught in this way, see [3].

We give very few lectures; most arise naturally from student questions or our critique of student work. Classroom time is spent in student presentations and our criticisms thereof. Students are expected to solve problems before class, not extemporaneously. Only rarely do we lead the class or an individual student through a problem as this tends to produce an unjustified illusion of success on the part of both teacher and student. Often we rewrite a student's presentation in a clearer way, explaining why our exposition is better. The emphasis here is on obtaining a presentation clear enough for anyone to judge its correctness in the absence of sample solutions with which to compare it.

CLASS ORGANIZATION. In order that all students be sufficiently challenged, we first call on those students whom we think likely to have solved a particular problem with the greatest effort. We also call on students who have not yet presented much, especially when we consider a problem relatively easy. For each student, we record those problems which he has solved correctly but make no record of his incorrect attempts. In graduate algebra, there are no tests, and grades are set according to the difficulty of the problems a student can work. In calculus, students are divided into groups of 5 or 6 and encouraged to work together. Points are assigned to the whole group when any member presents a correct solution. This not only enables us to handle 30 students but also encourages students to articulate and evaluate their ideas before presenting them in class. In calculus, we also have tests, but each test has fewer problems than normal because we never ask problems which have occurred before and each test covers the entire course to date. On tests students are encouraged to use their notes, as well as computational devices. Presently each student has an HP-28S calculator; we hope to add 30 PCs next year.

THE TECHNOLOGY NEEDED. Both students and teachers need technological support for such courses. Students attempting to solve a problem without a sample to mimic will often want to make computational experiments. Certainly our calculus students use their HP-28S calculators in this way and they could use more computing power. In addition, since this sort of teaching is slow, it is useful to streamline the syllabus. There are several topics in calculus which can be covered less intensively provided the computing tools used are always available to the students. For this reason inexpensive technology is especially interesting to us.

While we do not now use computational technology for our algebra course we think it would be useful. It should not be difficult to adapt even the HP-28S to operate on a large variety of algebraic structures.

An altogether different technological requirement of students in such courses involves the notes. Notes are essential because standard textbooks contain sample solutions and cannot be continuously adjusted to the progress of the students. We envision the notes as a problem solving tool, more like a computer manual than a textbook. For efficient use they should be concise and cross-indexed. In addition, they should be expandable so students can add sample solutions, proofs, comments, etc. from classwork without destroying the organization. We now write notes by hand on only one side of each page and number pages decimally, e.g., pages 7, 7.1, 7.11, 7.2, 8. We are careful that items such as theorems, examples, etc. not contain page breaks. We number items by pages, e.g., 7E5 is an example (hence, the E) which is the 5th item on page 7.

The details of the organization of the notes are important as these can effect the psychological accessibility of the information. This can be illustrated with a classroom experience. On the first try, none of our students could work: 37E2 *Show that* $f(x) = sin\ x$ *is continuous at all numbers, a.* On the other hand, adding / 22T2, 37T1 (refering to a list of trig identities and an alternate form of continuity) produced five or six volunteers, one of whom worked the problem correctly. (Here E refers to "example" and T refers to "theorem".) Well organized notes can help to gradually free students from the need for hints.

Ideally the notes should also be available to students in a hypertext version, i.e., presented on a computer in a way so that relevant definitions, theorems, etc. can be called up quickly when needed to help solve a problem [4].

The need to produce notes which correspond to the sophistication of the students and the style of the teacher calls for a technological aid, as the production of notes by hand is very labor intensive. What is required here might be called a *dynamic book*, i.e., a large master set of notes in TEX on disks from which the teacher could select on a weekly or daily basis. Such a dynamic book should also contain (1) a specialized editor to facilitate the teacher's selection and supplementation of the notes, (2) adequate cross-referencing to encourage the use of the notes as a problem solving tool, and (3) provision for insertion of student solutions as they are developed.

INITIAL STUDENT REACTION. We recently distributed a questionnaire to gauge initial student reaction to the calculus course. We found students to be generally pleased with the course. Most favored working in small groups. Some would prefer more explicit instruction on the use of the HP-28S. A few are computer oriented and know as much about the calculator as we do.

We have a graduate assistant assigned to the course whose job it is to observe the class and make a written transcript of what occurs. Not surprisingly, he has observed that some students take rather complete notes, whereas others take hardly any. Eventually we want to examine the relationship between student note-taking and grades, but it seems premature now as the course has just begun. Another observation the assistant made was that despite student presentations being central to the course, the teacher still spends half of the time talking. This includes time spent in course organization and explanations of calculator use, as well as actual critiques of student work and an occasional mini-lecture. This suggests that active student participation in the form of student questions and boardwork in a traditional lecture course must be rather small. It also suggests that the course is not such a radical departure from the traditional one as might be supposed.

REFERENCES

1. Schoenfeld, Alan H., *Mathematical Problem Solving*, Academic Press, Orlando, 1985.

2. Selden, John, Alice Mason, and Annie Selden, "Can Average Calculus Students Solve Nonroutine Problems?", preprint.

3. Selden, Annie, and John Selden, "Errors and Misconceptions in College Level Theorem Proving," *Proceedings of the Second International Seminar on Misconceptions and Educational Strategies in Science and Mathematics*, Cornell University (July 1987), 457-470. Available through ERIC.

4. "Special Issue: Hypertext," *Communications of the ACM*, Vol. 31, No. 7 (July 1988).

5. von Glasersfeld, Ernst, "Learning as a Constructive Activity," *Problems of Representation in the Teaching and Learning of Mathematics*, Lawrence Erlbaum Associates (1987), 3-17.

THE EMERGENCE OF A
MATHEMATICS LABORATORY

Lester Senechal

Department of Mathematics,
Statistics, and Computation
Mount Holyoke College

1. Beginnings. Three years ago my department was little involved with computation, although some of us were teaching computer science outside of the Department. This was the situation in spite of heroic efforts on the part of a few of my colleagues to increase our involvement, first with a connection to Dartmouth's Kiewit Center in the 1960's, later to the CYBER at the University of Massachusetts. Most of us, I think, were quite put off by the frequent exaggerations and misrepresentations concerning the uses of computers in classroom teaching.

For reasons which are difficult to recall, much less to document, a sudden change of attitude occured about four years ago. Certainly the conference in computer algebra, which was held at the Courant Institute in the spring of 1984, was influential. At about the same time, one of my colleagues produced a commercially distributed calculus package for classroom demonstrations and student experimentation. In any case, in the annual report to the President of the College, dated 10 July 1985, there appears the following:

> The Department as a whole and especially its more computer science oriented members would like to see computers become the laboratory instruments of mathematics.

On the surface, there is nothing very revolutionary in such an idea. Computers have been, and will continue to be, used extensively for mathematical computation. Our notion, however, is a pedagogical one which seems not to have been tried before: we would like all our students to be involved in experimentation to the same extent that a chemistry student might be. For our majors, and especially our thesis students, we would like to run a summer program which emphasizes computer experimentation.

In the course of just three years, these ideas have come to fruition, with a degree of success which we couldn't have anticipated. The _ability_ to accomplish it derived from a wonderful diversity unusual in a mathematics department. Its faculty includes two applied statisticians, a physicist, and a mathematician who serves as chairman of the computer science program. It is also a department which is both research-active and innovative in its teaching, even at the most elementary level.

2. The Quantitative Reasoning Laboratory. "Case Studies in Quantitative Reasoning" was the first course to focus on computer experimentation. Developed with support from the Sloan Foundation through its New Liberal Arts program, it is an interdisciplinary course in data analysis and mathematical modelling. It was taught for the first time in the spring of

1987 and has been taught each semester since by a group of faculty whose membership changes from semester to semester. Presently, the course consists of of three modules:

I. "Salem Village Witchcraft" concentrates on data analysis in exploring the relationships between wealth and power as reflected in historical records from the seventeenth century.

II. "Aptitude and Achievement" concentrates on hypothesis testing, experimental design, and modelling. It uses blind data on recent graduates of Mount Holyoke, including aptitude and achievement test scores, grade point averages, and choices of majors.

III. "Population, Resources, and Disease" treats case studies drawn from the sciences, particulary ecology.

All students attend a large common lecture, smaller discussion sessions, and a weekly two-hour laboratory. The laboratory room houses ten MacIntosh Plus's, arranged along the walls of a modest-sized room. We have found this arrangement of the computers to be good for allowing for minimum distraction when students are using the computers, but forcing students to turn away from the machines when the lab instructor wants to have the attention of the class.

The first two units of the course use StatView software, which is an excellent package for hypothesis testing and statistical inference. The data sets are fairly large and are kept in files which can be pulled down from the software menu. Typical questions for study in these units are: "In New England in the period 1630-90, were women more likely than men to be convicted of witchcraft?" and "Is the mathematics SAT score a good predictor of performance in the freshmen year by students who subsequently elect science majors?" The fact that the data sets are real and lend themselves to questions in which our students have a natural interest contributes to the excitement of the course.

For the third unit, which deals with dynamical systems, we use Stella software. Here population growth is studied under a variety of assumptions. Typical instructions for the laboratory are: "Double the rate of food production and run the simulation. What happens to the carrying capacity of the environment? Change the death rate so that it rises more rapidly as food per individual declines and run the simulation. What happens to the population in the long term?"

The course serves a wide variety of students, most of whom will not subsequently study calculus and many of whom would otherwise have avoided coming to grips with quantitative issues. More than half are freshmen. Some will subsequently do further work in statistics, their motivations and abilities strengthened through the experience of this course. As is the case in many situations where much learning occurs, students consider the work of the course very demanding. However, some of the students who have completed the course regard it as being a turning point in their educational experience. One student commented: "It made all the difference in my study of biology and economics."

3. "Calculus in Context". This is an NSF-funded effort to improve the teaching of calculus. The project is in its infancy, so it would be premature to try to spell it out in much detail. I will simply quote from the Project Summary, as it appeared in the proposal:

> We will develop a calculus sequence in which the major mathematical concepts grow out of attempts to explore significant problems from the social, life, and physical sciences. Throughout the courses we will stress four mathematical themes: optimization, estimation and approximation, differential equations, and functions of several variables. The computer will enter as a basic conceptual device for structuring the way we think about problems and what it means to solve them.

Here we use software we created ourselves to meet our needs efficiently and precisely. As do StatView and Stella, it provides a set of tools which, like laboratory instuments, must be manipulated by the students in order to obtain measurements. It is menu-driven and runs on IBM-PC's with CGA graphics. But here, far from having any physical facility which could properly be called a laboratory, we are presently dependent on ten machines scattered across campus, two of which are in faculty offices.

Students begin the semester by obtaining plots of functions and estimating the location of local extrema. The next set of computer problems involves arc length calculuations, which is gives a nice introduction to the idea of approximation and paves the way technically for integration problems, which comprise the next topic.

I will try to give some idea of how integration works. We have control over two points, a "kept" point (X0,Y0) and a "variable" point (X,Y). Initially these coincide, but the location of the variable point is controlled by "L" (left) and "R" (right) keystrokes, which decrease or increase X by an amount H, which is also under our control. A "K" keystroke brings the kept point to the location of the variable point.

To integrate by the trapezoidal rule, we define an operator

$$0.5*(Y+Y0)*(X-X0) + AC,$$

where AC represents an "accumulator". An "O" keystroke performs the operation. Beginning with both X and X0 equal to the left endpoint of the interval of integration, we perform the succession of keystrokes R-O-K sufficiently many times to traverse the interval of integration. There is a also a programming feature which permits the succession of keystrokes to be done automatically.

Next we make estimates of derivatives, using central differences, and the inverse relationships between integration and derivation are explored, all the while computationally and visually.

The final topic in the first semester is differential equations, for which we have also developed software capable of handling systems of up to four equations. In particular, we are easily able to treat the logistic equation and the Lotka-Volterra equations in the first semester.

4. The Laboratory in Mathematical Experimentation. Another NSF-funded project, the Laboratory is intended to encourage students to come to grips with mathematical ideas through experiment, simulation, and conjecture. Subsidiary objectives are to ensure that students are competent in the use of computers; to foster precise thinking and writing; to introduce students to the rich interplay between technology and theory; and to interest more students in continuing their study of mathematics by awakening them to the intellectual liveliness and utility of the discipline.

The course is intended primarily for freshmen and sophomores and will be offered for the first time in Spring, 1989. Novices will be expected to learn enough Pascal to implement the algorithms. In the first offering, we plan to deal with the following topics: chromatic polynomials; the generation and the detection of randomness; distributions of primes; iteration of maps; the Mandelbrot set.

5. A Seminar in Software. In order to stimulate further innovation in computer experimentation, we are presently designing a weekly seminar, which will be organized cooperatively by Amherst, Hampshire, Holy Cross, and Mount Holyoke Colleges. Additional colleges and universities will participate. There will be weekly meetings and an annual conference with national participation, much in the spirit of the Nonlinear Dynamics Seminar at the University of Texas.

We will invite to the weekly meetings people who are doing interesting things in the way of curricular and software development. It is clear from our experience that there is no substitute for <u>seeing</u> what a software package can do, in the presence of someone who knows its capabilities. The seminar will provide a forum for the exchange of ideas on how computers can profitably be incorporated into the undergraduate curriculum, with high priority given to ways of making more mathematics accessible to students, especially at the most elementary level.

6. Why Experiment? One reason is that it's fun. Another is that the computer laboratory offers teachers of mathematics a nice way of bridging the enormous gulf between their training as mathematicians and the mathematical level of their students as they emerge from the high schools. Our experience so far seems to indicate the possibility of teaching mathematics to a vastly larger proportion of the college population. Professor James Yorke of the University of Maryland even foresees the possibility of a ten-fold increase in the number of students who study calculus.

At Mount Holyoke, we are in competition for majors with the other sciences. We found that the lack of a laboratory put us at a considerable disadvantage in certain important respects. For instance, laboratories are places where students spend time and where they socialize.

Also, computer laboratories provide a viable way of extending science education to a broader segment of our students, giving them a much better chance of being able to handle the mathematics which is necessary for science. Most important, computer laboratories are places where students can become actively involved in doing mathematics.

COMPUTATION ERRORS AND THE CALCULUS STUDENT
by Peter Shenkin
John Jay College of Criminal Justice

Calculus is most often taught in what we shall call the 'classical mathematical manner.' By this we mean that a theory is constructed in which the existence of solutions (or lack of solutions) to certain general types of problems is proved or demonstrated. Examples generally illustrate the given result almost exactly with 'closed form' solutions the rule. It seems that for the most part both the topics taught and the methods of teaching have not changed much for the past 20 (50) years. The advent of handheld calculators and personal computers present an opportunity to modernize the ways in which the concepts of calculus are taught and possibly to modify the core concepts. We feel that these opportunities have not yet been adequately exploited. It is true that many calculus texts have added problems which can be solved by calculator or computer, but for the most part these problems are not integrated into the course itself. The prospect for change is hopeful. Several authors have already written software packages for calculus courses and there are several independently written commercial packages such as those written by John Kemeny at True Basic, Inc. This paper discusses results from several programs and software packages developed at the John Jay College and currently being used in calculus and precalculus sections. In particular, we shall give special emphasis to errors which may occur when computers are used to 'prove' and evaluate results.

An obvious application where computers may be used is in the estimation of limits. It is quite easy, for example, to write a program in BASIC, PASCAL or some higher level language which will permit the student to guess (prove) what the limit of a certain function at a certain point is by evaluating the function at several arguments getting nearer and nearer to the desired point (or just getting larger and larger if the limit is at infinity.) The student may see that the value seems to be approaching some limiting value. Even though this does not actually show that a limit exists we feel that a computer demonstration of how a sequence seems to approach some value is a very valuable way of demonstrating just what a limit is.

There are problems with this numerical approach, however. The particular type of problem we would like to talk about in this paper involves various types of errors which are generated when using computers for computations in calculus course. If these errors are not considered then computer results can be misleading and even entirely incorrect.

A typical example is the evaluation of the limit of $(1+1/x)^x$ as x goes to infinity. From classical calculus we know that the limit is e or about 2.718281828459045. The results from the programs ESINGS.BAS, and ESINGERR.BAS displayed in the appendix were calculated using Microsoft's QUIKBASIC 4.0 and show an interesting result. If we keep x below about 2500 then it seems that as x gets large our function is approaching 2.71.... So the student in effect 'proves' what the limit is by looking at a finite number of terms. This is certainly not correct in the classical mathematical sense but to most students this seems like going to the limit. In this particular case a seemingly (to the student) strange thing happens x is increased further and further. At first it seems that there is no limit. Then we reach a new limit. This new limit seems to be 1. How did this occur?

In actuality, BASIC was only holding 7 significant digits of every number. $1+1/x$ eventually becomes 1 (in the computer) as x increases and 1 to any power is 1. Thus just about all significance was lost. Pity the poor student who 'proved' that the limit is 1 via this example. Notice that in ESINGERR.BAS we managed to display more than 7 digits even though BASIC only supports 7 in its 'single precision' mode. Even the values of $1 + 1/x$ are incorrect in ESINGERR.BAS as the $1 + 1/x$ ERROR column indicates. Note, also from ESINGERR.BAS that even though single precision gives 7 digits of accuracy our estimate seems to be most accurate when x has only 4 digits (x = 2560).

EDOUB.BAS is the essentially the same program as ESINGERR.BAS with the exception that all computations are done in 'double precision' mode. This means that there are 15 or 16 digits of accuracy in values and computations. Notice how much closer our results come to the 'true' limit in this case. However, we do give a printout which shows that even with double precision, if we let x increase enough we get a 'limit' of 1.

The graphs of $(1 + 1/x)^x$ shown were generated by a graphics program described in "A Computer Companion for Undergraduate Mathematics" by Wieschenberg and Shenkin and graphically display the results shown in EDOUB.BAS. The graphics program was written in True BASIC. These graphs seem to indicate that $(1 + 1/x)^x$ approaches zero very rapidly (in the computer) for a value of x in the neighborhood of 9E15.

When the computer is used to numerically solve calculus type problems several types of errors may occur:

Roundoff error - Due to the 'discreteness' of the computer's number system. e.g. using QUIKBASIC single precision only 7 digits matter.

Truncation error - Occurs when a process requiring an infinite number of steps is terminated after a finite number of steps, e.g. stop the above process for $x = 2560$ or truncate all Taylor series terms after the fifth term.

Propagated error - Error caused by error in some initial input, e.g. approximate PI.

Significance error - number of meaningful digits (significant digits) in an answer is less than expected.

Overflow and Underflow Error - Error caused when calculations, including intermediate calculations, get larger than the computer's infinity or closer to 0 than the smallest computer non-zero value.

We will look at the definite integral of $\exp(-x^2)$ to examine some of these types of error.

The program ETOXDBL.BAS (see Appendix) computes the definite integral of $\exp(-x^2)$ between 0 and some value input during the run of the program. The computations are done in double precision. A companion program, ETOX.BAS computes the same results using single precision arithmetic This program is not shown in the Appendix These programs use Taylor series approximations with a number of terms also chosen during the run of the program. The series for $\exp(-x^2)$ and the resulting integral approximation are convergent alternating series so the last term gives a good idea of the size of the error at least if enough terms are taken. If we look at some runs of the programs we see several things regarding error.

1. Truncation error can be a major factor. In fact look at the display showing an integral with limits from 0 to 3 with 13 terms has a value of over 70 for the last retained term. However, if the upper limit equals 1 the last term is on the order of 10^{-11}.

2. The classical truncation error bound (1st neglected term with alternating series of decreasing terms) is meaningless in single precision when these bounds yield values on the order of 10^{-9} while we are working with integral of order 1 with six or seven significant digits.

3. By comparing single and double precision printouts we see that roundoff error isn't severe but single precision is not necessarily accurate to 7 significant digits. Of course in the limiting process for e mentioned previously we saw that roundoff and significance error might be severe.

The results from the programs RTSSNG, RTSDBL compare the rectangular, trapezoidal and Simpsons rule for various interval widths. Note some of the following results:

1. The example printout from RTSSNG.BAS with $\exp(-t^2)$ as the integrand shows that, at least in the single precision case, increasing the number of subintervals does not necessarily improve the approximation.

2. This is again shown in the example where x^3 is integrated. Theoretically Simpson's rule should be exact here. However, look at the results as the number of subintervals increases.

3. We also show some results using the rectangular rule, the trapezoidal rule and Simpson's rule where we approximated e by 2.7. The results showed small changes in value all in the direction expected. This is an example of propagation error and does not seem to be serious in the integration problems we are tackling.

In conclusion, we feel that the student should be exposed to examples such as those described to aid them in results gotten by using a simple numerical rule as implemented on a computer may be more in error than theory as usually given in beginning calculus courses would indicate. The hands on, especially the interactive approach would probably be most useful to the greatest number of students.

```
1 REM etoxdbl.bas
2 REM PROGRAM BY A. WIESCHENBERG
3 CLS
4 PRINT
5 PRINT "THIS PROGRAM IS TO FIND THE "
5 PRINT "DEFINITE INTEGRAL FROM 0 TO X OF E TO THE NEGATIVE T SQUARED"
7 PRINT
10 REM I = NUMBER OF TERMS DESIRED
20 REM X# = TO FIND THE INTEGRAL FROM 0 TO THIS VALUE X
30 INPUT "NUMBER OF TERMS DESIRED"; I
40 INPUT "FIND THE VALUE OF THE INTEGRAL FROM 0 TO:"; X#
50 T# = X#
60 S# = X#
70 K# = 0#
80 PRINT
90 PRINT "TERM          TERM VALUE              SUM OF TERMS"
100 PRINT "-------------------------------------------------------"
110 PRINT K#; TAB(13); T#; TAB(40); X#
120 FOR N = 1 TO I - 1
130    K# = K# + 1#
140    TN# = (T# * X# ^ 2# * (2# * K# - 1#)) / ((2# * K# + 1#) * K#) * (-1# ^ K#)
150    S# = S# + TN#
160    T# = TN#
170    PRINT K#; TAB(13); T#; TAB(40); S#
180 NEXT N
190 END
```

```
                  THIS PROGRAM IS TO FIND THE
DEFINITE INTEGRAL FROM 0 TO X OF E TO THE NEGATIVE T SQUARED
NUMBER OF TERMS DESIRED  13
FIND THE VALUE OF THE INTEGRAL FROM 0 TO:  1

TERM        TERM VALUE              SUM OF TERMS
-------------------------------------------------------
0        1                         1
1         -.3333333333333333        .6666666666666667
2         .1                        .7666666666666667
3         -2.380952380952381D-02    .7428571428571429
4         4.629629629629629D-03     .7474867724867725
5         -7.575757575757576D-04    .7467291967291968
6         1.068376068376068D-04     .7468360343360344
7         -1.322751322751323D-05    .7468228068228069
8         1.458916900093371D-06     .7468242657397069
9         -1.450385222315047D-07    .7468241207011848
10        1.312253296380280-08      .7468241338237177
11        -1.089222103714857D-09    .7468241327344955
12        8.350702795147238D-11     .7468241328180025
```

```
                  THIS PROGRAM IS TO FIND THE
DEFINITE INTEGRAL FROM 0 TO X OF E TO THE NEGATIVE T SQUARED
NUMBER OF TERMS DESIRED  13
FIND THE VALUE OF THE INTEGRAL FROM 0 TO:  3

TERM        TERM VALUE              SUM OF TERMS
-------------------------------------------------------
0        3                         3
1         -9                        -6
2         24.3                      18.3
3         -52.07142857142857        -33.77142857142857
4         91.125                    57.35357142857143
5         -134.2022727272727        -76.84870129870129
6         170.3336538461538         93.48495254745254
7         -189.8003571428571        -96.31540459540654
8         188.4047662815126         92.08936168610802
9         -168.5726856203007        -76.4833297541927
10        137.2663297193877         60.78300578519503
11        -102.5428312923489        -41.75982550715389
12        70.75455359172075         28.99472808456686
```

RESULTS FROM RTSSNG.BAS
THESE ARE SINGLE PRECISION COMPUTATIONS

```
INTEGRAL OF EXP(-X^2) FROM 0 TO  1
NUMBER OF SUBINTERVALS =          10
LENGTH OF ONE SUBINTERVAL         .1

INTEGRAL BY RECTANGULAR RULE = .7778168
INTEGRAL BY TRAPEZOIDAL RULE = .7462108
INTEGRAL BY SIMPSON'S RULE   = .746825
```

```
INTEGRAL OF EXP(-X^2) FROM 0 TO  1
NUMBER OF SUBINTERVALS =          100
LENGTH OF ONE SUBINTERVAL         .01

INTEGRAL BY RECTANGULAR RULE = .7499784
INTEGRAL BY TRAPEZOIDAL RULE = .7468178
INTEGRAL BY SIMPSON'S RULE   = .7468241
```

```
INTEGRAL OF EXP(-X^2) FROM 0 TO  1
NUMBER OF SUBINTERVALS =          1000
LENGTH OF ONE SUBINTERVAL         .001

INTEGRAL BY RECTANGULAR RULE = .7471396
INTEGRAL BY TRAPEZOIDAL RULE = .7468236
INTEGRAL BY SIMPSON'S RULE   = .7468238
```

```
INTEGRAL OF EXP(-X^2) FROM 0 TO  1
NUMBER OF SUBINTERVALS =          2000
LENGTH OF ONE SUBINTERVAL         .0005

INTEGRAL BY RECTANGULAR RULE = .7469828
INTEGRAL BY TRAPEZOIDAL RULE = .7468247
INTEGRAL BY SIMPSON'S RULE   = .7468238
```

```
INTEGRAL OF EXP(-X^2) FROM 0 TO  1
NUMBER OF SUBINTERVALS =          4000
LENGTH OF ONE SUBINTERVAL         .00025

INTEGRAL BY RECTANGULAR RULE = .746903
INTEGRAL BY TRAPEZOIDAL RULE = .746824
INTEGRAL BY SIMPSON'S RULE   = .7468243
```

```
REM RTSDBL.BAS
REM RECTANGULAR, TRAPEZOIDAL, SIMPSONS'S RULE
REM DOUBLE PRECISION COMPUTATIONS
REM SHENKIN, WIESCHENBERG  01/06/88
DEFINT N
DEFDBL M, R-Z
DIM V(5000)
DEF FNF# (X) = EXP(-X ^ 2)
CLS
PRINT "INTEGRAL OF EXP(-X^2)"
PRINT "USING RECTANGULAR, TRAPEZOIDAL, & SIMPSON'S RULE "
PRINT
INPUT "DO YOU DESIRE PRINTED OUTPUT(Y/N) "; PRTS
INPUT "ENTER NUMBER OF SUBINTERVALS DESIRED (EVEN PLEASE) "; N
INPUT "THE INTEGRAL WILL BE FROM 0 TO ?          "; X

M = X / N        'M = STEP SIZE

FOR I = 0 TO N
    V(I) = FNF#(I * X / N)
NEXT I

FOR I = 0 TO N
    IF I <> N THEN
        RN = RN + V(I)
    END IF

    IF I = 0 OR I = N THEN
        TN = TN + V(I) / 2
    ELSE
        TN = TN + V(I)
    END IF

    IF I = 0 OR I = N THEN
        SN = SN + V(I)
    ELSEIF I / 2 <> INT(I / 2) THEN
        SN = SN + 4 * V(I)
        ELSEIF I / 2 = INT(I / 2) THEN
            SN = SN + 2 * V(I)
    END IF
NEXT I
RN = RN * M
TN = TN * M
SN = SN * M / 3

PRINT
PRINT "INTEGRAL BY RECTANGULAR RULE = "; RN
PRINT "INTEGRAL BY TRAPEZOIDAL RULE = "; TN
PRINT "INTEGRAL BY SIMPSON'S RULE   = "; SN

IF UCASES(PRTS) = "Y" THEN
    LPRINT "INTEGRAL OF EXP(-T^2) FROM 0 TO "; X
    LPRINT "NUMBER OF SUBINTERVALS = "; N
    LPRINT "LENGTH OF ONE SUBINTERVAL       "; M

    LPRINT
    LPRINT "INTEGRAL BY RECTANGULAR RULE = "; RN
    LPRINT "INTEGRAL BY TRAPEZOIDAL RULE = "; TN
    LPRINT "INTEGRAL BY SIMPSON'S RULE   = "; SN
    LPRINT
    LPRINT
    LPRINT
END IF
```

RESULTS ARE FROM RTSDBL.BAS
THESE ARE DOUBLE PRECISION COMPUTATIONS

```
INTEGRAL OF EXP(-T^2) FROM 0 TO  1
NUMBER OF SUBINTERVALS =          10
LENGTH OF ONE SUBINTERVAL         .1

INTEGRAL BY RECTANGULAR RULE = .7778168240731773
INTEGRAL BY TRAPEZOIDAL RULE = .7462107961317495
INTEGRAL BY SIMPSON'S RULE   = .7468249482544435
```

```
INTEGRAL OF EXP(-T^2) FROM 0 TO  1
NUMBER OF SUBINTERVALS =          100
LENGTH OF ONE SUBINTERVAL         .01

INTEGRAL BY RECTANGULAR RULE = .7499786042621125
INTEGRAL BY TRAPEZOIDAL RULE = .7468180014679697
INTEGRAL BY SIMPSON'S RULE   = .7468241328941758
```

```
INTEGRAL OF EXP(-T^2) FROM 0 TO  1
NUMBER OF SUBINTERVALS =          1000
LENGTH OF ONE SUBINTERVAL         .001

INTEGRAL BY RECTANGULAR RULE = .7471401317785986
INTEGRAL BY TRAPEZOIDAL RULE = .7468240714991844
INTEGRAL BY SIMPSON'S RULE   = .7468241328124359
```

```
INTEGRAL OF EXP(-T^2) FROM 0 TO  1
NUMBER OF SUBINTERVALS =          2000
LENGTH OF ONE SUBINTERVAL         .0005

INTEGRAL BY RECTANGULAR RULE = .7469821476238258
INTEGRAL BY TRAPEZOIDAL RULE = .7468241174841186
INTEGRAL BY SIMPSON'S RULE   = .7468241328124275
```

```
INTEGRAL OF EXP(-T^2) FROM 0 TO  1
NUMBER OF SUBINTERVALS =          4000
LENGTH OF ONE SUBINTERVAL         .00025

INTEGRAL BY RECTANGULAR RULE = .7469031440502019
INTEGRAL BY TRAPEZOIDAL RULE = .7468241289803483
INTEGRAL BY SIMPSON'S RULE   = .7468241328124258
```

```
INTEGRAL OF X^3 FROM 0 TO  1
NUMBER OF SUBINTERVALS =          10
LENGTH OF ONE SUBINTERVAL         .1

INTEGRAL BY RECTANGULAR RULE = .2025
INTEGRAL BY TRAPEZOIDAL RULE = .2525
INTEGRAL BY SIMPSON'S RULE   = .25
```

```
INTEGRAL OF X^3 FROM 0 TO  1
NUMBER OF SUBINTERVALS =          100
LENGTH OF ONE SUBINTERVAL         .01

INTEGRAL BY RECTANGULAR RULE = .245025
INTEGRAL BY TRAPEZOIDAL RULE = .250025
INTEGRAL BY SIMPSON'S RULE   = .25
```

```
INTEGRAL OF X^3 FROM 0 TO  1
NUMBER OF SUBINTERVALS =          1000
LENGTH OF ONE SUBINTERVAL         .001

INTEGRAL BY RECTANGULAR RULE = .2495003
INTEGRAL BY TRAPEZOIDAL RULE = .2500003
INTEGRAL BY SIMPSON'S RULE   = .2500001
```

```
INTEGRAL OF X^3 FROM 0 TO  1
NUMBER OF SUBINTERVALS =          2000
LENGTH OF ONE SUBINTERVAL         .0005

INTEGRAL BY RECTANGULAR RULE = .2497501
INTEGRAL BY TRAPEZOIDAL RULE = .2500001
INTEGRAL BY SIMPSON'S RULE   = .25
```

```
INTEGRAL OF X^3 FROM 0 TO  1
NUMBER OF SUBINTERVALS =          4000
LENGTH OF ONE SUBINTERVAL         .00025

INTEGRAL BY RECTANGULAR RULE = .2498751
INTEGRAL BY TRAPEZOIDAL RULE = .2500001
INTEGRAL BY SIMPSON'S RULE   = .2500002
```

```
REM ESINGS.BAS

LET h$ = "            x                     1 + 1/x           (1 + 1/x)^x"
LET f$ = " ###,###,###,### "
LET xmax = 1E+08

CLS
LPRINT "Problem: Evaluate e using (1+1/x)^x for various x"
LPRINT "         Calculate in single precision "
LPRINT

LPRINT h$
x = 10
DO WHILE x < xmax
   LET t1 = 1 + 1 / x
   LET t2 = t1 ^ x
   LPRINT USING f$; x; : LPRINT TAB(25); t1; TAB(47); t2
   x = 2 * x
LOOP
LPRINT CHR$(12)
```

```
REM ESINGERR.BAS

WIDTH "lpt1:", 132
LET h0$ = "
LET h$ = "            x                1 + 1/x              (1 + 1/x)^x              ERROR              ERROR    "
LET f$ = " ###,###,###,###      #.###############      #.###############      ##.##############      ##.##############"
LET xmax = 1E+08

CLS
LPRINT "Problem: Evaluate e using (1+1/x)^x for various x"
LPRINT "         Calculate in single precision but display excess characters"
LPRINT "         Machine double precision value of e equals ", EXP(1#)
LPRINT

LPRINT h0$
LPRINT h$
x = 10
x# = 10#
DO WHILE x < xmax
   LET t1 = 1 + 1 / x
   LET t2 = t1 ^ x
   LET t1# = 1# + 1# / x#
   LPRINT USING f$; x, t1, t2, t1 - t1#, t2 - EXP(1#)
   x = 2 * x
   x# = 2# * x#
LOOP
LPRINT CHR$(12)
```

Problem: Evaluate e using (1+1/x)^x for various x
    Calculate in single precision

x	1 + 1/x	(1 + 1/x)^x
10	1.1	2.593743
20	1.05	2.653295
40	1.025	2.685061
80	1.0125	2.701495
160	1.00625	2.709846
320	1.003125	2.714005
640	1.001562	2.71612
1,280	1.000781	2.717386
2,560	1.000391	2.717917
5,120	1.000195	2.717353
10,240	1.000098	2.717486
20,480	1.000049	2.720871
40,960	1.000024	2.720904
81,920	1.000012	2.707668
163,840	1.000006	2.707676
327,680	1.000003	2.761084
655,360	1.000002	2.761086
1,310,720	1.000001	2.553589
2,621,440	1	2.553589
5,242,880	1	3.490342
10,485,760	1	3.490343
20,971,520	1	1
41,943,040	1	1
83,886,080	1	1

Problem: Evaluate e using (1+1/x)^x for various x
    Calculate in single precision but display excess characters
    Machine double precision value of e equals   2.718281828459045

x	1 + 1/x	(1 + 1/x)^x	1 + 1/x ERROR	(1 + 1/x)^x ERROR
10	1.100000023841858	2.5937430858612060	0.0000002384185782	-0.1245387425978390
20	1.049999952316284	2.6532952785491940	-0.0000000476837158	-0.0649865499098508
40	1.024999976158142	2.6850614547729490	0.0000000236371586	-0.0332203736860959
80	1.012500047683716	2.7014951705932620	0.0000000476837158	-0.0167886578657834
160	1.006250023841858	2.7098457813262940	0.0000002384185782	-0.0084360471327511
320	1.003124952316284	2.7140054702758790	-0.0000000476837158	-0.0042763581831662
640	1.001562476158142	2.7161197662353520	-0.0000000476837158	-0.0021620622236935
1,280	1.000781297683716	2.7173864841461180	0.0000000476837158	-0.0008953443129269
2,560	1.000390648841858	2.7179169654846190	0.0000000236371586	-0.0003648629744260
5,120	1.000195264816284	2.7173531055450440	-0.0000000476837158	-0.0009287229140011
10,240	1.000097632408142	2.7174856662750240	-0.0000000236371586	-0.0007961621840207
20,480	1.000048875808716	2.7208712100982670	0.0000004768371586	0.0025893816392215
40,960	1.000024437904358	2.7209043502807620	0.0000000236371586	0.0026225218217166
81,920	1.000012159347534	2.7076678276062010	-0.0000000476837158	-0.0106140008528439
163,840	1.000006079753767	2.7076761722564700	-0.0000000236371586	-0.0106056562025754
327,680	1.000003099641528	2.7610840797424320	0.0000002384185782	0.0428022512833866
655,360	1.000001549720764	2.7610864639282230	0.0000000236371586	0.0428046354691776
1,310,720	1.000000715255737	2.5535886287689210	-0.0000000476837158	-0.1646931996901242
2,621,440	1.000000357627869	2.5535891056060790	-0.0000000236371586	-0.1646927228529660
5,242,880	1.000000238418579	3.4903423786163330	0.0000000476837158	0.7720605501572879
10,485,760	1.000000119209290	3.4903426170349120	0.0000000236371586	0.7720607885758670
20,971,520	1.000000000000000	1.0000000000000000	-0.0000000476837158	-1.7182818284590450
41,943,040	1.000000000000000	1.0000000000000000	-0.0000002384185782	-1.7182818284590450
83,886,080	1.000000000000000	1.0000000000000000	-0.0000001192092891	-1.718281828459045

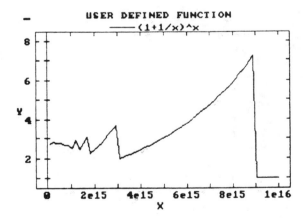

USER DEFINED FUNCTION

—— (1+1/x)^x

```
REM EDOUB.BAS

WIDTH "lpt1:", 132
LET h0$ = "
LET h$ = "            x                     1 + 1/x              (1 + 1/x)^x              ERROR    "
LET f$ = " ##,###,###,###,###,###      #.###############      #.###############      ##.###############"
LET xmax = 9E+16

CLS
LPRINT "Problem: Evaluate e using (1+1/x)^x for various x"
LPRINT "         Calculate in double precision"
LPRINT "         Machine double precision value of e equals "; EXP(1#)
LPRINT

LPRINT h$
x# = 10#
DO WHILE x# < xmax
   LET t1# = 1# + 1# / x#
   LET t2# = t1# ^ x#
   LPRINT USING f$; x#, t1#, t2#, t2# - EXP(1#)
   x# = 4# * x#
LOOP
LPRINT CHR$(12)
```

Problem: Evaluate e using (1+1/x)^x for various x
    Calculate in double precision
    Machine double precision value of e equals 2.718281828459045

x	1 + 1/x	(1 + 1/x)^x	ERROR
10	1.100000000000000	2.5937424601000020	-0.1245393683590428
40	1.025000000000000	2.6850638383899630	-0.0332179900690819
160	1.006250000000000	2.7098355763078150	-0.0084462521512298
640	1.001562500000000	2.7161612079478540	-0.0021206205111914
2,560	1.000390625000000	2.7177511040752590	-0.0005307243837862
10,240	1.000097656250000	2.7181491117321900	-0.0001327167268550
40,960	1.000024414062500	2.7182484670602920	-0.0000331813987531
163,840	1.000006103515625	2.7182735329280950	-0.0000082955309497
655,360	1.000001525878906	2.7182795474357190	-0.0000020737233264
2,621,440	1.000000381469726	2.7182813093552150	-0.0000005191038306
10,485,760	1.000000095367432	2.7182817013728750	-0.0000001270861700
41,943,040	1.000000023841858	2.7182817859282130	-0.0000000425308322
167,772,160	1.000000005960465	2.7182818608634900	0.0000000324064449
671,088,640	1.000000001490116	2.7182816644115500	-0.0000001640474951
2,684,354,560	1.000000000372529	2.7182824760416940	0.0000006475826488
10,737,418,240	1.000000000093132	2.7182792359781370	-0.0000025924809077
42,949,672,960	1.000000000023283	2.7182921978694360	0.0000103694103912
171,798,691,840	1.000000000005821	2.7182403510785500	-0.0000414773804955
687,194,767,360	1.000000000001455	2.7184477442764960	0.0001659158174512
2,748,779,069,440	1.000000000000364	2.7176182664385440	-0.0006635620205011
10,995,116,277,760	1.000000000000091	2.7209376971568440	0.0026558668697709
43,980,465,111,040	1.000000000000023	2.7076842519337600	-0.0105975765252855
175,921,860,444,160	1.000000000000006	2.7610885385500930	0.0428067100910483
703,687,441,776,640	1.000000000000001	2.5535894580629250	-0.1646923703961196
2,814,749,767,106,560	1.000000000000000	3.4903429574618610	0.7720611290027754
11,258,999,068,426,240	1.000000000000000	1.0000000000000000	-1.7182818284590450
45,035,996,273,704,960	1.000000000000000	1.0000000000000000	-1.718281828459045

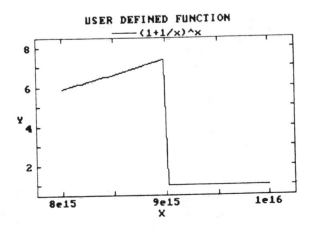

USER DEFINED FUNCTION

—— (1+1/x)^x

CAN WE "BOOT-UP" MATHEMATICS AND REMAIN PROFESSIONALLY HONEST?

LARRY E. SHERWOOD

Computer technology has worked its way into the calculus. Well, at least the majority of calculus texts have sections dealing with computer applications. Certainly, those sections can be skipped as we do many other sections we don't have time for or the inclination to include. However, it is no longer justifiable to skip Newton's Method, the Trapezoidal Rule or other numerical methods. One need not read journals or reports, nor try to stay abreast of the thinkers and doers to witness this intrusion into the traditional curriculum; it's staring us in the face. Therefore, the "new technology" will find its way into the curriculum in a kind of natural way and we can hope it evolves into a meaningful influence. So, relax!

Relax, only if you don't read the literature, review the software and distinguish the learning process from the extensions and applications of what is learned. Even the most casual reading of the literature can confuse. When is the computer an aid to instruction and/or its management and when is it an aid in problem solving? This confusion can be very legitimate. For example, iterative processes, tabulation, etc. can serve both purposes quite well. Assuming this confusion has substance beyond confession, can we incorporate the computer into the traditional mathematics curriculum without delineating these roles?

Even if I am the only one confused, don't reject what follows as a description of just a first step towards integration of this technology, for there may be something of value I am not aware of. The attempt to determine what my first step should be forced the realities of teaching at a community college to raise its challenging head. These realities are not as important as the awareness that I must make my bibliography clear: namely, (1) the general literature, (2) the extremes of the subject matter I teach – Basic Arithmetic through Differential Equations – in the same day, (3) the varied backgrounds, learning styles, expectations and epistemologies of my students at each curricular level and in each class, (4) myself – as a primary resource that I must contend with and satisfy in terms of my traditions and perceptions of mathematics education, and most importantly (5) my students! It was the latter which directed me to the first step in answering how I can incorporate computers into the traditional curriculum – other than applying for a grant and duplicating someone else's ideas, research and efforts.

I designed a course whose objective is to interact with the computer as it applies to the traditional calculus. The purpose was to be free from the presentation of reflexive and manipulative skills; discuss and solve the problems therein and see how the computer can facilitate, enhance, direct and redirect one's thought. Most important was my interest in the response of the student to this type of encounter. I embedded myself in this objective and this purpose, not by design, but necessity. I knew I could deal with the mathematics, but had less than minimal computer expertise. Guts or stupidity? I solicited and found seven calculus students - the best we had. The initial course offering had two Calc I students and five Calc II students. Three of the five Calc II students enrolled for the second half of the course concurrent to their enrollment in Calc III. The two Calc I students had a variety of reasons for not continuing. One reason was they felt unable to compete - not justified by their performance. One of these wants to continue, but I advise him to find a cohort, because I will maintain the structure of student presentation of the course and thus he will need someone to help in the struggle, not someone who knows the answers. Kind of R.L. Mooreish, but effective.

To prepare for my interlocutorship I purchased an IBM clone, researched the materials available and found a paperback by C.H. Edwards, Jr. entitled Calculus and the Personal Computer, which met my objectives. For two intensive weeks at fourteen hours per day, I worked through this book compiling a syllabus. I had located a good DOS manual, BASIC manual and help from the Data Processing Department.

The students understood we were in this together and are going to see how the computer fits into the traditional calculus - as a tool. Some had never turned on a computer. Some had more computer expertise than I, but it didn't take long for them to realize that magic fingers and fancy menus won't solve problems. This is a mathematics course with clear objectives. Why are you turning on the computer and what for? What does the program do and how do you know it is addressing the problem? What assumptions are you making? Do you see the convergence of discrete and indiscrete Mathematics? Do you even understand the problem? Can you modify the program to solve bigger problems? It went on and on. They responded superbly, reinforced their mathematical experiences, saw mathematics they had never seen, got frustrated and in spite of all this, I still ended up seeing some fancy menus and some programming tricks!

As an aside, it is of interest to note that I invited the Physics instructor to help us through some projectile problems, resulting in a lengthy discussion with the students. We argued at great length on what the author intended and our collective confusion. A true learning experience. To get this close to what education must be was exhilarating. In addition, I invited a member of the English department to instruct them on writing a term paper on their experiences. This experience with writing across the curriculum, (WAC), raised more questions than it answered - reason to pursue! These students had never experienced mathematics education in this diverse a way even though they came from all corners of the earth - the Middle East, the Far East and Middle America.

Two major conclusions surfaced from this experience: (1) to maintain this forum for A and B students enrolled in calculus as a legitimate problems course and testing ground for student reactions to educational innovation; and (2) to let these students take their experience back to the traditional classroom as the second step to incorporation.

The third and final step will be to convince the powers that be that all mathematics courses must become computer laboratory-type courses in which the students have guided assistance to the problems assigned. Even the best students are under the illusion that if you look long enough you can find a program which, at one keystroke, will solve the problem. As the technology evolves this is certainly going to be the case. For example, to find the derivative of a function or to find the tangent line to a curve which passes through a point not on the curve can both be reduced to this technological solution. The latter problem, I feel, is the first non-trivial problem the student encounters in calculus, for no other reason than that they will solve the problem as if the point lies on the curve and will generate an answer, submit their solution in great confidence and think they have been duped when they receive no credit. The mathematics involved here is difficult for the student and must not be hidden by technology. As a mathematics educator I feel obliged to devote my time to the interaction of the mathematics and the technology used to solve such problems and let economy and efficiency of solution be discovered by the student rather than presented by the instruction. Yet, the instructor must maintain a watchful eye that this discovery is eventually made. This is the "art" of teaching.

A final word of caution to those convinced of the viability of the above course. No matter how 'good' your students are they will complain of the time required by such a course and may not continue in light of their required studies. This is why you must choose these students and not open it up to all comers. Furthermore, don't let the powers that be force you into making it a bigger and popular course. The bottom line in such a course is for you, as an instructor of traditional courses, to discover student responses which will facilitate incorporation of what is learned into the traditional course.

In conclusion, bring on the new technology, provided each of us mathematics educators have a clear understanding of the interaction of the technology and what we feel are the important ideas to be learned and the problems to be solved. The above is but one way to achieve this clarification.

```
  A     B     C     D     E     F     G     H
1
2
3
4          THE ELECTRONIC SPREADSHEET IN CALCULUS
5
6                             by
7
8
9
10                     Samuel W. Spero
11                 Department of Mathematics
12                 Cuyahoga Community College
13               2900 Community College Avenue
14                  Cleveland, Ohio 44115
15                     (216) 987-4561
16
17
18
19
20
  A1
Width:  9  Memory:363 Last Col/Row:A1     ? for HELP
F1 = Help; F2 = Erase Line/Return to Spreadsheet; F9 = Plot; F10 = View
```

I.   INTRODUCTION - GRAPHING FUNCTIONS IN CALCULUS

One of the most important applications of the computer in the study of Calculus is in the area of GRAPHING FUNCTIONS. From the very first chapter of almost every Calculus text, and then throughout the text, students are asked to graph functions. For example in the exercises in the text by Purcell and Varberg (published by Prentice-Hall) of the 126 sections in the text, in 56 of them - 44% of the total - the student is asked to graph functions or supply a sketch. In all of these exercises, sketching the graph does increase understanding of the underlying concepts of that particular section. Without computational help, no more than a few of the exercises can realistically be completed. On the other hand with computational help, a much more substantive learning experience can take place.

In this paper we will discuss the computer approach that we suggest for GRAPHING FUNCTIONS in the Calculus. This approach uses the electronic spreadsheet, in particular SUPERCALC 4 published by Computer Associates, for GRAPHING FUNCTIONS.

II. GRAPHING FUNCTIONS AND THE COMPUTER

To graph a function by hand a Table of Values must be
computed. In the first column of this Table the values for the
independent variable
are set down. In the
adjoining column the
corresponding values
of the function are
then calculated. De-
pending on the func-
tion to be graphed,
5 or 10 points or
more are calculated.
If more than one
function is being
graphed, or para-
metric or polar func-
tions are being eval-
uated more columns may be required.

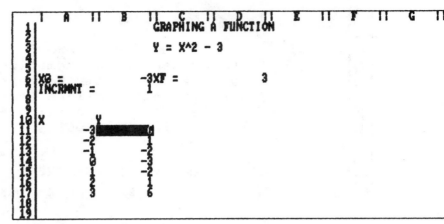

Before the graph can be drawn
a set of axes and an appropriate
scale must be established. Once
the scale has been established
the points calculated in the Table
can be graphed. In most cases the
points should be connected in a
smooth curve to complete the graph
of the function. If the graph is
to be handed in, the graph and the
axes should be properly labeled.

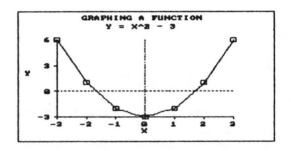

The computer is an ideal tool to be used for graphing
functions in the study of Calculus. The computer can be used
to help set up the Table of Values, as well as graphing the
values in the Table which includes scaling the axes, plotting and
connecting the points, and providing the headings and labels. The
computer also provides "hardcopy" of both the Table and the
graph.

While there are many aspects of the above process that can
be automated using a computer or calculator, it is this author's
opinion that the computational method of choice should retain the
above features of the "process" of graphing but should eliminate
the "busywork." The electronic spreadsheet does that.

III. THE ELECTRONIC SPREADSHEET AS A GRAPHING TOOL

There are electronic spreadsheets available for all popular
computers. In the present discussion we will be discussing
software that runs on IBM PC´s and compatibles which are readily
available both on- and off-campus. Our approach is to have each
student buy their own copy of the software which they then can

use anywhere and at anytime.

This is possible because electronic spreadsheet packages are reasonable in price. Almost all of the popular spreadsheet packages now offer student editions which sell for between $20 and $40. In fact Borland sells the complete QUATTRO package (an outstanding spreadsheet) to educators for that price.

Learning to use a spreadsheet is not just learning to graph. Spreadsheet skills can also be used in a broad spectrum of other courses as well as in one´s personal life and in the workplace. For example, the author uses SUPERCALC 4 for his gradebook, for generating problem sets for students, and even for calculating his mortgage payments. The electronic spreadsheet is really a powerful computing language that is very easy to use.

The graphics - which we consider the most important application for our Calculus students - are especially strong and versatile. To prepare a graph the student specifies the function and its limits and sets up the procedure in a very intuitive manner. There is not a lot of typing involved. Especially nice are the printouts of both the Table and the graphs. Furthermore the flexibility is there to deal with parametric equations, polar equations, and in the newer packages even three-dimensional graphics.

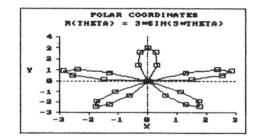

Aside from graphics the electronic spreadsheet can be used in Calculus in the study of the solution of non-linear equations (Newton´s Method), numerical integration, the numerical solution of differential equations, and the solution of simultaneous equations.

Because the electronic spreadsheet is being used very extensively in industry, it is now, and will continue to be, well supported. There is also extensive documentation available both from the manufacturer and from other sources.

IV. AN EXAMPLE

The various stages of graphing a function are easily accomplished with an electronic spreadsheet.

The electronic spreadsheet - in what follows we will be referring to SUPERCALC 4, or SC4, by Computer Associates - is a large electronic "sheet of paper" on which you can write using the keyboard as your electronic "pencil".

To type text into the spreadsheet, you "point" to where you

want to place the text and just type it in. No special command is needed. The same is true for numbers, you just "point" and type in the number. Even complex formulas can be typed in in a similar fashion.

To perform the tedious tasks associated with the process of computation and graphing there are simple commands.

For example, in the preparation of the Table of Values, the student has only to set up the calculation of the function at the first point in the domain, an SC4 command copies that formula and evaluates the function at all of the other points of the domain.

In the preparation of the graph, the student specifies the type of graph required and where on the spreadsheet the data to be graphed is. SC4 automatically sets up the axes and a scale appropriate to the points defined, graphs the function and connects the points in a smooth curve. Headings for the graph and labels for the axes are easily specified and pressing one key prints out the graph on the printer.

The spreadsheet for the function $Y = X^2 - 3$ and its graph are shown above. The spreadsheet with the equations in place instead of the numbers can be displayed with a simple command.

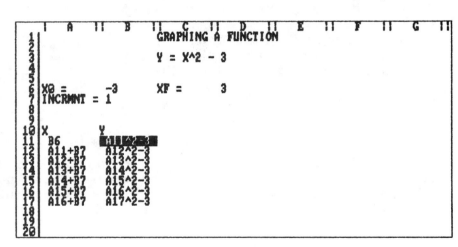

Other graphing applications such as polar coordinates and parametric equations are no more difficult. For the function

$$R(\theta) = 3SIN(5\theta)$$

The graph has been displayed earlier in this paper. The spreadsheet is to the right.

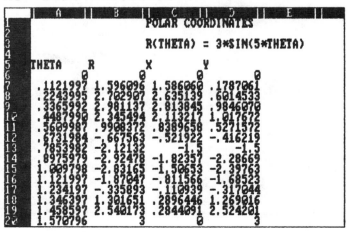

The author has described this process in greater detail in several articles[1] and also in a tutorial for students entitled GRAPHING FUNCTIONS which may be ordered from the author.

V. EXPERIENCES

The author has been using electronic spreadsheets - specifically the SUPERCALC series - in teaching Calculus since 1981.

SUPERCALC 2 was orginally selected for use with an 8-bit CPM computer because it was the only software available. When IBM technology was introduced, SUPERCALC 3 was selected because it was available in inexpensive student editions (SUPERCALC 3 by Lamont R. Lockwood, IBM PC Apprentice Personal Computer Learning Series, Prentice Hall,1984 and LEARNING TO USE SUPERCALC 3: AN INTRODUCTION by Gary Shelly and Thomas J. Cashman, Boyd and Fraser Publishing Company, 1986). Each student purchased their own software. This was very cost effective for the school, as well being very satisfying to the students since they could take the software to any IBM compatible machine.

Students need at least a two-hour orientation to SC4 which is usually sufficient to teach them to draw graphs. We have designed the above-mentioned tutorial on GRAPHING FUNCTIONS specifically for this two-hour orientation. Following this the students are able to undertake assignments from the Calculus.

The approach has been very successful. While we have not formally evaluated the use of the electronic spreadsheet, the students have indicated that they are more motivated when they can use SC4. The students are definitely completing more of those assignments which include graphing of functions. They have also acquired a new skill which they are using in their other courses.

FOOTNOTES:

1. a) The Electronic Spreadsheet as a Problem Solving Tool in Mathematics Instruction, paper presented at Fourth Annual Conference Applying New Technology in Higher Education, March 4-5, 1985, Orlando, Florida. Published in the PROCEEDINGS OF THE FOURTH ANNUAL CONFERENCE APPLYING NEW TECHNOLOGY IN HIGHER EDUCATION, National Issues in Higher Education, Kansas State University.

    b) Spreadsheets in the Classroom: Using SUPERCALC to teach Algorithms in Mathematics, CREATIVE COMPUTING, Volume 11, Number 10, October, 1985.

    c) The Electronic Spreadsheet as a Problem Solving Tool in Mathematics Instruction, paper presented at International Conference on Courseware Design and Evaluation, Ramat Gan, Israel, April 8-13, 1986. Published in the PROCEEDINGS OF THE INTERNATIONAL CONFERENCE ON COURSEWARE DESIGN AND EVALUATION, Israel Association for Computers in Education.

    d) The Electronic Spreadsheet as a Problem Solving Tool in Vocational Education, paper presented at the 24th National Conference on Technical Education of the American Technical Education Association, March 25-28, 1987, Cleveland, Ohio.

"Zooming In" on a Graphing Calculator

Alan Stickney
Department of Mathematics, Wittenberg University

Introduction

Whether one is using a computer or a calculator to display graphs, the ability to choose a new, smaller viewing rectangle is a very useful pedagogical tool, since it allows the instructor or student to "zoom in" on points of particular interest (roots, extrema, intersections, etc.). While there are several schemes for doing this, probably the most convenient method is for the user to indicate the corners of a new viewing rectangle in an interactive mode while the graph is displayed.

The purpose of this paper is to show how this "zoom-in" feature can be implemented on the Casio fx-7000/8000G series of calculators. Related programming topics on the Casio, and an implementation of the "zoom-in" feature for the HP-28S are also considered.

A Program for the Casio

To implement a "zoom-in" feature on the Casio, enter the following program as "Prog 9" in the program memory:

```
Cls
Lbl 1: Prog 0
Plot ◢
X → A
Y → D
Plot ◢
X → B
Y → E
( B - A ) ÷ 12 → C
( E - D ) ÷ 8 → F
Range A, B, C, D, E, F
Goto 1
```

"Prog 0" should contain one or more "Graph" commands which will generate the desired graph. For example:

```
Graph Y = X x^Y 5 - 3 X x^Y 3 + 2
Graph Y = 2 X - 1
```

The program is used as follows:

1. Enter the range information for the first graph as usual.

2. Begin execution of "Prog 9".

3. Following the appearance of the graph on the screen, the calculator enters "plot" mode. A <u>small</u> blinking dot will appear at the center of the screen, and its x-coordinate is displayed at the bottom of the screen.

4. Use the arrow keys to move the blinking dot to the <u>lower-left</u> corner of the desired viewing rectangle.

5. Press the "EXE" button. The lower-left corner will now be marked by a non-blinking dot, and the calculator will again enter "plot" mode.

6. Use the arrow keys to move the new blinking dot to the <u>upper-right</u> corner of the desired viewing rectangle.

7. Press the "EXE" button, and the graph will be redrawn in the new viewing rectangle.

8. Repeat steps 3-7 until done. Pressing the "AC" button at any time will terminate execution of the program.

Once the program has been terminated, it can be restarted with the last viewing rectangle that was displayed unless the range information has been changed.

Possible Enhancements

There are several options which can enhance the usefulness of the above program, although they all require additional space in program memory, a resource which is rather limited on the fx-7000G. A partial list includes:

1. <u>Last Graph Recall</u>. To do this, use additional variables to store the range information contained in A-F before it's changed. If an unsatisfactory graph is obtained, the user can stop the program with "AC", restore the range data for the previous graph (using a "Range" command in a separate program), and restart the zoom-in program.

2. <u>Improved Marking of Lower-Left Corner</u>. The program above leaves only a small "dot" to mark the lower-left corner of the new viewing rectangle. It is possible, but complicated, to plot several points at that corner in any shape, perhaps an "+", using the "Plot" function. This would improve the visibility of the lower-left corner of the rectangle.

3. Improved Scale Markings. The program as given has a nice
    number of scale markings on each screen, but they occur at
    "strange" numerical values. If scale marks at "even" values
    (e.g. 2.343, 2.344, 2.345, etc.) are desired, the scale value
    can be calculated from the values of A, B, D, and E. For
    example, the following does a reasonably good job of choosing
    the x-scale if the x-range is less than one:

$$10 \ x^y \ \text{Int} \ (\log \ (B-A) \ - \ 1.3 \ ) \ \rightarrow \ C$$

4. Zoom-In with Other Graphing Routines. The program can be
    used with a wide variety of graphing routines. Any routine
    that generates a graph without changing the range values can
    be stored as "Prog 0" and used with the zoom-in program. For
    example, a polar graphing procedure can be stored as "Prog 0".
    In this case, the user should be allowed to re-specify the
    range of theta values before each new view is obtained. If
    this is not done, the calculator might waste a great deal of
    time attempting to plot points which are not in the viewing
    rectangle.

Modularity and Subroutines on the Casio

The Casio is designed to allow programming with subroutines
(one program calling another), and that is a distinct advantage
when using the calculator in a classroom setting. The "zoom-in"
program given above can be entered as "Prog 9" just once at the
beginning of a course, and it never needs to be changed. For
different graphing applications, students only need to enter new
graphing commands in "Prog 0". This reduces programming errors
and student frustration with procedures that are used repeatedly
in different applications.

Polar and parametric plotters, as well as other procedures
for student use, can all be written in a similar fashion. The
idea is to isolate those parts of the program which will change
from one application to another (functions, equations, etc.) by
putting them in a separate subroutine in the program memory. The
general procedure is then entered just once, and students only
change the isolated subroutine as they move from one application
to another.

On the other hand, it should be noted that a maximum of 10
programs can be stored in the program memory at one time. This
tends to limit the number of separate subroutines that should be
used in programming.

How to Zoom-In on an HP-28S

It is also easy to mark a new viewing rectangle for a graph on the HP-28S. In fact, it can be done using only the built-in functions of the calculator and doesn't require any programming. Here's how:

1. Press "PLOT" so that the Plot Menu shows in the display and enter the function/equation using "STEQ".

2. Set the initial range values using "PMIN" and "PMAX".

3. Press "DRAW" to draw the graph.

4. Use the arrow keys to move the cursor to the lower-left corner of the desired viewing rectangle.

5. Press "INS" to store the coordinates of that point on the stack. Note that this does not place a "mark" at the point.

6. Use the arrow keys to move the cursor to the upper-right corner of the desired viewing rectangle.

7. Press "INS" again to store the coordinates.

8. Press "ON" to return to the text screen.

9. Press "PMAX" followed by "PMIN" to store the new range information.

10. Repeat steps 3-9 until done.

Summary

A "zoom-in" feature is now readily accessible on at least two different graphing calculators. The implementations are easy, and students should find them convenient to use. The only drawback is that the outline of the new viewing rectangle does not appear on the screen while the cursor is being moved to the second corner, as it does in some microcomputer software. This is unfortunate, but not a significant problem.

Marvin Stotz
23968 Maple Ridge Road
North Olmsted, OH   44070-1351
September 12, 1988

ABSTRACT:

ONE HIGH SCHOOL MATHEMATICS TEACHER'S PERSONAL ATTEMPTS TO INTEGRATE COMPUTER GRAPHING TECHNOLOGY INTO THE ALGEBRA I THRU A. P. CALCULUS CURRICULUM WITHOUT A CLASSROOM SET OF COMPUTERS OR GRAPHING CALCULATORS.

I call my application the M. A. R. V. Method which stands for M athematics A chievement R ealized thru V iewing!

During the 1987-88 school year, this teacher of Algebra I, Advanced Algebra II, Precalculus, and A. P. Calculus, introduced students at Fairview High School to the graphing software of Bert Waits and Frank Demana of OSU (Now Addison Wesley Publishing Co.).

The main thrust of my use of the computer graphing technology was to look at specific topics and determine how computer technology could give students meaningful mathematical experiences. Next I designed complete step-by-step handout instructions for students to follow in each "experiment" or assignment. I also had my A. P. Calculus students design and field test individual lessons!

Thus each lesson was a stand alone "mathematical experience" to achieve a certain objective. One illustration in Algebra I was to examine the slope intercept form of a line, $y = mx + b$. Students were to determine the effect of changing the values of b (e.g. from -4 to 4 step 2) while holding m constant. Next the values of m were changed and b was held constant. Another example in trigonometry involved the graph of $f(x) = A*\sin(Bx + C)$. Alternately changing the values of A (from -4 to 4 step 2) then B, and finally C, illustrated the concepts of amplitude, period, and phase shift.

A. P. Calculus used the graphing package mostly for limits of a function. The Zoom In feature, and Zoom Out feature were most helpful!

Each particular software package has it's own unique strong points. One program allowed entire functions to be entered by striking a single key (e.g. "S" gave you the sine function.)

In summary, the effort and time in rounding up hardware and overhead projection devices was well worth the educational advantages in preparing my students for future school and job opportunities.

M. A. R. V.   Method

for

enhancing high school mathematics

(M athematics   A chievement   R ealized   thru   V iewing)

Sources
of
Computer Software

Franklin Demana or Bert K. Waits
Department of Mathematics
The Ohio State University
231 West 18 th Avenue
Columbus, OH 43210

(The Computer Graphing Laboratory Manual and PreCalculus
Mathematics A Graphing Approach by Franklin Demana and Bert K.
Waits is now available from Addison Wesley Publishing Co.)

Margaret Raub
Mathematics Teacher
Strongsville High School
20025 Lunn Road
Strongsville, OH 44136

Marvin M. Stotz
Fairview High School
West 213 th Street
Fairview Park, OH 44126

Many textbook publishers offer computer graphics packages
Example:
        Addison-Wesley Publishing Company
        Calculus and Analytic Geometry, Sixth Edition
        George B. Thomas, Jr. and Ross L. Finney

        Toolkit programs (Derivative Grapher, Super*Grapher,
etc.)
    Have fun exploring the possibilities of this new technology!
Share these ideas with a fellow teacher and let me know how you
might use this computer software in the classroom or as a special
project!

Marvin M. Stotz
23968 Maple Ridge Road
North Olmsted, OH 44070-1351

COMPOSITION OF FUNCTIONS

Sketch the following composition functions f(g(x)) giving the domain and range. You may use a computer graphing program to assist you. There are 100 composition functions in all! Your evaluation will be three of these composition functions. Mastery will be two or more correct.

g(x) =   1.   2.   3.   4.   5.   6.   7.   8.   9.   10.

f(x) =

1.   2*x-5

2.   x^2

3.   x^2 - 4

4.   abs(x)

5.   sgn(x)

6.   sin(x)

7.   cos(x)

8.   exp(x)

9.   ln(x)

10. INT(x) or [x]

Name _____
Date _____

A Computer Assisted Activity

Load a computer graphing program into your computer. By
using your knowledge of domain of a function (and trial and
error), determine the domain, estimate the range, and sketch the
given curve. Your evaluation will be to sketch five (5) related
graphs, giving the domain and range. Mastery level is 80% or 4
correct.

FUNCTION	DOMAIN	RANGE

A. absolute value   ABS(X)

1. f(x) = abs(x)
2. f(x) = abs(x) + 2
3. f(x) = abs(x+3)
4. f(x) = abs(x) - 4
5. f(x) = abs(x-5)
6. f(x) = 2*abs(x)

B. greatest integer function   INT(X)

1. g(x) = [x]
2. g(x) = [x+3]
3. g(x) = [x-1]
4. g(x) = [x] + 2
5. g(x) = [x] - 5
6. g(x) = 3*[x]

C. signum function   SGN(X)

1. h(x) = sgn(x)
2. h(x) = sgn(x+2)
3. h(x) = sgn(x-3)
4. h(x) = sgn(x) + 4
5. h(x) = sgn(x) - 5
6. h(x) = 4*sgn(x)

D. linear function   y = m*x+b

1. i(x) = 3*x-4
2. i(x) = x
3. i(x) = -x
4. i(x) = 4
5. i(x) = -2x + 3

E. exponential function   EXP(X)

1. j(x) = e^(x)

THE EL-5200 GRAPHIC SCIENTIFIC CALCULATOR:
PRECALCULUS-TEACHER-GENERATED CURRICULUM IDEAS

by

David L. Stout

Department of Math/Stat

The University of West Florida

During the Summer semester, 1988, fourteen (14) students in a MAT program at the University of West Florida participated in a two-week long seminar in which the impact of an enhanced scientific calculator with graphics capabilities (the Sharp EL-5200 graphic scientific calculator) on the pre-calculus mathematics curriculum was explored and discussed.

The seminar consisted of ten four-hour days. Approximately the first three days were used to familiarize the participants with the EL-5200, six days were used to explore and highlight its capabilities in the context of the standard pre-calculus curriculum, and one day was used to summarize our findings and discussions. The main objective of the seminar was to cause the participants to think about possible curriculum changes and how the EL-5200 could be effectively implemented in the classroom. As part of the seminar requirements, each participant was to produce a mini-curriculum project which dealt with a specific area of the pre-calculus curriculum. The participants were told to assume every student had access to an EL-5200 and that all students had been taught how to use it. With these assumptions in mind, the participants produced projects which addressed both computational and graphical aspects of the curriculum. Some of the projects used the programming capabilities of the EL-5200 as a springboard for applying pre-calculus mathematics. Other projects combined the matrix-handling and graphing capabilities of the EL-5200 to alter the way we teach the solving of systems of equations; while others simply took another look at how we teach the concept of function. And still others addressed how conic sections could be effectively taught using the EL-5200. It is the teaching of conic sections, specifically the parabola, and concerns of the participants on which the remainder of this paper will focus.

Using the EL-5200 in the teaching of the properties of the parabola can be accomplished in any number of ways; however, many of the participants suggested the following:

(1) Use the EL-5200 to graph a significant number of parabolas in a given form;

(2) Use the trace and zoom features of the EL-5200 to explore and analyze properties of the given parabola (i.e., find values of the vertex, axis of symmetry, etc.);

(3) Provide the definition of the parabola; and

(4) Derive equations of parabolas.

Some participants suggested that the directrix and focus of the parabola also be graphed using the LINE and PLOT features of the EL-5200.

An example will serve to illustrate the progression outlined above. After graphing several parabolas whose equations are all of the form $y = ax^2$, students would analyze the graphs in order to understand the significance of the value of "a". Next, students would be asked to graph several parabolas of the form $y = ax^2 + k$ in order to determine the effect the constant k has on the parabola. At this point, students would be asked to use the trace and zoom features in order to find the coordinates of the vertex. Next, students would graph several parabolas of the form $y = a(x - h)^2$ to gain an understanding of the effect the value of h has on the parabola. Again students would be asked to use the trace and zoom features to determine the coordinates of the vertex of each parabola. Finally, students would graph parabolas of the form $y = a(x - h)^2 + k$ and use the trace and zoom features to find that the coordinates of the vertex are (h,k). Thus, after such a learning sequence, students would have an understanding of the significance of the constants found in the standard form of the equation of a parabola. It should be noted that parabolas of the form $y^2 = 4px$ could also be graphed simply by transforming the equation into $y = \pm\sqrt{4px}$. Even if p < 0, the correct graph can easily be obtained since the EL-5200 will ignore any value of x in the given range which give non-real values. After students have explored and analyzed many parabolas written in standard form, they would be asked to (1) graph some in the form $y = ax^2 + bx + c$, with $a \neq 0$; (2) use the trace and zoom features to find the coordinates of the vertex; and (3) transform the given general equation into the standard form $y = a(x - h)^2 + k$. This activity would further strengthen their understanding of the significance of the constants a, h, and k.

Max-min problems which give rise to quadratic functions can be solved quite efficiently by graphing the quadratic function involved and using the trace and zoom features to determine the coordinates of the vertex. Consider an example: "Suppose we have 200 feet of fencing and want to enclose a rectangular region of maximum area. Find the dimensions of the rectanglar region." A solution might progress in the following manner: (1) write an equation which describes the situation; (2) decide on a sensible range for each of the variables involved [note: this step is usually overlooked by students]; (3) graph the function (equation) which models the problem; and (4) use the calculator to find the required extreme value. Thus, the graphic calculator allows students to focus on the solution process and not be concerned with the algebraic and/or computational aspects of the situation.

In conclusion, I would like to cite some issues or concerns expressed by the seminar participants. Initially, there was some concern that students would not be willing to purchase a graphic scientific calculator. However, this concern moderated somewhat once the usefulness of the calculator was demonstrated. Another concern was one of competence. That is, participants asked "how proficient must I be with such a device?" As the seminar came to a close, most felt that there must be a basic level of proficiency in order for them to be able to demonstrate and use the machine in its various modes. Probably the biggest concern surrounded the unrestricted use of a graphic scientific calculator by students. There seemed to be a consensus that the teacher should actively guide students in its intelligent use and seek to highlight traditional topics which could be best taught and learned with the aid of a graphic scientific calculator. Indeed, all of the participants felt that the precalculus curriculum could be enhanced by introducing students to the many capabilities of a graphic scientific calculator.

Paul Thompson

Story Problem Analysis System

Examinations in mathematically-based curiculae inevitably require the ubiquitous story problem. While multiple-choice problems are easy to grade, they are presented in a manner which performs much of the important understanding process for the student, leaving them with the much easier task of merely computing some quantity. Such problems require that the student:

- select the appropriate formulae;
- select the correct information from the problem;
- solve the equations using the information;
- indicate the correct answer;
- make a decision.

The solutions for grading purposes of such problems require that many contingencies be pre-planned.

While the problem designer has no difficulty implementing a conditional grading strategy, the author's experience with graders has not been uniformly positive. Either unforseen contingencies arise, or the grading algorithm is defective. The professor frequently must regrade problems for consistency.

A solution to this problem is currently being created. The Story Problem Analysis System (SPAS) consists of:

- a course-wide equation list, for all equations discussed in the class;
- a problem generator, capable of the generation of both constant text and random values for each individual (both data and summary quantities);
- the solution to the problem, stated in equations and decision inequalities;
- a low-level decision-making engine. This section, the heart of the system, will consist of comparisons of the answers from the student to the correct answers. The system will self-correct to ensure that partial credit is fully obtained.

The SPAS system, when completed, will allow the teacher to administer and grade such long-answer problems, and handle certain broad classes of errors. This will facilitate both consistency in grading, and research into student preparation and learning. Finally, once a problem is created, the time previously spent grading may be spent considering why the problem did not work like it was supposed to!

Conceptual vrs. Procedural Knowledge in Introductory Calculus - Programming Effects

Fredric W. Tufte
University of Wisconsin-Platteville

Student Knowledge. Evidence abounds that many calculus students are learning to execute tasks in a routine, algorithmic fashion while lacking very basic conceptual understanding of the topics being studied. For example, Table I gives results of several final exam questions given to approximately 200 students who were completing an introductory calculus course. The examples in this table demonstrate that we are having remarkable success on algorithmic problems. Some would say that with a little more practice we might insure a short life span for the HP-28C.

TABLE I
Points on Selected Drill and Template Problems

Question	Percent of Total
1. Find dy/dx: $e^{x+y} = \sin y$	86%
2. Find dy/dx: $y = x^{\pi} \cos(\pi x)$	92%
3. Integrate: $\int x^2 e^x \, dx$	89%
4. Integrate: $\int \dfrac{3}{x(x^2 + 1)} \, dx$	73%
5. Find the first three terms of the Taylor series for $f(x) = \sqrt[4]{x}$ about 1.	90%
6. Find the area of the region bounded by $y = 4x - x^2$ and $y = x^2$.	86%

The same student subjects reported on in Table I were given a Calculus Concept Test during the final week of the semester. Three of the problems on this test are reported in Table II. The results on these problems suggest that many students do not recognize the derivative presented in symbolic form, and that they are unable to formulate or interpret geometric representations of fundamental concepts. Not only did students have conceptual difficulty with fundamental concepts of calculus, but problem 2(a) indicates that almost 40% of these students had conceptual difficulty at an even more basic level. These students were unable to correctly interpret the relationship between a function and its graph. This suggests that many of the intuitive explanations that instructors use in the classroom may be of little benefit for many students. The results on problem 3 indicate that when students had no

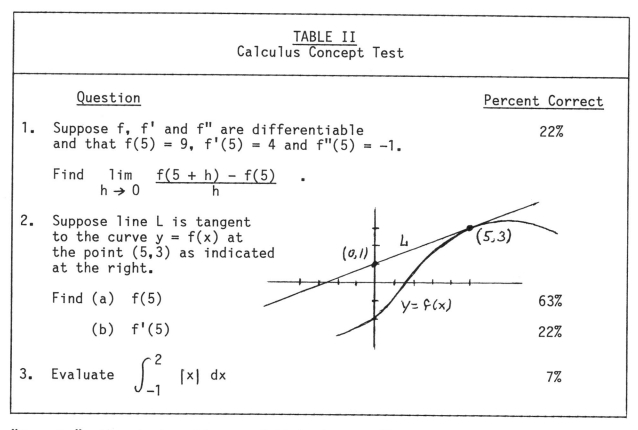

TABLE II
Calculus Concept Test

Question	Percent Correct
1. Suppose f, f' and f" are differentiable and that $f(5) = 9$, $f'(5) = 4$ and $f''(5) = -1$. Find $\lim\limits_{h \to 0} \dfrac{f(5 + h) - f(5)}{h}$.	22%
2. Suppose line L is tangent to the curve $y = f(x)$ at the point $(5,3)$ as indicated at the right. Find (a) $f(5)$	63%
(b) $f'(5)$	22%
3. Evaluate $\int_{-1}^{2} \lfloor x \rfloor \, dx$	7%

"formula", they had nothing to fall back on. Their concept of the Riemann integral did not allow them any alternative approach to the problem.

The examples that have been examined tend to confirm the contention that even when students are emersed in the study of calculus, our instructional emphasis results in conceptual knowledge that is indeed minimal. The evidence may also suggest that many students have difficulty in relating geometric explanations to the corresponding algebraic or symbolic representations of fundamental concepts.

A Programming Example. Because many students are now entering colleges and universities with experience in a programming language, we believe that having students write computer programs could be helpful in developing conceptual schemas related to the fundamental ideas of the introductory calculus course. We have instituted a one-credit supplemental course in computational calculus where students write their own programs to find limits, right and left hand derivatives, Riemann integrals, solutions to equations, etc. These programs, used in conjunction with graphics utilities (which students do not write) are used by the students to investigate several questions, among them the existence of derivatives and integrals for various functions at specified points or on specified intervals. Other types of questions are also investigated. It is quite easy, for example, to consider functions with discontinuities and examine the limit of Riemann sums when using left endpoints, right endpoints, midpoints or randomly selected points in the subintervals. By running a few examples, and examining the numerical output, students are easily convinced that the integral is independent of the points selected.

Table III lists several questions that were given on the Calculus Concept Test to those same students referred to earlier in this paper, and also to a group of 24 students who completed the computational calculus course. As can be seen from these results, students who were enrolled in the supplemental course appear to have developed richer schema related to both the derivative and integral concepts.

Justification of programming. Three seemingly logical arguments are often advanced for justifying the use of "canned programs" in calculus instruction. One argument draws a parallel between the discrete nature of computers and students' experiences with discrete situations. Discrete numerical output belongs to the students' world, and is more likely to be properly interpreted than are the symbolic representations of continuous functions. Second, the graphics capability, and in particular the ability to

TABLE III

Concept Test (Percent Correct) Question	Regular Class	Regular + Programming Class
1. Suppose f, f' and f" are differentiable and that $f(5) = 9$, $f'(5) = 4$ and $f''(5) = -1$. Find $\lim_{h \to 0} \dfrac{f(5+h) - f(5)}{h}$	22%	63%
2. Supose line L is tangent to the curve $y = f(x)$ at the point (5,3) as indicated at the right. Find		
(a) $f(5)$	63%	88%
(b) $f'(5)$	22%	46%
3. Evaluate $\displaystyle\int_{-1}^{2} |x|\, dx$	7%	50%
4. If $f(x) = (x + 1)^{10}$, find find $\lim_{h \to 0} \dfrac{f(h) - f(0)}{h}$	8%	42%
5. What is the maximum slope of the curve $f(x) = -x^3 + 3x^2 + 9x - 27$	6%	46%

6. Suppose P is a partition of $[0, \pi/2]$ into   10%   67%
   n subintervals, and u_i is an arbitrary

   point in the ith subinterval $[x_{i-1}, x_i]$.

   Explain why $\displaystyle\lim_{\|P\| \to 0} \sum_{i=1}^{n} (\cos u_i) \, \Delta x_i = 1.$

---

produce dynamic representations, provides a new dimension that was not previously available. The human mind appears to be quite adept at processing visual information of this type. And finally, the use of these software packages in an experimental mode creates an environment that is more conducive for learning. Detailed elaborations of these arguments can be found in the literature, and there is beginning to appear a body of supporting research.

But what about the effects of programming itself. We believe the following arguments lend support to the hypothesis that writing computer algorithms can enhance the understanding of fundamental concepts and help in developing mathematical maturity.

1. In constructing their computer programs to find limits, left and right hand derivatives, Riemann integrals, etc., students are forced to deal with the definitions at a more basic level, at a step beyond what is required when merely examining numerical or graphical output from "canned programs". They must pay attention to the language of mathematics, and deal with the association between that language and their own conceptions.

2. The language used in programming can help bridge the gap between natural language and the formal language of mathematics. Understanding a sequence of programming code may enhance understanding of the associated mathematical symbolism. Particularly when working cooperatively in a laboratory situation, students have available another means for communication, another language with which they may feel more comfortable.

3. When writing a computer program, students are put into a situation in which they are doing the teaching. They are teaching the computer what to do. As most teachers will attest, teaching a topic to someone else requires an increased precision in ones knowledge of the topic, in terms of both comprehension and expression.

4. When students use instruments (programs) of their own creation, their mathematical investigations become intrinsically more interesting and exciting. There is a creation of conflict when output does not conform to expectations, whether from naive judgements, analytic or graphical analysis, or by an instructors edict. Might there be added incentive to reconcile this conflict when ones own creation is in error?

5. Creation of a satisfactory computer algorithm requires a degree of planning, and creation of a sequence of logically developed steps that lead to some intelligible results. Consequently its creation is not unlike developing a formal mathematical proof. Since many Calculus instructors have given up on requiring students to produce proofs, the development of computer algorithms may be the next best thing.

It may be that the production and examination of computer algorithms can put a little more excitement and joy into both teaching and learning elementary calculus. And though we have nothing against the use of "canned programs" (they can be very beneficial), we believe that when students can develop their own algorithms, their efforts are more akin to "doing mathematics."

Research on Reading and Interpreting Computer Generated Graphs Using Eye-Tracking Technology

Charles Vonder Embse
Central Michigan University

One of the most frequent and valuable uses of computers in the mathematics classroom is for computer generated graphs of mathematical functions and relationships. New advances in calculator technology include interactive graphics capability on hand-held machines. Computers and graphing calculators can generate a wide variety of graphs of functions and relationships. But, more importantly, these machines provide speed and flexibility to produce multiple graphs and to manipulate these graphs in various ways. A student can graph more functions in a three-hour homework session using this technology than would normally be done in an entire year of mathematics instruction. Students are free to manipulate factors such as scale, rotation, translation, and viewing area in the plane. With the increasing availability of micro computers and inexpensive graphing calculators, this use of technology in the mathematics classroom will increase in the future. This paper provides research evidence that suggests increased use of technology can enhance students graph reading ability in the same way that practice in reading text improves text reading ability.

Graphs are an important teaching and learning tool which allow students to visualize the relationship between variables in a mathematical expression. Recent assessments of mathematical skills (National Assessment of Educational Progress; Carpenter,1975, 1980, 1983; Second International Mathematics Study; Travers, 1985) have shown students were able to read simple graphs but could not perform related skills such as interpreting, generalizing, integrating, or extending the information in a graph. There is little empirical evidence about how the human visual system decodes the symbol system we call a graph. With the research evidence concerning students' skill levels and the increasing use of computer and calculator generated graphs in learning and teaching mathematics, it seems appropriate to ask basic questions about how people read and interpret mathematical graphs. This information will be helpful in designing new curriculum materials and teaching methods which will make better use of the technology in the classroom.

This research study used eye-tracking technology to investigate how subjects read polynomial graphs generated by a computer. Eye-tracking technology allowed precise measurement of the location and duration of the subjects' gaze while viewing the graph on the computer screen. The eye-tracking computer used an inferred sensitive television camera to measure the angular rotation of the eye. This angular rotation was mapped to a location in the viewing plane. Data on the location of the viewing position was updated 60 times per second. This was three times as fast as the eye could establish a focal point and transmit visual information to the brain. The duration of a single visual fixation was then the sum of the 1/60 second readings when the eye was focused on a given point. Typical fixation durations while reading were from 150 to 400 milliseconds with the average fixation duration of about 250 milliseconds.

The study attempted to define the differences and similarities between novices and experts on the variables of fixation duration and total time spent viewing polynomial graphs generated on a computer screen. For the analysis of viewing patterns, the graphs were blocked in a grid pattern (see figure 1) and grid squares were categorized by the graphical features contained within them. Important areas of the graphs were grid squares which contained features critical to interpretation of the graphs such as scale, intercepts, and maximum, and minimum points. Fixation duration was defined as the average length of a fixation when viewing areas of the graph.

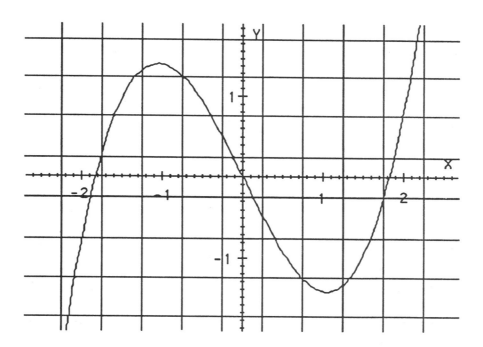

Figure #1 Graph showing blocking used to define important and unimportant area of the graph.

Each subject was allowed to view a graph for up to one minute. Subjects could self-select to stop viewing at any time up to the one minute. Since each subject's total viewing time could vary, the total viewing time variable was stated as a percent of total time spent. For the important areas of the graph, the percent of the total time spent in each important area was compared. For example, a subject may have viewed a certain graph for 45 seconds and of that total time spent 5.4 seconds, or 12% of the total time, looking at the grid area containing the y-intercept of the graph. Another subject may have spent 7.2 seconds of a total viewing time of 60 seconds looking at the same grid area. This would also represent of 12% viewing time in that grid area.

Twenty-five expert and twenty-five novice subjects participated in the study. Novice subjects were students enrolled in remedial mathematics courses at a large Midwestern university. Expert subjects were graduate students and professors in mathematics and mathematics education. Each subject viewed five different polynomial graphs and completed a memory task immediately after viewing each graph.

Data was analyzed using MANOVA, Pearson Correlation, and Spearman Rank Correlations procedures. Multivariate analysis of data from important blocks of the graphs indicated that the fixation durations were significantly different ($p < .05$) between experts and novices for 4 of 5 graphs The percent of total time variable did not differ significantly for the two groups of subjects. For unimportant blocks of the graphs, there was no significant difference between experts and novices for either fixation duration or percent of total viewing time. Correlation analysis indicated that subjects were consistent between variables and within themselves across all the viewed graphs.

Results of the study indicated that novices and experts were able to locate the important areas of the graph and spent approximately the same amount of time viewing those areas. However, experts had significantly longer average fixations when viewing important areas of the graphs. This contradicted the notion that familiarity with graphs would cause experts to be faster in processing graphical information. When an expert fixated on an important feature of a graph, the average length of each fixation was significantly longer than that of a novice. Experts were processing the visual information is a different way than novices. When experts viewed unimportant areas of the graph, they did not differ from novices on the two variables assessed.

Results of the study imply that a graph is an effective symbol system for drawing a viewer's gaze to the important information. Once viewing the important information, experts spent more time during each fixation before moving their eyes. Longer fixation durations imply more complete cognitive processing. Just and Carpenter (1980) contend that fixation duration is a direct measure of cognitive processing of the information being viewed at the present time. Experts were able to control their viewing strategy in response to the type of information being viewed at the moment. Novices, however, were not able to shift to the longer fixations when viewing important information even though they spent as much total time as the experts viewing the important information. Shebilske and Fisher (1981) found fixation durations to be longer in areas of text passages containing more important ideas than in other less important areas. This observation also seems to be true for expert graph readers, but not novices.

What factor taught experts to shift their viewing strategy for the important information? More experience using graphs by experts seems to be one probable explanation for the differences between the two groups. Like reading text, the more experience a reader has, the more sophisticated he/she becomes in reading techniques. This seems to be true for reading graphs or text. While both groups were drawn to the important information, experience probably taught experts that longer fixations are needed to process more completely the important information. If experience reading and interpreting graphs is an important factor in improving graph reading skills, then increased use of technology to generate graphs would be appropriate in mathematics classrooms. Producing accurate graphs by plotting points is an inefficient way to give students experience with graphs. Using graphing technology in classroom for all types of activities, including lectures, demonstrations, work sessions, homework, explorations, concept development, problem solving, and testing, will provide students with significantly more experiences using graphs.

References

Carpenter, T., et al. (1975, October). Results and implications of the NAEP mathematics assessment: secondary school. *Mathematics Teacher, 68*: 453-470.

Carpenter, T., et al. (1980, May). Results of the second NAEP mathematics assessment: secondary school. *Mathematics Teacher, 73*: 329-338.

Carpenter, T., et al. (1983, December). Results of the third NAEP mathematics assessment: secondary school. *Mathematics Teacher, 76*: 652-659.

Just, M., & Carpenter, P. (1980). A theory of reading: From eye fixations to comprehension. *Psychological Review, 87*, 329-354.

Shebilske, W., & Fisher, D. (1981). Eye movements reveal components of flexible reading strategies. In M. L. Kamil (Ed.), *Directions in Reading: Research and Instruction*, 30th Yearbook of the National Reading Conference.

Travers, K. (1985). *Second International Mathematics Study, Detailed National Report: United States*. Champaign, IL: Stripes Publishing Company.

Vonder Embse, C. (1987). An eye fixation study of time factors comparing experts and novices when reading and interpreting mathematical graphs. (Doctoral dissertation, The Ohio State University, 1987). (ERIC Document Reproduction Service No. ED 283 672)

cT – Not Just Another Language

Russell C. Walker
Department of Mathematics
Carnegie Mellon University
Pittsburgh, PA 15213

Introduction. Modern computer applications emphasize powerful features such as high quality graphics, mouse interactions, a variety of text fonts, windows, and pop-up menus. While the end user finds these applications appealing and easy to use, only exceptionally skilled programmers can produce such applications with today's programming tools. The cT[1] language is intended to make development of programs employing such features possible for those with far more limited programming experience.

The historical antecedents of cT are the TUTOR and MicroTutor languages developed by the Computer-based Education Research Laboratory for the PLATO System[2] at the University of Illinois at Urbana-Champaign. The TUTOR language was implemented on Control Data Corporation time sharing systems, and MicroTutor was designed to run on a variety of personal computers.

cT was first implemented on the Andrew environment of networked workstations at Carnegie Mellon University to provide students and faculty with a language to easily take advantage of the power of that system. Building on the efforts of the developers of MicroTutor to produce a language that would run on a variety of personal computers, cT has been implemented on several machines. cT programs can be run without change on the Macintosh, the IBM PC[3], and the IBM PS/2[4] series and the IBM RT PC, Sun 2, Sun 3, and VAXstation II models of advanced function workstations.

Further, because the main motivation of cT's developers was the production of educational software, cT includes features to facilitate handling of student responses and the presentation of text and graphics. In order to be used by instructors with a minimum of assistance from professional programmers, cT includes a highly supportive development environment, including an on-line reference manual and informative error diagnostics. As the language has become better known, it has also begun to be used in the development of research tools.[5]

cT language features. cT includes the usual structures of loop, reloop, outloop including a unified treatment of until and while; case; and if, elseif, endif. In addition, it contains a number of commands specifically designed for educational software development.

A set of **graphing** commands generates axes and appropriate scaling for graphs. Tick marks and labels on the axes may be set by the programmer. Rotatable displays, pattern filled polygons and disks, and animated icons are also available.

Multi-font text including bold, italics, bigger, smaller, and centered are available. The display rectangle for text can be adjusted during execution.

Portability between computers is instant. Rescaling features of cT make it possible for the programmer to minimize the effort necessary to make displays adapt to differing screen sizes. Because of incompatibilities in fonts, one does encounter difficulties in porting the program if special symbols are used.

The **menu** command places a prompt associated with a routine in the program on a menu. The type of the menu generated depends on the machine, i.e. pop-up menus on a workstation, pull-down on a Mac.

On machines that permit several programs to run simultaneously, the **execute** command allows a cT program to initiate a Lisp, Fortran, or C program.

Response handling commands permit an entry from the keyboard to be examined for mathematical accuracy, for correct spelling, inclusion of keywords, or the inclusion of mathematical operators.

The usual **sequencing** commands are supplemented by menu entries for the next routine or the previous routine to facilitate moving through a sequence of lessons based upon a student's progress.

The cT development environment. Because cT is designed to allow inexperienced programmers to develop sophisticated applications with speed and ease, a number of supportive features are built into the development environment. Several of the features are illustrated in the figure below from a Macintosh screen.

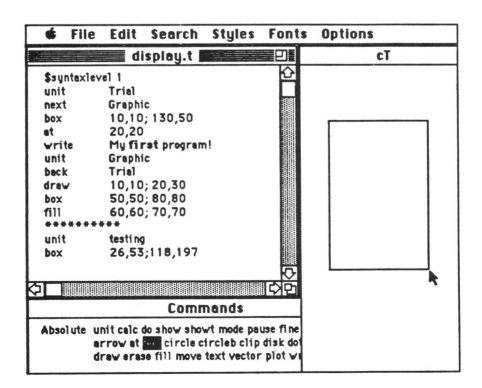

The **commands window** contains a list of cT commands. Clicking on a command causes that command to be entered in your program at the position of the insertion pointer. In the figure, the "box" command appears in reverse video in the commands window, indicating that that command was entered by a click in the commands window.

Graphics editing permits screen locations to be entered in the source code by means of mouse clicks. In the figure, the coordinates for the box command have been entered by clicks at the top left and the bottom right of the box. The requested box has then been drawn.

An on-line **reference manual** provides a tree-structured guide to the cT language. The manual is accessed by mouse clicks and includes executable examples that may be copied into a program.

cT programs are **incrementally compiled**, i.e. when a program is executed, the procedures needed first are compiled first. When changes are made, only the changed procedures are recompiled. This gives the speed of a compiled language, with a quick modify, test, re-modify cycle. A binary version of the program can be requested.

Useful **error diagnostics** are provided during both compilation and execution. Source code is scrolled to the location of the error, a message is displayed, and the insertion pointer placed at the position of the error.

Some sample cT applications.

Simplex Algorithm Mastery (Russell Walker, Mathematics Department, Carnegie Mellon University)

This program guides a student through the simplex algorithm to solve a linear program. Computations are done by the program, but the student is required to make all decisions regarding selection of pivots, formation of the dual, use of artificial variables, etc. Errors are detected, their consequences indicated, and the opportunity to correct the error is presented. For problems involving two or three variables, the progress toward solution is indicated on a graph of the set of feasible solutions.

Graphs & Tracks (David Trowbridge, Center for Development of Educational Computing, Carnegie Mellon University)

Graphs & Tracks I and II are aimed at the difficulties students have in making connections between observations of motion and graphs of that motion.

In part I, students are presented with a graph and must create a motion that matches the graph by setting up inclined tracks, positioning a ball, and starting the ball with a particular initial velocity. Part II goes the other way: the student views a motion on inclined tracks and must sketch a graph corresponding to the motion. There is abundant feedback to the student. These programs recently won EDUCOM/NCRIPTAL awards.

Handling of lab data (Robert Schumacher, Physics Department, Carnegie Mellon)

The ability of cT to present graphics, manipulate strings, and initiate processes in other languages makes it useful in the laboratory. In this application, a cT program accepts a binary data file, prepares it for processing by a FORTRAN program, and submits the FORTRAN program to a CRAY supercomputer. When the output is received, it strips off extraneous material and generates a graphic display of the result. This facility is used both to process research data and data from a modern physics experiment investigating chaos.

Fourier (Brad Keister and Harry Stumpf, Physics Department, Carnegie Mellon University)

Designed to help students visualize summation and convergence properties of Fourier series, this program begins with a tutorial which steps through the basic sequence of entering parameters and viewing results. The program calls for a functional form for the Fourier sine and cosine coefficients, along with the end points which define the period interval, and the function which the series is to represent. A histogram illustrates the relative weights of the coefficients. The user then enters the maximum number of terms to be summed, and the resulting partial sum is plotted alongside the anticipated function. Fourier is especially useful for examining convergence properties, Gibbs phenomena, etc.

----

1. cT is a service mark of Carnegie Mellon University.

2. The PLATO System is a development of the University of Illinois. PLATO is a registered trademark of Control Data Corporation.

3. IBM is a registered trademark. PC and PS/2 are trademarks of the International Business Machines Corporation. Macintosh is a registered trademark of Apple Computers, Inc.

4. Announcement of PC family implementation expected in early 1989.

5. For a more extensive discussion of cT, see "The cT Language and Its Uses: A Modern Programming Tool" by B. A. Sherwood and J. N. Sherwood, to appear in the Proceedings of the Conference on Computers in Physics Instruction, North Carolina State University, August 1-5, 1988.

Gerald L. White
RR #2
Colchester, Il 62326

(309) 776-4592

Mathematics Dept.
Western Illinois Univ
Macomb, Il 61455

(309) 298-1383

Testing and Evaluation

USING MICRO COMPUTER SUPPORT FOR A CURRICULUM MANAGEMENT PLAN
IN A JTPA BASIC EDUCATION PROGRAM.

An Apple IIe microcomputer system and a Scrantron 1300 Optical
Card Reader were used in conjunction with CDM, a curriculum
management plan, to provide immediate listings of individual
weaknesses in arithmetic skills and continual updating of
records of progress in mastery of objectives in a three week
JTPA Basic Education Program.

Forty-two youth between sixteen and twenty years old were
enrolled in a Job Training Partnership Act (JTPA) Basic Skills
Program in a three week residential experience at Western
Illinois University. They needed to be separated into three
classroom groups and spend six hours each day developing
arithmetic, problem solving and calculators skills. An
additional hour each day was scheduled in an Apple Computer
Lab for investigations with LOGO.

A list of thirty-three objectives involving arithmetic skills
were established. Stanford Diagnostic Mathematics Tests, Blue
Level, were used for pre and post testing. The curriculum
management system and a microcomputer made it possible to
quickly form the required groups based upon objectives not yet
mastered as shown by the pretest. By continually updating the
data base, we were able to share with each individual a
printout of the number of objectives that they had mastered.
Also each instructor could be provided a matrix showing
objectives mastered and not yet mastered for their group each
day.

Post test scores showed the mastery of eight new objectives
overall and an average increase of three years in grade level
equivalences.

COMPUTER GRAPH ANALYSIS

Agnes Wieschenberg
John Jay College of Criminal Justice - CUNY

In a Pre-Calculus and Calculus course a considerable amount of class-time is spent on functions and the graphing of functions. To be familiar with the shape and characteristic properties of a graph means better understanding of the function that generated it. Therefore, the development of the visual imagery of our students can mean a great deal when it comes to grasping some of the fundamental concepts in Calculus.

New technologies, namely calculators and micro computers, that are available today at most Colleges and Universities can greatly enhance our struggle for achieving better understanding of the calculus material. As we use these new tools, however, some caution must be exercised since these sophisticated tools work for us in different ways than the earlier more primitive pencil and paper methods. One **must** be familiar with the "old" methods to know what questions to ask as we incorporate the new techniques.

In the spirit of the "Calculus for a New Century" movement, Professor Peter Shenkin and I have spent a great deal of time in developing software to be used in all Pre-Calculus and Calculus courses at the John Jay College of Criminal Justice. (Actually we developed software to be used by every required mathematics course at the College which includes courses in Algebra and Finite Mathematics.) Each class spends one week in our microcomputer laboratory, which is equipped with IBM XT and AT computers, for a hands-on computer experience. The software is "user friendly" so that students need no prior computer experience to use it. Modules include finding slopes and intercepts of linear equations, systems of linear equations, and quadratic equations. All of these modules offer algebraic solutions as well as graphs at the fingertip of the user. Other modules include Linear Programming (using the Simplex method), Probability, Statistics, Limits, Graphs and Integration using numerical methods. [1]

The focus of this paper is the use of the Graphs Module. The package will run on IBM PCs and compatibles (and is being converted to Macintosh capability), is an easy to use graphics package. The student may enter some pre-programmed functions that appear on the menu such as

```
Exp(x) and Exp(2*x)
Log(x)
Sin(x) and Cos(x)
Sin(x) and Sin(x+1)
Sin(x), 2*sin(x) and Sin(2*x)
```

or interactively enter the function definition and the domain to graph. Both the function and the domain may be easily modified. The graph rapidly appears on the monitor. Upon examination of the graph, points of relative maxima and minima of the function as well as points of inflection and intervals of increase and decrease are easily estimated. The user can also enter two functions to be graphed on the same coordinate system and find intersections of the functions by estimating. The accuracy of this estimate depends on how far the user is willing to go to "close in" on the point of intersection.

The first example will illuminate the above points. Let us take the graph of two familiar functions, $f(x)=\sin(x)$ and $g(x)=\cos(x)$ and graph them on the same coordinates. We will try to estimate the value of y at the intersection of these functions in the interval between 0 and π. The questions that we may ask are: what does this point represent and how can we find an answer to three point accuracy? The answer to these questions may be simple to some but even this group will find it interesting to "zero in" on an estimated answer that can be more accurate than some trigonometric tables. Changing the parameters we can easily approach the desired number (in this case three decimal point accuracy). Now the student can check the tables and see that indeed the number that we found, .707 is corresponding to a 45° angle where the values of sine and cose are indeed the same. Thus the student learned not only the periodic nature of the sine and cosine graphs and that they have common points of intersections, but the nature of infinities between distances that otherwise appear finite. (Figure 1.)

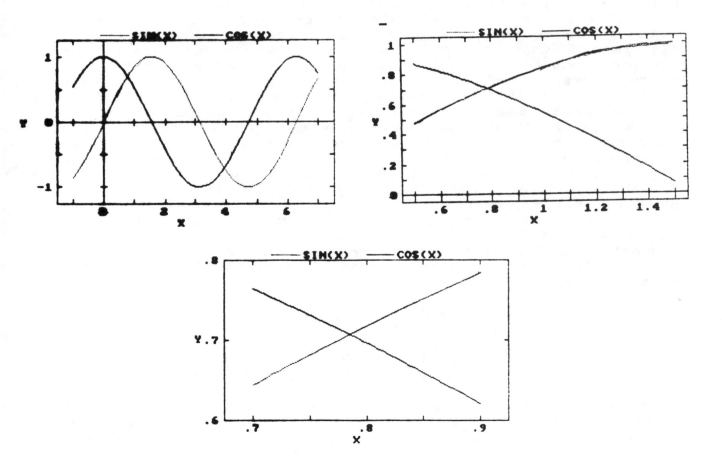

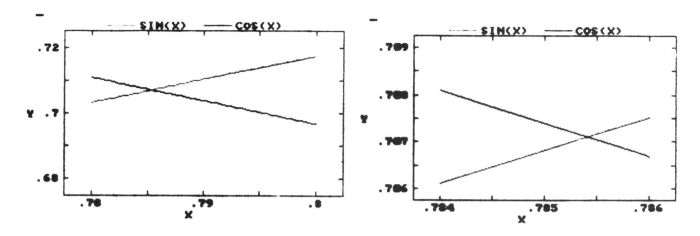

Figure 1.

Another lesson which may be learned is the fact that computations are very rapid on the computer but as we increase the difficulty level of the calculations, the computer will also need more time to produce an answer. Students may also learn from other examples as they experiment freely with the computer that computer errors can be significant in the calculation of certain formulas depending on the operations that are performed. Some students may investigate this further to develop ways to circumvent the problem.

Now let us can look at an example that is not so familiar. Let us select two functions: $f(x)=\sin(\sin(x))$ and $g(x)=\cos(\sin(x^2))$, and look at them at an interval from 0 to 6. After generating the graphs, all we have to do is to enter the functions and go from x=-1 to x=6, we immediately will see that there are four intersections in this interval. Using the above method we now can zero in on any one of those points and get an accurate estimate of its location. The same approach may be used for approximating maxima and minima, points of inflection and intervals of increase and decrease. (Figure 2)

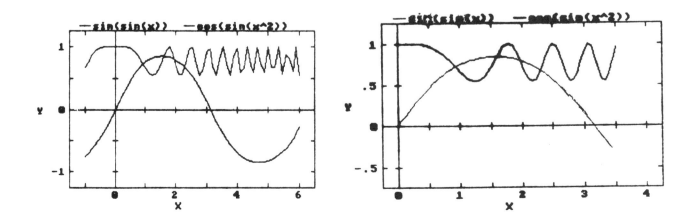

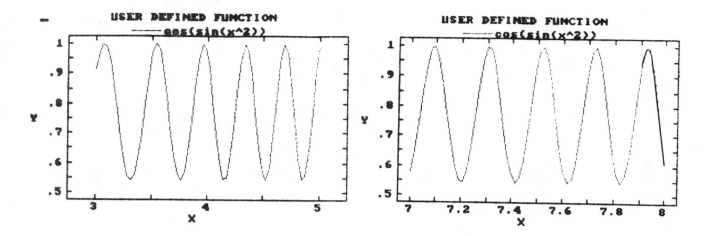

Figure 2.

If we want to investigate the graph of these functions as we are increasing the value of x, say maybe around x=30, we can change the parameters to values in this neighborhood. The user will soon discover that going from x=0 to x=30 will not result in a useful graph. To investigate how the the function cos(sin(x^2)) behaves as we increase the value of x, we must take smaller intervals for inspection. Since the range of the function is selected by the computer, some adjustments are sometimes needed in this selection. Thus easy experimentation becomes possible which in turn makes analyses accessible and more fun. Going on the wrong track will not result in frustration but rather learning of what steps will lead to fruitful results.

For functions such as f(x)=1/x, where there is a vertical asymptote, the computer will exhibit unusual behavior near the asymptote. The algorithm used in this module plots a value of zero in cases like this. The rest of the graph, however, is accurate.

For intervals where the value of the function is a complex number, the graph generated will behave irregularly indicating to the user that something is wrong.

Since the module allows free experimentation and rapid results students at all level of mathematical development can benefit greatly.

References

[1] Wieschenberg, Agnes and Shenkin, Peter. A Computer Companion For Undergraduate Mathematics Simon & Schuster Higher Education Publishing Group, Ginn Press, Needham Heights, Massachusetts, 1988.

TEACHING NUMERICAL ANALYSIS

IN A SMALL COLLEGE ENVIRONMENT

Daniel S. Yates
Randolph-Macon College

The purpose of this paper is to share my experience teaching a one-semester, undergraduate course in numerical analysis at a small, four-year private college, Randolph-Macon College, in Ashland, Virginia. I should say at the outset that I am not an expert at teaching numerical analysis; to the contrary, I am a novice. I brought several handicaps which I describe below. Yet I am encouraged by my first experience. The person I hope to reach with my commentary is that person who anticipates teaching a numerical analysis course for the first time and who might avoid some of the mistakes I made and be reassured by some of my successes.

Randolph-Macon College has an enrollment of approximately 1000 students and a mathematics faculty of seven. In the past half dozen years, there has been an average of about 10 math majors each year. Numerical analysis is not required for the mathematics major but may be elected by mathematics and computer science majors to satisfy their requirement for hours in the major field.

The Instructor   I joined the R-MC faculty between terms during the 1987-88 school year when there was an unexpected need to identify an instructor for the numerical analysis course and several introductory statistics sections. For fifteen years previously, I had been a mathematics and computer resource teacher for public schools in the Richmond, Virginia area. My previous college teaching experience consisted of three years as an instructor at Virginia Tech in the nineteen sixties. I did have considerable experience in recent years teaching programming in BASIC and Logo at the public school and college levels.

My training in mathematics, twenty years ago, consisted entirely of pure mathematics, with two terms of statistics and no other experience in any area of applied mathematics. I had neither taught nor even taken a course on numerical analysis prior to the spring 1988 term, although I have, in recent years, explored some numerical techniques, such as writing a program to approximate the sum of a series and using the bisection method to obtain roots of an equation on a spreadsheet. I wanted to teach this course for my own professional development and because of my interest in using the computer to enhance the teaching of mathematics and to make learning more efficient (and more appealing to students).

The Course   I consulted with several numerical analysis instructors at nearby universities to get different perspectives on the most prominent texts. I selected the Burden and Faires text (Prindle-Webber) because it is a respected text, but also because the authors take what seemed to be a sensible, middle of the road position on the role of the computer in such a course. That is,

they provide algorithms (without program listings) that are easily
converted into computer code.

For the one-semester course, taught during the spring term, I
arranged to move the class into the college's Computer Literacy Lab
in order to have access to a demonstration computer with LCD
overhead projection device and multiple computers and printers (all
aging IBM's).

At Randolph-Macon the fall and spring semesters are short,
lasting only 13 weeks, with a four-week January term in the middle.
Because of the relatively short spring term, the topics to be
covered in the course had to be chosen carefully. The course
sequence reflected my personal preference:

Chapter 1	Mathematical Preliminaries
	Review of Calculus
	Round-off errors and computer arithmetic
	Algorithms and convergence
Chapter 2	Solutions of Equations in One Variable
	The bisection algorithm
	Fixed point iteration
	The Newton-Raphson Method
	Error analysis for iterative methods
Hour Test 1	
Chapter 2	Accelerating convergence
	Zeros of real polynomials
Chapter 3	Interpolation and Polynomial Approximation
	The Taylor polynomials
	Interpolation and the Lagrange polynomial
	Iterated interpolation.
Hour Test 2	
Chapter 6	Direct Methods for Solving Linear Equations
	Linear systems of equations
	Gaussian elimination and backward substitution
	Linear algebra and matrix inversion
	The determinant of a matrix
	Pivoting strategies
	Special types of matrices
	Direct factorization of matrices
Final Exam	(3 hours) comprehensive, but emphasized Chapter 6 material

<u>Programming</u> The course description in the catalog states that a
prerequisite for the course is a "working knowledge of a computer
programming language." I polled the students to determine their
language of choice and was surprised that all 14 enrolled students
indicated BASIC, although basic computer science courses at the
college provide instruction only in Pascal. My conclusion was that
the students were most comfortable with BASIC from their high school
experience. I indicated that a continuing requirement in the course
would be to translate the provided algorithms into programs that
executed, and to use these programs to solve selected problems that
included tedious computations.

Things that went well   The use of the computer definitely motivated
study of the numerical methods and added an extra dimension by
eliminating much of the tedious computations by hand. In fact, I
can´t imagine teaching this course in any practical way without
having access to the computer, both for in-class demonstrations and
for students to code and use on non-routine problems and
applications. I am convinced that the students perceived the
computer as an indispensable tool and one that freed them to
concentrate on the various approximating methods.

The demonstration computer, in particular, was a most important
asset for it allowed me to follow theoretic discussions of
approximating techniques and algorithms with quick, efficient
solutions to sample problems. Being able to quickly display program
listings, or to graph a complex function made the instructional
process more efficient and demonstrated, over and over, the special
symbiosis between man and computer.

Although I had some reservations about endorsing the use of
BASIC as a programming language in the course (because of the
negative press BASIC has received recently and the emerging role of
Pascal in college computing curricula), I found that the algorithms
translate directly into BASIC without the opportunity for "spaghetti
programming" that detractors say BASIC permits. Non-structured
programming by students was not a factor because of the nature of
the algorithms presented in the text.

I elected to use a function graphing utility on several
occasions when the objective was to approximate roots to specified
degrees of accuracy. Most numerical routines to do this require
that one begin with an interval that brackets a root. Yet we
encountered occasional equations that were sufficiently complex that
the general location of the graph was not easily determined. In
these instances, I pointed out to the students that one could make
use of the built-in mathematical functions in BASIC to quickly get a
general feel for the characteristics of a function and its graph.
The two steps are to enter the function into computer memory with a
one-line program such as:

        10 DEF FNF(X) = 16*X^4 - 40*X^3 + 5*X^2 + 20*X + 6

and then type, in immediate execution mode:

        FOR X = -10 TO 10 : PRINT X, FNF(X) : NEXT

to produce a table of ordered pairs, (x,f(x)). In most cases, one
can determine by inspection where the function is increasing or
decreasing and the general location of the roots (although not for
this example--here, there are two roots between 1 and 2). Of
course, the interval and increment can be modified to either check
another interval or to obtain a refinement within the previous
interval. The students picked up on this technique immediately and
made good use of it thereafter.

Things that did not work well   I made the mistake of not providing a
review of BASIC programming syntax prior to the first programming

assignment, and found that about one-third of the students had programming skills that were weak at best. Subsequently, I departed from the text to provide an overview of those techniques that they would need most often. My second mistake was in not providing practice programming exercises before returning to the text. Those students with weak skills continued to have trouble with their coding throughout the course. Next time, I will definitely devote more time early in the course to insuring that students can quickly translate the algorithms into code.

Although the move to the computer lab did provide for a demonstration computer and screen, the other computers in the room and the fact that each student was facing a computer that is kept running continuously provided yet another problem. Students, facing the front of the room, could surreptitiously develop their own agenda while giving the impression that they were paying attention to the instruction at the front of the room. Suffice it to say that I will avoid this situation in the future.

I also found that the majority of my students had a somewhat less than satisfactory recall of important results from calculus (consistent with the "Nothing transfers" hypothesis). But I suspect that this will always be a source of irritation for teachers regardless of the role of the computer.

<u>Conclusions and recommendations</u> Having considered the above, I have made several decisions for the next time I teach this course (spring 1989), and offer them for your consideration.

1) Retain the text; it is rigorous but accessible. The algorithms provided strike a good balance between the mathematics of the continuous processes that are the principle object of numerical analysis and the aspects of computers and codes that enable numerical solutions.

2) Classroom environment: restrict the equipment to a single, demonstration computer with projection capability or linked to large screen monitor(s).

3) Provide a review of programming language syntax prior to the first coding assignment.

4) Emphasize to faculty advisors the prerequisite of a working knowledge of a programming language.

<u>Postscript</u> I would be happy to share my tests and examination and/or a 5 1/4" disk of programs in BASIC for the IBM with any prospective numerical analysis instructor who would be interested. Write to Dr. Dan Yates, Mathematics Department, Randolph-Macon College, Ashland, VA 23005.

The subject of ordinary differential equations is an old and beautiful one. Indeed, one of the first and most important differential equations expresses Newton's Second Law: F=ma. Of course in the three hundred years since this early beginning in classical mechanics, differential equations have been used to model an increasingly broad and rich selection of inherently dynamic phenomena. Unfortunately, a vast number of our introductory courses fail to convey this dynamic nature. Rather they are viewed as boring "techniques" courses in which assorted formulas for solutions are matched to a variety of seemingly unrelated equations. Students leave the course with the misleading impression that all ordinary differential equations have explicit solutions, when in fact quite the opposite is true. The numerical and qualitative analysis of equations receives little attention, particularly the latter topic. This inattention is regrettable, since in many "real-world" modeling situations, the most important questions often center on: how a solution changes in response to various parameters that appear in the equation? or how does the solution behave with respect to its initial conditions? We would be remiss in our educational mission, if we did not afford our students exposure to this manner of analysis.

Recently, however, Hüseyin Koçak [2] published PHASER, a remarkable piece of software that holds great potential to make this type of analysis, particularly the qualitative theory, more accessible to our students. The package produces sophisticated graphical displays of solutions to differential and difference equations and for a detailed description of the many features of PHASER, we cite the excellent review article of Bridger [1]. Here we simply note the features that make the package so attractive for the classroom. All of the equations that are built into the software appear in a very general format, allowing for parameter values to be changed and multiple initial conditions to be considered. This also applies to any "external" equation that a user might enter in for study. A variety of graphical displays (including 3-D images with perspective) is also available, and different numerical approximation schemes can be used to compute the solutions, so that comparisons of algorithmic efficiency and accuracy can be made. Since the package is entirely menu-driven and no programming skills are required, it is particularly well suited to serve as a "laboratory" environment for student self-discovery and exploration. For the instructor, there is a utility that allows an entire sequence of screen images to be saved to a file so that a "slide show" demonstrating a key concept or idea can be displayed in class. Note, however, that each screen image is actually redrawn at the time of display. Hence students can watch an individual solution evolve in time, as well as watching a sequence of such solutions, resulting from a sequence of changes in the underlying equation. In this way PHASER animates what are normally perceived as inanimate models.

At the University of New Hampshire we have been working to incorporate PHASER into our introductory sophomore level course since the fall of 1986. Support for this work has come primarily from the University's computer-aided instructional initiative, Project DISCovery, or Directions in Scholarly Computing. We have used three vehicles for this effort: large lecture demonstrations, student performance of specific experiments or "computer labs" designed to explore concepts, and individual self-instruction. Since the course is organized around large (125+ students) lecture sections, supplemented by twice weekly recitation sections, most of our efforts have involved the development of sets of in-class demonstrations. These have been displayed by using a number of large-screen projection systems. In the two parts of the project involving "hands on" use of PHASER, exercises are intended to be a paradigm for part of an applied mathematician's work. A computer experiment is performed and the resulting data (both numerical and graphical) suggest certain patterns. These form the basis of a conjecture, which must then be verified through rigorous mathematical analysis. Here we have had to rely on a self-selecting group of students each semester to work on the special "computer laboratory" assignments. Before discussing the natural issues of implementation that are raised here, we note that in the fall of 1989 we will be offering a special small section of ordinary differential equations in which PHASER will be used from the very beginning of the course and will in effect serve as the "text". For a more in depth description of both the in-class demonstrations and the "laboratory" assignments, see Zia [3].

Regardless of the software package being used, some type of projection system is absolutely crucial for in-class demonstrations, unless the class size is so small as to permit everyone to see a monitor image without strain. Our experience has been that a permanently mounted system complete with the appropriate computer hardware is the ideal situation, particularly for a large lecture section, as anything short of that poses an obstacle to the less technologically inclined instructor. A number of portable systems which use an LED display board with a standard overhead projector are also available and these can prove effective for the smaller class sizes. Since PHASER produces color displays, a color projector is recommended but not essential. Another logistical issue involves providing students access to machines. While the large demonstrations are useful, they are still only a <u>passive</u> learning experience. Ideally we would like to see an entire class <u>actively</u> <u>engaged</u> in the use of PHASER. Certainly as more micro-computer clusters are created and as more students obtain their own machines, we can move in this direction. In the interim we have designated some of our recitation sections as "computer oriented" so as to give hands on experience with PHASER to perhaps not all students in the course, but at least a larger group than one obtained by self-selection. If this avenue is taken, we do recommend that computer exercises <u>replace</u> rather than overload what is normally required for a recitation, as this approach helps to maintain student motivation.

Beyond these logistical challenges there is the question of the impact that the use of PHASER has on the material in a standard introductory course. Since graphs of any dependent variable versus time can be displayed, PHASER is entirely suitable for the analysis of autonomous scalar equations, which is the typical starting point of most courses. But rather than considering higher order equations next as is frequently done, we found it useful to consider first order linear systems and develop the necessary linear algebraic concepts, thus setting the stage for using PHASER to illustrate the techniques of phase plane analysis for linear and non-linear systems. (These would of course include non-autonomous scalar equations). Having discussed first order systems of equations, we could then move directly to higher order scalar equations and recast them as first order systems in the usual way. We found this particularly useful in presenting the classical damped and forced harmonic oscillator.

Having advocated the use of PHASER to include the study of numerical and qualitative analysis in the introductory differential equations course, we must be prepared to omit an area that is normally discussed and we submit that the typical two to three week section on power series solutions can be dropped. While the basic idea of approximating a solution by an infinite linear combination of "simple" functions, namely powers of x, illustrates a powerful theme in mathematics, the tedious algebraic manipulations required to compute the coefficients in the typical expansion really only serve to obscure the picture. In fact, we suggest that if an instructor deemed it absolutely necessary to present the technique of power series expansions, the use of any symbolic manipulation package or calculator would be extremely appropriate for these computations and would streamline the entire presentation.

Finally, we note that the "responsible" use of the computer as a tool for understanding must be encouraged. One way that this goal can be accomplished is through the creation of "computer counterexamples" and PHASER easily provides a wonderful example. The user simply calls up the equation for the undamped and unforced harmonic oscillator which in theory produces periodic solutions that look like concentric circles about the origin in the phase plane. But if the standard Euler method (which is not "self-correcting") is used to compute the numerical solutions rather than the default Runge-Kutta algorithm, the numerical orbits do not close up and in fact they spiral away from the origin! Thus an experiment with the simplest two-dimensional linear system, if performed without care, can lead to seriously erroneous data.

References

1. Bridger, M. 1986. Review of <u>Differential and Difference Equations Through Computer Experiments</u>. BYTE The Small Systems Journal, 11(7), 63-66.

2. Koçak, H. 1986. <u>Differential and Difference Equations Through Computer Experiments</u>. Springer-Verlag, New York.

3. Zia, L. L. 1988. Computer aided instruction in ordinary differential equations. Journal of Research on Computers in Education, <u>in press</u>.

Lee L. Zia

Department of Mathematics

University of New Hampshire

Durham, NH 03824